Methods in Molecular Biology

Series Editor
John M. Walker
School of Life and Medical Sciences
University of Hertfordshire
Hatfield, Hertfordshire, AL10 9AB, UK

For further volumes:
http://www.springer.com/series/7651

Sorghum

Methods and Protocols

Edited by

Zuo-Yu Zhao

Corteva Agriscience™, Agriculture Division of DowDuPont™, Johnston, IA, USA

Jeff Dahlberg

UC-ANR Kearney Agricultural Research and Extension (KARE) Center, Parlier, CA, USA

Humana Press

Editors
Zuo-Yu Zhao
Corteva Agriscience™
Agriculture Division of DowDuPont™
Johnston, IA, USA

Jeff Dahlberg
UC-ANR Kearney Agricultural
Research and Extension (KARE) Center
Parlier, CA, USA

ISSN 1064-3745 ISSN 1940-6029 (electronic)
Methods in Molecular Biology
ISBN 978-1-4939-9038-2 ISBN 978-1-4939-9039-9 (eBook)
https://doi.org/10.1007/978-1-4939-9039-9

Library of Congress Control Number: 2018966334

This Humana Press imprint is published by the registered company Springer Science+Business Media, LLC, part of Springer Nature.
The registered company address is: 233 Spring Street, New York, NY 10013, U.S.A.

Preface

Sorghum [*Sorghum bicolor* (L.) Moench] was domesticated about 8,000 years ago in northeastern Africa and then spread to other parts of the world. Sorghum, a diploid (2n=20) C_4 plant, belongs to the Poaceae family with a genome size of 730 Mb. It is a self-pollinated crop (less than 15% outcrossing) with high photosynthetic efficiency and high tolerance to abiotic stresses, having adapted to varied agroclimatic environments throughout its early domestication and movement. It has germplasm adapted to drought, high temperature, and infertile soils, making it highly adaptable to climate changes. These characteristics made it among the top five most important cereal crops that feed the world. It is grown in more than 90 countries, on about 40 million hectares, and used as the dietary staple for over 500 million people, mostly in developing countries in Africa and Asia. In addition to human food, it has been recognized as a multipurpose crop with a wide range of uses, which include animal feed and fodder, fermented foods and beverages, and renewable industrial materials and biofuels. The world production of sorghum grains in the recent decade is around 60 million metric tons annually. The United States is the world's leading producer of sorghum, with a production of about 11.5 million metric tons.

For centuries, the efforts for improving sorghum relied on farmer selections and more recently on plant breeding efforts from both public and private sectors. Breeding for improved sorghums took an evolutionary change in 1954 with the discovery of the cytoplasm male sterility system. This discovery triggered the rapid growth of the hybrid sorghum industry in the United States with the first commercial hybrid sorghum sold in 1956. Rapid adoption by farmers followed, and by the early 1960s, 95% of all US sorghum fields were planted with hybrid sorghums. The success of genetic transformation in sorghum with biolistic bombardment was reported in 1993 and with *Agrobacterium* in 2000. Genetic maps began to appear in the 1990s, and the sequence and annotation of the entire sorghum genome was reported in 2009. Other revolutionary discoveries in sorghum breeding, haploid inducers SMH01 and SMH02, and the first CRISPR/Cas9-mediated gene (*Sb-CENH3*) knockout were achieved in 2018.

This book contains 19 chapters reviewing and outlining sorghum breeding technologies, grain compounds, nutrition and digestibility, biotechnology methods, broad renewable applications, and an economic study. These 19 chapters are categorized as 5 review chapters, 5 case study chapters, and 9 protocol chapters made up of contributions by 44 authors. The information provided in these chapters cover up-to-date comprehensive reviews, new study results, or state-of-the-art protocols.

This book targets a broad range of readership interested in sorghum from traditional breeding, modern breeding technologies, sorghum nutrition and compounds related to human health, and renewable biofuels to advanced biotechnologies, biofortification, and genomic editing. We hope this book can provide useful information and tools to an array of readers looking to research and utilize sorghum.

Johnston, IA, USA *Zuo-Yu Zhao*
Parlier, CA, USA *Jeff Dahlberg*

Contents

Contributors

MARC ALBERTSEN • *Corteva Agriscience™, Agriculture Division of DowDuPont™, Johnston, IA, USA*
AJITH ANAND • *Corteva Agriscience™, Agriculture Division of DowDuPont™, Johnston, IA, USA*
DOUG BICE • *United Sorghum Checkoff Program, Lubbock, TX, USA*
JOHN BURKE • *Plant Stress and Germplasm Development Unit, Cropping Systems Research Laboratory, USDA-ARS, Lubbock, TX, USA*
GLORIA BUROW • *Plant Stress and Germplasm Development Unit, Cropping Systems Research Laboratory, USDA-ARS, Lubbock, TX, USA*
PING CHE • *Corteva Agriscience™, Agriculture Division of DowDuPont™, Johnston, IA, USA*
JUNPING CHEN • *Plant Stress and Germplasm Development Unit, USDA-ARS, Lubbock, TX, USA*
RATAN CHOPRA • *Plant Stress and Germplasm Development Unit, Cropping Systems Research Laboratory, USDA-ARS, Lubbock, TX, USA; Department of Agronomy and Plant Genetics, University of Minnesota, St. Paul, MN, USA*
JEFF DAHLBERG • *UC-ANR Kearney Agricultural Research and Extension (KARE) Center, Parlier, CA, USA*
J. ANITA DILLE • *Department of Agronomy, Kansas State University, Manhattan, KS, USA*
JOHN DUFF • *National Sorghum Producers, Lubbock, TX, USA*
KWAKU G. DUODU • *Department of Consumer and Food Sciences and Institute for Food, Nutrition and Well-being, University of Pretoria, Pretoria, South Africa*
LINDA DYKES • *Cereal Crops Research Unit, Red River Valley Agricultural Research Center, USDA-ARS, Fargo, ND, USA*
MARIO G. FERRUZZI • *Department of Food, Bioprocessing and Nutrition Sciences, Plants for Human Health Institute, North Carolina State University, Kannapolis, NC, USA*
CLEVE FRANKS • *Corteva Agriscience™, Agriculture Division of DowDuPont™, Johnston, IA, USA*
KIMBERLY GLASSMAN • *Corteva Agriscience™, Agriculture Division of DowDuPont™, Johnston, IA, USA*
IAN D. GODWIN • *School of Agriculture and Food Sciences, The University of Queensland, Brisbane, QLD, Australia*
KAREN R. HARRIS-SHULTZ • *Crop Genetics and Breeding Research Unit, USDA-ARS, Tifton, GA, USA*
CHAD M. HAYES • *Plant Stress and Germplasm Development Research, USDA-ARS, Lubbock, TX, USA*
MARK HINDS • *Corteva Agriscience™, Agriculture Division of DowDuPont™, Johnston, IA, USA*
IAN HOEFFNER • *National Sorghum Producers, Lubbock, TX, USA*
HALEE HUGHES • *Plant Stress and Germplasm Development Unit, Cropping Systems Research Laboratory, USDA-ARS, Lubbock, TX, USA*
TANVEER HUSSAIN • *Corteva Agriscience™, Agriculture Division of DowDuPont™, Johnston, IA, USA*

TODD J. JONES • *Corteva Agriscience™, Agriculture Division of DowDuPont™, Johnston, IA, USA*
JOSEPH E. KNOLL • *Crop Genetics and Breeding Research Unit, USDA-ARS, Tifton, GA, USA*
LARRY LAMBRIGHT • *National Sorghum Foundation, Lubbock, TX, USA*
JIEQING LI • *College of Agriculture, Anhui Science and Technology University, Fengyang, China*
GUOQUAN LIU • *School of Agriculture and Food Sciences, The University of Queensland, Brisbane, QLD, Australia*
ABABACAR NDOYE • *Institute of Food Technology, Dakar, Senegal*
DARWIN ORTIZ • *Department of Food Science, Purdue University, West Lafayette, IN, USA; Department of Agronomy, Purdue University, West Lafayette, IN, USA*
BOTOROU OUENDEBA • *3N Program, Niamey, Niger*
COURTNEY PAYNE • *The National Renewable Energy Laboratory, Golden, CO, USA*
KRISTEN RINEHART • *Corteva Agriscience™, Agriculture Division of DowDuPont™, Johnston, IA, USA*
JEFFRY D. SANDER • *Corteva Agriscience™, Agriculture Division of DowDuPont™, Johnston, IA, USA*
JOHN H. SANDERS • *Agricultural Economics Department, Purdue University, West Lafayette, IN, USA*
JUSTIN SLUITER • *The National Renewable Energy Laboratory, Golden, CO, USA*
NIABA TÉMÉ • *Institute of Agricultural Economy, Bamako, Mali*
SOUNGALO TRAORE • *Federal Extension Service, Mopti, Mali*
AUSTIN VINCENT • *National Sorghum Producers, Lubbock, TX, USA*
ED WOLFRUM • *The National Renewable Energy Laboratory, Golden, CO, USA*
EMILY WU • *Corteva Agriscience™, Agriculture Division of DowDuPont™, Johnston, IA, USA*
ZHANGUO XIN • *Plant Stress and Germplasm Development Unit, USDA-ARS, Lubbock, TX, USA*
ZUO-YU ZHAO • *Corteva Agriscience™, Agriculture Division of DowDuPont™, Johnston, IA, USA*
GUIHUA ZOU • *Plant Stress and Germplasm Development Unit, USDA-ARS, Lubbock, TX, USA; Institute of Crop and Nuclear Technology Utilization, Zhejiang Academy of Agricultural Sciences, Hangzhou, Zhejiang Province, China*

Part I

Breeding Technologies

Chapter 1

Hybrid Sorghum Product Development and Production

Larry Lambright

Abstract

There are many moving parts involved in developing and taking a new commercial hybrid to market. Sorghum [*Sorghum* bicolor (L.) Moench] hybrid development involves development of parental lines based on a cytoplasmic male sterile system including the pollen parent (R-line) and seed parents (A- and B-lines). New parental lines are developed by recombining existing elite parental lines to create new breeding populations or by adding specific traits of interest to existing parental lines by crossing elite lines with donor parents. Molecular markers are utilized to identify plants with particular traits of interest during parental line development. Newly developed parental lines are crossed together or with other elite parental lines to create new hybrid combinations. New hybrid combinations are evaluated in target geographies for improved yield, good agronomics, and specific traits of interest. Multi-year, multi-location evaluations are used to identify hybrid entries with improved yield, stable performance, and good agronomics. Evaluation of the parental lines involved in these new hybrids helps establish the produce ability and potential cost of goods which have direct impact on the potential commercial release of new hybrid products.

Key words Parental lines, A, B, R-lines, Elite, Hybrids, Molecular markers, Multi-year, Multi-location, Yield, Performance

1 Introduction

Developing and releasing a new sorghum hybrid into the market is a process that involves a variety of different stakeholders within a sorghum seed company. Sorghum breeders develop new parental lines which are evaluated as hybrid combinations in breeder trials. When potential new commercial products have been identified using data from breeder trials, breeder seed is provided to seedstock personnel for pilot hybrid seed production and parental line increase. Seed generated from these pilot production plantings is then provided to company agronomists and district sales managers for evaluation in field scale strip trials. Those strip trial entries with strong performance and good agronomics are in the next generation, produced by seed production staff for release into the market. Prior to release, company marketing personnel, using both breeding data and data from strip trials, position the new hybrid for

Zuo-Yu Zhao and Jeff Dahlberg (eds.), *Sorghum: Methods and Protocols*, Methods in Molecular Biology, vol. 1931, https://doi.org/10.1007/978-1-4939-9039-9_1,

release into a defined geography. Finally, sales personnel in the targeted release area introduce the new product to seed dealers, distributors, and customers.

The total process may take 10 to 12 generations from start to finish. By efficiently utilizing contra-season nurseries as well as greenhouse crops, which result in multiple generations per calendar year in the early line development stage, as well as conducting concurrent operations in the final hybrid evaluation and seed production stages, significant time can be shaved from this process. Access to genetic marker technology as well as technologies new to sorghum, e.g., double haploid, can also provide significant speed to market improvements within the process as described.

2 Materials

The first step in developing a new sorghum hybrid is parental line development. Sorghum has a perfect flower, as such, before entering into a description of hybrid sorghum product development; it is important to understand the mechanics of the cytoplasmic male sterile system utilized in the development and production of hybrid sorghums. Sorghum's cytoplasmic male sterile system is made up of three parental lines. The seed parent (female) has both sterile, A-line, and fertile, B-line, versions. The pollen parent (male), R-line, has restorer genes that restore fertility in F_1 hybrid combinations. A-lines, which are sterile, are propagated by crossing an A-line with a B-line of the same genotype. The B-line, sometimes referred to as the maintainer, is fertile and sheds pollen, but does not have restorer genes; thus progeny of A-line x B-line crosses is sterile. Both B-lines and R-lines are propagated by self-pollination. Commercial F_1 hybrids are generated by crossing A-lines x R-lines. The resulting progeny of an A x R cross is fertile due to the restorer genes provided by the R-line (pollen parent) in a hybrid combination.

3 Methods

3.1 Parental Line Development

The developmental process of seed parental lines (B-lines) and pollen parental lines (R-lines) is identical through the F_3 generation and begins with the process, whereby R-lines and B-lines are crossed/recombined with other R-lines or B-lines. As a general rule, R-lines are only crossed with R-lines and B-lines are only crossed with B-lines. In crops utilizing sterile systems, it is generally best to avoid mixing restorer lines and maintainer lines during parental line development. Elite R- and B-lines are chosen to be included in the recombination process based on their strong performance in hybrid combinations. On occasion, a particular parental line's inclusion into this recombination process might be the addition of a particular trait(s).

As both B-lines and R-lines that are to be crossed in the recombination process are fertile, it is necessary to emasculate one of the lines included in a combination in order to secure recombination from the other parent which provides the pollen. Emasculation is normally accomplished in one of two ways. One of the techniques called hand emasculation involves removing the anthers from 40 to 50 sorghum flowers prior to bloom by hand utilizing forceps or other small tools. This operation is quite tedious as the sorghum flowers are small and very delicate. Thus it is very easy to do damage to the flower if not done properly. The second technique involves using a plastic bag to cover a portion of the sorghum head prior to blooming. The plastic bag causes condensation under the bag and the moisture destroys the pollen when the anthers emerge during bloom. If done correctly, the hand emasculation method yields very few to no self-pollinations while the plastic bag method normally generates a higher frequency of self-pollinations. In both emasculation methods, pollen is provided by the other line involved in the combination and the resulting seed are a F_1 breeding population.

The F_1 seed generated following the emasculation process is then selfed. Care is taken to avoid harvesting any suspected self-pollinations which may have occurred during the emasculation process. Seed harvested from selfed F_1's is planted in F_2 nurseries at sites which are representative of target geographies. Available genetic markers are utilized to identify individual plants with specific traits of interest. From this subgroup which are positive for the markers of interest, selections are made by the breeder based on visual appearance and agronomic characteristics exhibited by individual plants. Population size in F_2 nurseries may vary depending on the trait(s) targeted, the number of genes involved, and the inherence of the trait(s). Minimum population size should be in the neighborhood of 1000 plants/segregating population, but a larger population size may be necessary depending on the elements involved in a particular population.

Selections made from segregating F_2 populations are planted in F_3 nurseries the following generation. Normally, a single row containing approximately 100 plants is planted from each selection. The presence of genetic markers for traits of interest is confirmed in each F_3 row. Breeder selections are then made at harvest from rows confirmed to be positive for markers of interest, and that have acceptable agronomic traits and good uniformity. Maturity is also an important consideration. Rows of elite parents are scattered in F_3 nursery plantings for the purpose of maturity comparisons.

3.2 R- and A-Line Development

While R-line and B-line populations are handled identically in generations F_1 through F_3, beginning with the F_4 generation selections planted in crossing nurseries are handled differently. The R-line F_4

selections are testcrossed onto two or three elite seed parents to generate F_1 hybrids which are utilized to evaluate the potential combinability of the new male parental lines (R-lines).

B-line F_4 selections are crossed on to elite-related females (A-lines) to begin the sterilization process ($CMS_{0\rightarrow1}$). By using a related female for the initial sterilization cross, testcrossing can normally occur with fewer sterile backcrosses. The following generation, CMS_1 A-line crosses are paired with the F_5 selections of the new B-line for an additional cross into sterile cytoplasm ($CMS_{1\rightarrow2}$). CMS_2 A-line crosses are paired with the F_6 selections of the new B-line in the next generation to provide a $CMS_{2\rightarrow3}$ version of the new A-line. Utilizing the same row of CMS_2 plants, testcrosses are also made with two or three elite pollen parents to generate F_1 hybrids which will be used to evaluate the potential combining ability of the new seed parents (A-lines). During the CMS process, sterility evaluations of potential new seed parents are monitored at each generation. Those entries with questionable sterility are discarded accordingly.

3.3 Testcross and Hybrid Evaluation

The testcross generation is the final step in parental line development. As discussed, there are quite a few generations required to get to the testcross stage. Contra season nurseries and greenhouse plantings are utilized to accomplish two to three generations per year during the early stages of parental line development as well as B-line sterilization during the CMS process.

After new parental lines are developed and testcrossed, the next step in the process is hybrid evaluation. New parental lines are initially evaluated as testcrosses. As mentioned previously, each new parental line, whether pollen parent (R-line) or seed parent (A-line), is testcrossed with elite tester lines. New pollen parents are crossed with two or three elite seed parents while new seed parents are crossed with two or three elite pollen parents. These testcrosses are evaluated in single rep trials at a minimum of three to four testing sites in target geographies. Adapted, maturity appropriate commercial check hybrids are included in these testcross trials. Based on yield performance relative to commercial check hybrids as well as maturity and agronomics, normally between 5 and 10% of the testcross entries are advanced to more intensive testing. New parental lines, both A-lines and R-lines, which have exhibited strong performance on multiple testers are flagged as potential general combiners (GCA). New parental lines, whether seed parents or pollen parents, which have shown GCA tendencies are crossed with a panel of additional elite lines to generate new F_1 hybrids. These hybrids by-pass the testcross evaluation stage and move directly into more intensive testing. Additionally, new parental lines flagged as possible GCA's move into a pool for recombination in order to generate new F_1 breeding populations for parental line development.

3.4 Advanced Hybrid Testing

Following the completion of testcross evaluation, in the next generation the 5–10% of testcross entries flagged for further consideration are advanced into more intensive testing and are evaluated in multi-rep, multi-location (5–6 sites) hybrid trials in target geographies. A panel of five to six well-adapted, maturity appropriate commercially available check hybrids, both internal and external, are included in each trial. Overall yield performance, maturity, stalk and root strength, as well as general agronomics are considered. The best hybrids relative to checks are moved into advanced level testing the next generation. Advance testing involves multi-rep, multi-location (10–12 sites) hybrid trials scattered over the target geographies. A panel of eight to ten maturity appropriate commercially available check hybrids both internal and external are included in each trial. A high priority is placed on total yield performance, which includes yield stability, yield for maturity, harvestable yield, as well as general agronomics.

Yield stability provides stable performance relative to competing products across environments and differing weather conditions which will vary from year to year. Generally those products that exhibit excellent yield stability tend to have wider adaptation. Yield for maturity identifies those hybrids that are able to generate the most yield with the earliest maturity in each maturity classification. A simple calculation of total yield divided by the number of days to mid-bloom is an excellent indicator of yield for maturity (yield/day). Harvestable yield takes into account all of the biotic and abiotic stressors which detract from the genetic yield potential that exists in a hybrid product. Insects, plant diseases, weak stalks or roots, drought stress, heat stress, and soil pH are all examples of stressors that can and do negatively impact harvestable yield.

3.5 Cost-of-Goods Evaluation

Concurrent with advanced testing, Cost-of-Goods (COG) evaluation of parental lines involved in those hybrids included in advance trials is initiated. Primary characteristics considered are seed parent (A-line) yield potential, pollen parent (R-line) pollen volume, and synchronization of bloom of the two parents involved in the hybrid. COG's evaluations are handled in replicated trials made up of parental lines utilized in advance trial entries. COG's trials are evaluated in seed production areas with planting dates commensurate with normal seed production plans. Seed parent yield potential has an obvious impact on cost of goods, but pollen volume and synchronization of bloom, commonly referred to as split, also impact green seed yields, as well as play a significant role in the genetic purity of the seed produced.

Pollen volume available during the period of female bloom is key to maintaining genetic purity and minimizing outcross potential. Low pollen volume R-lines are objectionable and once identified are discarded. Synchronization of bloom is controlled by split planting or sometimes referred to as delay plantings. The larger the difference in bloom timing between the seed parent and the pollen parent, the greater the risk of genetic purity issues. Split differences which require longer planting delays may be subject to weather issues which can alter planting plans. The goal of split planting is to have a large load of pollen available at the beginning of female bloom through the bloom period until the completion of bloom. It is common to make three male plantings to help insure good volumes of pollen availability during female bloom. The type of split plantings also has impact on risk of synchronization of bloom. Female delay plantings to achieve desired splits are generally accepted to be of higher risk than male delays.

3.6 Breeder's Seed Increase

During the same generation that advance testing occurs and COG's evaluation begins, breeder's seed increases of all new parental lines involved in hybrids evaluated as advance trial entries are initiated. Target quantities of these nursery breeders' seed increases normally are in the 5–10 lb. range. In the case of new seed parents, both A-line and B-line versions of the seed parent will need to be increased. In addition to seed of the A-line for pilot hybrid seed production, an increase in the A-line, as an AxB increase planted in one to two acres, will also be needed in order to be ready for a larger production of the new hybrid if warranted. The A-line and R-line increases are needed in order to be prepared for pilot seed production of those advance trial entries, which, based on performance, show promise as potential commercial releases. Seedstock personnel who grow these pilot hybrid seed production plantings (1 to 3 acres) normally have a target of 2500 to 5000 lbs. per hybrid. This seed will be utilized by company agronomists and district sales managers for field scale testing in strip trials the following growing season.

3.7 Strip Trials

Field scale strip trials with producers are scattered over the anticipated target release area. Strip trial entries include experimental hybrids as well as maturity appropriate commercially available hybrids for comparison. Data including yield, grain moisture, plant height, stalk quality, root strength, disease or insect response differences, as well as other differences that may manifest are collected and entered into a database that also includes small plot data collected from all breeding program evaluations involving this set of hybrids.

Experimental hybrids that exhibit yield potential improvement of 5% or greater vs. other maturity like products in the commercial hybrid product portfolio, stable performance relative to check hybrids across locations, good agronomic traits, and acceptable Cost-of-Goods are advanced to commercial status. These new offerings are normally targeted to replace a current product in the portfolio. The growing season prior to the planned sales of a new hybrid, seed production (500 to 1500 units) is planted to generate a targeted amount of the new hybrid product.

3.8 Seedstock Development

Seedstock of new parental lines which will be utilized by seed production staff is sourced from either new pollen parents harvested from pilot productions or in the case of new seed parents, seed produced in the one to two acre AxB increases mentioned previously. Planting split data from COG's evaluations as well as information generated from small pilot productions are used to determine planting splits for this introductory seed production. As mentioned, normally three planting dates of the pollen parent are utilized to help insure good pollen coverage during the seed parent bloom period.

3.9 Hybrid Seed Production

Hybrid sorghum seed production requires isolation from other grain sorghums, forage sorghums, sudangrass, and wild sorghum species including johnsongrass [*S. halepence* (L) Pers.] and shattercane [*S. bicolor* (L.) Moench]. Isolations can be generated with distance or with time utilizing delayed plantings. Following harvest of the seed production in the fall, growout samples drawn from harvested seed are included in winter growout plantings in Mexico, the Caribbean, or Central America to be evaluated for genetic purity prior to commercial sales. Following growout readings, those seed lots that are within company standards for genetic purity and germination are then conditioned and bagged.

4 Notes

Utilizing performance data generated from both breeder trials as well as field scale strip trial plantings, company marketing staffers position the new hybrid for release into a defined geography. As the final step in this process, sales personnel introduce the new product to seed dealers, distributors, and customers in the defined geography. The new product, which based on company-wide data is expected to have very strong performance in the targeted release area, is then anticipated to expand into other regions in subsequent growing seasons.

References

No references were utilized in preparing this chapter. Information in the chapter titled Hybrid Sorghum Development and Production was based on my reflections and observations during 40+ years of working in the private seed sector as a commercial plant breeder.

Chapter 2

Mapping QTLs and Identification of Genes Associated with Drought Resistance in Sorghum

Karen R. Harris-Shultz, Chad M. Hayes, and Joseph E. Knoll

Abstract

Water limits global agricultural production. Increases in global aridity, a growing human population, and the depletion of aquifers will only increase the scarcity of water for agriculture. Water is essential for plant growth and in areas that are prone to drought, the use of drought-resistant crops is a long-term solution for growing more food for more people with less water. Sorghum is well adapted to hot and dry environments and has been used as a dietary staple for millions of people. Increasing the drought resistance in sorghum hybrids with no impact on yield is a continual objective for sorghum breeders. In this review, we describe the loci, quantitative trait loci (QTLs), or genes that have been identified for traits involved in drought avoidance (water-use efficiency, cuticular wax synthesis, trichome development and morphology, root system architecture) and drought tolerance (compatible solutes, pre- and post-flowering drought tolerance). Many of these identified genes and QTL regions have not been tested in hybrids and the effect of these genes, or their interactions, on yield must be understood in normal and drought-stressed conditions to understand the strength and weaknesses of their utility.

Key words Drought avoidance, Drought tolerance, Drought escape, Trichomes, Epicuticular wax, Stay-green, Root architecture, Water-use efficiency, Compatible solutes

1 Introduction of Drought Resistance

Drought, described as soil and/or atmospheric water deficit, has been a major cause of crop loss [1]. With increasing aridity in many areas of the globe and a growing human population, water will likely be a scarcer commodity [2]. Thus the utilization of crop cultivars adapted to water-stressed conditions is a long-term solution for stabilizing crop productivity [3]. Water is essential for plant growth and serves many roles such as a solvent, a reactant, and a maintainer of cellular turgidity [4]. Thus, water limitation causes plants to halt cell enlargement, leaf and stem elongation, and photosynthesis and to decrease respiration and turgor, which increases wilting [4]. Drought can occur at any plant growth stage and can affect crop productivity depending on the time of

Zuo-Yu Zhao and Jeff Dahlberg (eds.), *Sorghum: Methods and Protocols*, Methods in Molecular Biology, vol. 1931, https://doi.org/10.1007/978-1-4939-9039-9_2,

onset, duration, and intensity of drought stress [5]. Drought stress during the seedling or reproductive stages can result in stand losses or severe yield losses [6, 7].

Drought resistance is defined as, under the same drought intensity, certain genotypes of crop plants having higher relative yield as compared with others [8]. According to Levitt [9], plants that are resistant to drought physiologically use three strategies that include escape, avoidance, and tolerance strategies of which plants may combine these responses [10]. Plants that escape drought can complete their life cycle before physiological water deficits occur and exhibit developmental plasticity [2]. Desert plants exhibit high stomatal conductance for carbon dioxide and photosynthetic efficiency, rapid growth and development, and can complete their life cycles in a few weeks [11]. Additionally, in another example of drought escape, the use of early maturing genotypes or the timing of the planting of crops can be chosen, so that during the growing season, local seasonal or climatic droughts are avoided [10, 12].

Drought avoidance refers to plants that can endure drought by avoiding tissue dehydration [2]. This involves minimizing water loss from transpiration or the cuticle and/or maximizing water uptake by the roots [2]. Water loss is minimized by closing stomata [4], by the excessive deposition of epicuticular wax on the aerial surface of the plant that prevents cuticular transpiration [13], and by the presence of trichomes that entrap and retain surface water allowing foliar water uptake [14]. Other drought avoidance mechanisms include leaf abscission [15], dormancy [16], and leaf rolling to reduce water loss from transpiration [17]. Additionally, plants that use C_4 photosynthesis have less water loss from transpiration than C_3 plants [18]. Furthermore, plants possessing deep, large root systems that can penetrate hard soil layers are often associated with the ability to maintain water supply during low water conditions as well as the ability of the root system to respond to water stress [19]. Under water stress, increased root/shoot ratios are often seen, and the root/shoot ratios have been used as a criterion to identify drought resistance in plants [10]. The ability of some plants to store water using specific organs (tubers, etc.) is also a drought avoidance trait [20].

Drought tolerance mechanisms allow the plant to maintain metabolic activity under low tissue water potential by osmotic adjustment and increased antioxidant defense [3, 21, 22]. Plants can accumulate compatible solutes such as sugars and sugar alcohols (fructan, trehalose, mannitol, d-ononitol, and sorbitol), ammonium compounds (polyamines, glycinebetaine, *b*-alanine betaine, dimethyl-sulfonio propionate, and choline-*O*-sulfate), and amino acids (proline and ectoine) which accumulate in the cytosol in order to lower osmotic potential and maintain the turgor of both shoots and roots [23]. During drought, reactive oxygen species (ROS) generation is increased due to the disruption of the electron transport system [24]. ROS are damaging to cellular components and can cause lipid peroxidation or protein and nucleic acid modification [24]. Antioxidants allow plants to detoxify ROS by using antioxidant enzymes and free radical scavengers [25]. Antioxidant enzymes include superoxide dismutase (SOD), catalases, peroxidases, glutathione S-transferase, gluatathione reductase, monodehydro/dehydroascorbate reductase, alternative oxidase, etc. and the free radical scavengers are carotenoids, ascorbate, tocopherols, cytochrome f, flavanones, anthocyanins, glutathione, etc. [10, 25]. Thus, osmotic adjustment and the alleviation of drought damage are mechanisms of drought tolerance.

2 Crosstalk Between Abiotic and Biotic Stress Responses

In the past, the molecular mechanisms of each abiotic and biotic stress were examined independently but plants respond to abiotic and biotic stresses with common and overlapping pathways [26] which will be briefly reviewed here. Abiotic and biotic stresses activate the hormone signaling pathways regulated by abscisic acid (ABA), salicylic acid, jasmonic acid, and ethylene, generate ROS, and cause mitogen-activated protein kinase (MAPK) cascades [26]. To reduce the effects of the stress, plants use hormone signaling, ROS accumulation, and MAPK cascades to activate transcription factors that cause the expression of downstream response genes [27].

Plant hormone signaling controls abiotic and biotic stress responses [26]. Abiotic stress is largely controlled by ABA whereas biotic stress is controlled by the antagonism between salicylic acid and jasmonic acid/ethylene signaling pathways [27]. Additionally, ABA can increase the susceptibility of plants to pathogens and ABA insensitive lines have higher levels of pathogen resistance than wild type plants [28]. ABA can also have a positive effect on pathogen defense but the timing of infection and the type of pathogen affects the influence of ABA [29].

ROS accumulate constantly as by-products of metabolism but their concentrations increase to high levels after abiotic stress [27]. Additionally, plants under pathogen infection induce a hypersensitive reaction and/or systemic acquired resistance activation which leads to ROS generation [30]. The generation of ROS after abiotic and biotic stress then causes plants, in order to prevent damage to membrane phospholipids and fatty acids, to induce large sets of genes that encode ROS-scavenging enzymes or antioxidants [31].

MAPK cascades transduce the perception of the environmental and developmental cues into internal signaling pathways [32]. During biotic stress, plant transmembrane receptors detect pathogen-associated molecular patterns and trigger MAPK cascades to start pathogen response signaling [33]. Likewise, during abiotic stress, MAPK cascades are induced following activation of receptors [32]. Downstream targets of MAPK cascades include transcription factors, other kinases, enzymes, and hormone response factors [32, 34].

Transcription factors control a wide range of downstream events and many transcription factors have been found to be involved in the response to both biotic and abiotic stress [27]. The transcription factors Myelocytomatosis Oncogene2/Jasmonate Insenstitive1(*MYC2/JIN1*), the myeloblastosis (MYB) family of transcription factors; No Apical Meristem (NAM), Arabidopsis Transcription Activation Factor (ATAF), and Cup-shaped Cotyledon (CUC) of which these three are members of one of the largest families of plant-specific transcription factors and are abbreviated (NAC); APETALA2/Ethylene Responsive Factor (AP2/ERF), ABA-responsive Element-binding Factor (ABF), transcription factors defined by the conserved WRKYGQK sequence (WRKY), Zinc-fingers, and heat shock transcription factors all have been shown to have a role in the interaction between abiotic and biotic stress signaling pathways [27]. Heat shock transcription factors control many downstream stress-responsive genes as well as the expression of heat shock proteins which bind and stabilize proteins that have become denatured after abiotic and biotic stress [35]. Each stress may elicit activation of a unique combination of heat shock proteins [27]. Heat shock transcription factors may detect the presence of ROS and cause the expression of heat shock proteins as well as ROS scavenging enzymes [35].

Compilation of drought responses, specifically, in a range of plant species reveals that drought induces hormone signaling by ABA, a hormone involved in stomata closure to reduce transpiration [36]. Drought suppresses cell growth and photosynthesis efficiency, decreases respiration [4], and induces genes that are classified into two groups [36]. The first group includes proteins such as chaperones, late embryogenesis abundant (LEA) proteins, osmotin, mRNA-binding proteins, osmolyte biosynthesis, water channel proteins, metabolite transporters, detoxification enzymes, and proteases which all function in abiotic stress tolerance [36]. The second group involves proteins involved in signal transduction and stress-responsive gene expression and includes kinases, protein phosphatases, calmodulin-binding proteins, and enzymes involved in phospholipid metabolism [36].

3 Sorghum

Sorghum (*Sorghum bicolor* L. Moench) is a C_4 annual grass originating from Africa [37] that is used as a grain crop, for forage, and as a sweetener [38]. It also has potential to be used as a bioenergy feedstock [39]. Sorghum has drought and heat tolerance and is grown in Africa, Asia, the Americas, and Australia in regions that are typically too dry for maize (*Zea mays* L.) production [40, 41]. The drought resistance seen in sorghum is attributed to morphological and physiological factors such as a root system whose depth is dependent on the amount of water received [42], the ability to maintain stomata open and thus continue photosynthesis at low water potentials, leaf senescence, and the ability for osmotic adjustment [43]. Furthermore, high epicuticular wax deposition on the sorghum plant surface reflects visible and near-infrared radiation and reduces cuticular conductance of water vapor [44]. In comparison to maize, sorghum has a higher water-use efficiency [45]. Sorghum is consumed as a dietary staple in parts of Africa and Asia and is used predominantly as a poultry/animal feed, with only a small amount being used for human consumption in the United States and Australia [46]. Consumption of sorghum provides energy, vitamins, and minerals, and the grain has a protein content of near 9% [47].

In the United States from 2000 to 2016, the average acres of planted grain sorghum were 7.65×10^6 acres [47]. In the U.S., sorghum is grown primarily from South Dakota to Southern Texas (the Sorghum Belt) [48]. Most of the sorghum grown in the U.S., for the last few years, has been exported; during 2013–2014, China imported four million metric tons of sorghum from the U.S [49]. China has been importing sorghum as an alternative to corn as there is no import quota on sorghum [50]. This demand from China has a strong, positive impact on the world and U.S. grain sorghum prices and farm production [47].

4 Molecular Mechanisms of Drought Responses in Sorghum

With the crosstalk of the abiotic and biotic signaling pathways as a framework, we will now briefly cover the literature on the molecular mechanisms of drought responses in sorghum. Transcript-profiling studies are beginning to elucidate the roles that various genes play in adaptation to drought. As part of a larger sorghum transcriptome project, three drought-induced expressed sequence tag (EST) libraries were constructed [51]. For the first library-water-stress 1 (WS1), five-week-old whole greenhouse-grown plants (including roots) of inbred line BTx623 were harvested after water was withheld for 7–8 days. Two libraries were created from the leaves of greenhouse-grown plants at either pre-flowering drought stress (DSBF1) or post-flowering drought stress (DSAF1). The pre-flowering drought-tolerant line Tx7000 was used to create DSBF1, while the stay-green post-flowering drought-tolerant line B35 (a.k.a. BTx642) was used to create the DSAF1 library. The DSBF1 and DSAF1 libraries were subtracted using cDNA from unstressed sorghum leaves to enrich these libraries with transcripts preferentially expressed under drought stress [52]. Dehydrins, heat-shock proteins, and other drought-inducible proteins of unknown function are among the transcripts expressed in these libraries. Fifty-five of these transcripts were also found to be upregulated by ABA in a microarray experiment highlighting the role of ABA signaling in the drought response [53]. The software program Simple Sequence Repeat Identification Tools (SSRIT) [54] was used to search for simple sequence repeat (SSR) sequences within the ESTs from the DSAF1 and DSBF1 libraries [55]. From over 9800 ESTs, 109 SSR markers were developed. Using 168 recombinant inbred lines (RILs) from a cross between 298B and IS18551, 28 of these markers were placed on a linkage map, along with 100 genomic SSRs. Following the publication of the *Sorghum bicolor* genome sequence [56] and the development of high-throughput sequencing methods, RNA-Seq was used to re-examine the transcriptomes of sorghum plants exposed to osmotic stress induced by polyethylene glycol (PEG) or ABA [57]. They found that about 1650 genes were upregulated by osmotic stress, and about 700 genes were downregulated. Exogenous ABA caused upregulation of about 2300 genes and downregulation of about 2600 genes. Depending on the tissue examined (leaf or root), between 12 and 30% of these genes showed similar regulation by osmotic stress and ABA. Genes that were upregulated by osmotic stress and ABA included a dehydrin and a late-embryogenesis-abundant (LEA) protein. Downregulated genes included a gene with sugar transporter domains in the root tissue and a peroxidase-like gene in the shoot tissue. Pathway analysis revealed that genes involved in synthesis of sugars, amino acids, and other osmoprotectant molecules were also upregulated by osmotic stress and ABA. Several pathways involved in hormone biosynthesis or regulation were also affected.

A transcriptome study using microarrays was conducted on sorghum seedlings subjected to heat, drought, or heat and drought in combination, as for summer-grown crops the two stresses are often simultaneous [58]. For drought stress only, genes involved in water deprivation, response to ABA, regulation of photosynthesis, wax biosynthesis, fluid transport, and amino acid metabolism were induced. Examination of the promoter motifs for the 966 induced genes revealed the most highly represented promoter motif was similar to the abscisic acid response element (ABRE): (C/T) ACGTGTC. Examination of the genes induced by drought with a previous study that used PEG [57] revealed that 32% of the induced genes were shared between the two studies. For heat and drought stress, genes regulated by the combined stresses encode ion transporters, heat shock proteins, LEA proteins, signaling proteins, calcium binding proteins, MAP kinases, and transcription factors. Ontology analysis revealed an enrichment of genes involved in polyamine metabolism and, in particular, spermidine biosynthesis which may be involved in reactive oxygen scavenging and membrane protection [59]. The most enriched promoter motif was similar to the ABRE, which was also the promoter motif for drought stress only. This study reveals although many common pathways and genes are regulated by heat stress, PEG, drought stress, and a combination of heat and drought stress, there are genes, and thus pathways, that are uniquely altered by each particular stress.

The sorghum proteome responses to drought stress and recovery from drought have also been examined at the fifth leaf stage using matrix-assisted laser desorption/ionization-time-of-flight mass spectrometry (MALDI-TOF-MS) [36]; for the principles and technical aspects of MALDI-TOF-MS see Kasper et al. [60]. Proteins that were differentially regulated after drought and/or recovery were homologs potentially involved in (1) metabolism esp. methionine synthase, (2) energy esp. C_4 phosphoenolpyruvate carboxylase and chloroplast NADP-dependent malic enzyme, fructose-1,6-bisphosphate aldolase, and glyceraldehyde-3-phosphate dehydrogenase, (3) transcription esp. RNA-binding proteins, (4) protein synthesis of which one protein was differentially regulated, elongation factor alpha, and (5) chaperones esp. heat shock protein 60. The authors conclude that the drought-tolerant genotype of sorghum maintained protein stability, reallocated metabolites to the newly developed structures, and maintained protein synthesis while the drought susceptible genotype had more proteins involved in cell death and proteases [36].

5 QTL Mapping

Many agriculturally important traits such as yield and drought response are controlled by many genes and are referred to as quantitative traits [61]. QTL are sections of DNA, usually DNA markers, which correlate with variation in a phenotype. QTL are typically identified by bi-parental mapping where a linkage map, made up of ordered DNA markers, is created and the parents (which differ in the traits of interest) and progeny are phenotyped in multiple locations and/or years. QTL are identified using single marker analysis, simple interval mapping, and composite interval mapping, of which, composite interval mapping is currently the most frequently used, reviewed in detail by Collard et al. in 2005 [61]. An alternative approach to QTL mapping is association mapping, which uses largely unrelated individuals as the population and this population is phenotyped in multiple environments and years and is genotyped [62]. Correlations are identified between genetic markers and phenotypes. Association mapping has higher mapping resolution than linkage mapping due to the larger number of recombination events in a sample [62].

Many QTL studies have been performed in sorghum due to the diploid tractable nature of the genome, the ease of population development and obtaining germplasm, and the advent of high-throughput sequencing and genotyping-by-sequencing [63]. Furthermore, a reference genome sequence of sorghum [56] and a collection of mutants are also available [64].

6 Sorghum QTLs and Identification of Genes

Drought resistance in sorghum, like many plants, is controlled by numerous small effect loci and hundreds of genes that affect morphological and physiological responses to drought [5]. In this section, we will review the morphological and physiological traits associated with drought resistance that have been examined in sorghum and the QTL studies performed on sorghum to date that are involved in drought resistance as well as the candidate genes identified. The traits measured to assess drought resistance can be classified as drought avoidance-related, drought tolerance-related, and integrative indicators [10].

6.1 Drought Avoidance Traits

6.1.1 Transpiration and Water Use Efficiency

Water use efficiency (WUE) can be defined as the amount of biomass produced divided by the amount of water used, or transpired, by the plant [65]. An indirect way to measure WUE is to measure the CO_2 assimilation rate (A) and the transpiration rate (E), both of which can be measured simultaneously using a portable infrared gas analyzer, such as the Li-Cor LI-6400 Portable Photosynthesis System (Li-Cor, Lincoln, NE), or similar apparatus [66, 67]. The transpiration ratio (A:E) is then calculated as (A/E)/VPD, where VPD is the vapor pressure deficit of water [67]. Water use efficiency is often equated with drought tolerance, but Blum [65] cautions that this is often not true, and that selection for WUE may result in a shorter growing cycle and lower yields.

In two greenhouse studies, A, E, and A:E were measured in a population of 70 RILs (Tx430 × Tx7078) [67]. The soil moisture in the pots was maintained at either 40% or 80% field capacity (FC) in Experiment 1, and at 80% FC only in Experiment 2. A genetic map was constructed from 261 digital genotyping markers [68]. Composite interval mapping identified one QTL (SBI-05) for A at 40% FC, and one (SBI-01) at 80% FC in the first experiment, and a different QTL (SBI-09) for A at 80% FC in the second experiment (Table 1). Three QTLs were identified for E, two on SBI-01 at different locations, and another on SBI-07. Three QTLs for transpiration ratio at 80% FC were found on SBI-09 (2) and SBI-10 in Experiment 1, and another on SBI-06 at 80% FC in Experiment 2. No QTL for A:E were found for the 40% FC

Table 1
QTLs identified in *Sorghum bicolor* that are associated with drought resistance

Trait	Mapping population	No. of progeny	Chr./ LG[a]	Flanking markers	PV[b]	References
Drought avoidance						
Transpiration and water use efficiency						
CO_2 assimilation rate (A)	Tx430 × Tx7078 F_6	70	1	DGA18–DGA20	12	[67]
			5	DGA216–DGA219	15	
			9	DGA40–DGA41	11	
Transpiration rate (E)	Tx430 × Tx7078 F_6	70	1	DGA29–DGA30	12	[67]
			1	DGA20	15	
			7	DGA291–DGA293	13	
A:E ratio	Tx430 × Tx7078 F_6	70	6	DGA6–DGA47	12	[67]
			9	DGA326–DGA327	19	
			9	DGA329–DGA330	21	
			10	DGA395–DGA398	17	
Stomatal conductance	Early Hegari-Sart × BK7 $F_{3:4,\ 4:5}$	28,107	7	58,192–58,594 kb	4	[69]
			10	0–781 kb	1	
Stomatal density	Tx430 × Tx7078 F_6	70	2	DGA89	17	
			7	DGA293–DGA294	15	
The cuticle						
Cutin/epicuticular wax	*blmc* × BTx623 F_2	220	10	Xsbarslbk10.47, Xcup42	–[c]	[89]
Epicuticular wax	BTx623 × RS647 F_2/F_7	605/116	1	Sblm13/Sblm16	–	[81]
Trichomes						
Trichome density-upper	296B × IS18551 F_7	168	10	Xnhsbm1044-Xnhsbm1013	16	[102]
			10	Xnhsbm1043-Xgap1	33	
Trichome density-lower	296B × IS18551 F_7	168	1	Xtxp32-Xtxp88	9	[102]
			1	Dsenhsbm80-xtxp302	5	
			4	Xcup48-Ungnhsbm32	9	
			6	Xtxp317-Xtxp274	5	
			10	Xnhsbm1048-Xnhsbm1013	15	
			10	Xnhsbm1043-Xgap1	23	

(continued)

Table 1 (continued)

Trait	Mapping population	No. of progeny	Chr./ LG[a]	Flanking markers	PV[b]	References
Root systems						
Crown root angle, mature	Early Hegari-Sart × BK7 $F_{3:4,\ 4:5}$	28,107	3	4631–5166 kb	3	[69]
Nodal root angle, seedling	BR23296 × SC170-6-8 F_6	141	5	SPb-5892–SPb-4323	10	[108]
			5	SPb-6287–SPb-9490	30	
			8	SPb-4767–SPb-4432	7	
			10	SPb-1660–SPb-7058	12	
Nodes with brace roots	Sansui × Jiliang F_2	611	6	Xtxp127–Xtxp6	7	[111]
			7	Dsenhsbm7–Xcup70	53	
Root dry weight, seedling	BR23296 × SC170-6-8 F_6	141	2	SPb-4366–SPb-5544	13	[108]
			5	SPb-4086–SPb-6323	12	
			8	SPb-7889–SPb-6935	7	
Root length, mature	E36-1 × SPV570 F_8	184	4	Xsnp56–Xiabt194	8	[110]
Roots/plant, mature	E36-1 × SPV570 F_8	184	1	Xiabt210–Xiabt69	18	[110]
Root:shoot ratio, mature	E36-1 × SPV570 F_8	184	10	Xiabt489–Xiabt364	8	[110]
			10	Xiabt312–Xiabt178	8	
Root volume, mature	E36-1 × SPV570 F_8	184	1	Xiabt210–Xiabt69	14	[110]
			4	Xtxp51–Xtxp270	13	
Root weight, mature	E36-1 × SPV570 F_8	184	4	Xtxp51–Xtxp270	9	[110]
Drought tolerance						
Pre-flowering						
Pre-flowering DT[d] (height stability)	Tx7078 × B35 F_5	98	A	bB13/35	43[c]	[125]
Pre-flowering DT (Prf F)	SC56 × Tx7000 F_7	125	F	psb553-psb708	22.2	[127]

(continued)

Table 1 (continued)

Trait	Mapping population	No. of progeny	Chr./ LG[a]	Flanking markers	PV[b]	References
Pre-flowering DT (Prf E)			E	txs1075a-psb758	11.9	
Pre-flowering DT (Prf G)			G	txs645-hhu504	15	
Pre-flowering DT (seed stability)			E	b258/94	14.2[e]	
Pre-flowering DT (yield per se)			H	b374/75-t380/67	42.1[e]	
			D	UMC85	42.1[e]	
Pre-flowering DT (Yield stability)			F	tC13/150	38.7[e]	
Post-flowering (stay-green)						
Stay-green (*stg1*)	BTx642 × Tx7000 F_7	97	3	NPI1414-BNL15.20	17.1	[41, 147]
Stay-green (*stg2*)			3	Xtxs584-A12-420	24.5	
Stay-green (*stg3*)			2	TXS1307–UMC5	10.7	
Stay-green (*stg4*)			5	TXS713-rcb	11.2	
Stay-green (*stg5*)			1	CPY79A1-UGT85B1	10.8	[149]
% dry leaves after drought	Association panel	107	1	Xtxp149	19–20	[151]

[a]LG-linkage group
[b]PV-phenotypic variation
[c]Single nuclear recessive gene
[d]DT-drought tolerance
[e]PV was combined

treatment. Stomatal density was measured in Experiment 2, and a QTL for this trait was found on SBI-02. The correlation coefficient (R^2) values for these QTLs ranged between 11 and 21%. This same population was also planted in a field experiment, and stomatal density was again measured. This time a QTL was found on SBI-07, with an R^2 of 15%.

Two QTLs for stomatal conductance on SBI-07 and SBI-10 in an Early Hegari-Sart × BK7 population were identified [69] (Table 1). During the growing season, stomatal conductance (g_s) was measured using a Li-Cor LI-6400XT. Further investigation of the population contrasting for the QTL on SBI-07 showed that the difference in stomatal conductance was not due to stomata number or pore size. There was also no negative trade-off between reduced stomatal conductance and CO_2 assimilation rate (A). The authors hypothesized that this QTL may function in regulation of stomatal closure.

Aquaporins are membrane-bound proteins which allow water, and sometimes other small molecules, to pass through cell and organelle membranes [70]. In drought-stressed rice (*Oryza sativa* L.) seedlings, it was found that aquaporins are the main route through which water enters the plant, as opposed to the apoplastic route [71], and thus their role in conferring drought tolerance in sorghum merits investigation. The roles of four aquaporin genes in field-grown maize and sorghum under drought versus well-watered conditions were studied [72]. As expected, the drought treatment reduced shoot growth in both species, but had a much greater effect on maize. Relative water content of maize was also reduced by the drought treatment, but it was unaffected in sorghum. The aquaporin gene *PIP2;3* was more highly expressed in root tissue under drought stress than under control (well-watered) conditions in the predawn and early morning hours in sorghum. A similar pattern was observed for aquaporin gene *TIP1;2* in sorghum roots, except that expression in the control increased in the late afternoon. These results suggest that these genes play an important role in water uptake by sorghum roots under drought, though more studies are needed. Drought also appeared to induce expression of aquaporin gene *PIP1;5* in drought-stressed sorghum leaves; expression was also detected in roots but the pattern was inconsistent. In hydroponically grown sorghum seedlings, addition of PEG to the culture medium induced osmotic stress, and caused upregulation of several aquaporin genes in the roots [73]. Treatment with silicon (Na_2SiO_3) also caused upregulation of several aquaporins, suggesting a possible role of this element in signaling of osmotic stress.

6.1.2 The Cuticle

The plant cuticle covers the above ground organs of all land plants providing a hydrophobic surface layer or "protective skin" [74, 75]. The cuticle is composed of a framework of cutin and integrated and superimposed lipids; the integrated waxes in the cuticle are called intracuticular waxes while the superimposed lipids are called epicuticular waxes [74]. The cuticle serves primarily as a barrier to transpirational water loss, allowing gas exchange and transpiration to be controlled by the stomata (a mechanism of drought avoidance) but also has a role in signaling, plant defense, and development [74, 75]. Cuticles are not impermeable and depending on the plant species, have permeances for water [76], chemicals [77], and gases [78, 79].

Epicuticular waxes (EWs) coat the aerial surface of sorghum plants and appear as a white soft material which is prominent at the boot leaf stage on the leaf sheath and the lower part of the leaf blade

[80, 81]. EWs reduce cuticular conductance and reflect incoming radiation at the UV and 400–700 nm wavelengths to reduce leaf temperature and transpiration [65]. Epi- and intra-cuticular waxes from sorghum sheaths are made up of primarily fatty acids (96%), of which the C28 and C30 acids composed 77% and 20% of these fatty acids, respectively [82]. The very long fatty acid acyl chains that are wax precursors are activated by coenzyme A (CoA) and synthesized by elongases; these elongases use malonyl-CoA as the two-carbon donor [83]. Two epidermal cell types produce epi-cuticular wax crystals on sorghum leaf sheaths; epidermal cork cells and epidermal long cells produce tubular filament wax crystals and plate-like waxes, respectively [84]. In sorghum, epicuticular wax mutants are designated as bloomless (lacking visible epicuticular wax) or sparse bloom (reduced visible wax) [85]. Jordan et al. [86] estimated that in sorghum EW load of about 0.067 g m^{-2} could be effective in preventing water loss through the cuticle. To determine the precise contribution of single EW loci, diethyl sulfate (DES) and ethyl methanesulfonate (EMS) were used on two drought-resistant sorghum inbreds (P954035 and P898012) to generate 33 M_2 generation mutants with altered EW [85]. Scanning electron microscopy was used to group the lines into 14 classes based on altered EW structure [85]. Furthermore, 12 of these bloomless (*bm*) and sparse bloom (*h*) mutants were examined for the amount and composition of cuticular waxes on the sheaths of the mutants and were found to have significant reductions in the amount of C28 and C30 fatty acids as compared to wild type plants [82]. Using this collection of mutants and some mutants previously described [87], an allelism study was conducted, and 10 *bm* and 10 *h* loci were identified [80]. Sorghum lines with mutations in the *Bm2* locus lack EW and have thinner and lighter cuticles [84]. Furthermore, these *bm2* mutants have rapid leaf water loss rates, leakier epidermal layers, and higher susceptibility to a fungal pathogen [84, 88]. Genetic mapping of the *bm2* locus using the mutated line *blmc* [89] (Table 1), which has been renamed as *bm2-7* [90], has been found to be on sorghum chromosome 10. Identification of the *Bm2* gene will help elucidate the cuticle/EW pathways in sorghum and its role in improved abiotic and biotic stress tolerance.

Recently, a sorghum EW locus derived from a spontaneous mutant RS647 was identified [81]. Genetic mapping using BTx623 × RS647 F_2 and F_7 populations identified a locus *bm39* to a region on SBI-01 (Table 1). A gene glycine-aspartic (glutamic) serine leucine (GDSL)-like lipase/acylhydrolase was suggested as the candidate gene and mRNA levels were lower in the transcript of bloomless near isogenic lines (NILs) as compared to those having bloom [81]. Furthermore, a subgroup of GDSL lipases which has a motif consensus amino acid sequence of Ser, Gly, Asn, and His

(SGNH) hydrolase is associated with hydrolase ability in the leaf epidermis in *Agave americana* L. [91].

Additionally, in an unrelated study, a bloomless sorghum mutant, 21-400-2-11-3, was generated by gamma irradiation [92]. RNA sequencing was used to identify the genes that were differentially expressed between the wild type and mutant plants in the leaf sheaths, and the most differentially expressed gene, *Sb06g023280*, was similar to an ABC transporter responsible for wax secretion in *Arabidopsis*. Additionally, a 1.4 Mb genomic inversion was identified proximal to the *Sb06g023280* promoter region in the bloomless mutant [92]. Thus, the mutation may have prevented the secretion of wax and this study identifies a gene likely involved in wax secretion in sorghum.

6.1.3 Trichomes

Plants with increased drought resistance often have more epidermal trichomes [10]. Trichomes are specialized epidermal cells that include hairs and scales which protrude above the epidermis of which the hairs may be glandular and able to produce a secretion [93, 94]. Trichomes function in plants to control water and to prevent insect and herbivore damage [95–97]. In many plant species, dense hairs reflect sunlight and reduce the heat load over the leaf surface protecting the tissue in hot and dry environments [98]. Thus, less light absorption reduces leaf temperature and thus reduces the leaf transpiration rate [95]. Additionally, dense trichome canopies can hold droplets of water above the leaf surface to prevent the interference with stomatal pores [99]. Trichome density can be increased in plants after mechanical damage or water shortage [95, 100]. Having trichomes may come at a cost to plants. At low wind speeds, trichomes may increase the turbulence in the boundary layer and increase photosynthetic gas exchange, leading to increased transpiration [101].

Trichomes also serve as a mechanical barrier to insect movement and feeding [102]. Chemical secretions of terpenes, phenolics, alkaloids, or other substances from glandular trichomes can deter or be toxic to herbivores or insects [96, 97]. Hooked trichomes may impale adults or larvae of insects and stinging trichomes can deter large animals [96].

For most studies using sorghum, trichomes have been examined not for their control of water but rather in relation to their effects on pest feeding. Satish et al. [102] performed a QTL study examining trichomes and sorghum shoot fly resistance (*Atherigona soccata* (Rondani) using a 296B (susceptible) × IS18551 (resistant) RIL population. The authors noted that in sorghum the shoot fly-resistant lines often had pointed unicellular trichomes as compared to blunt bicellular trichomes. They mapped this morphological trait as the resistant parent, IS18551, had unicellular pointed trichomes and the susceptible parent had bicellular blunt trichomes. The gene segregating for trichome morphology mapped as a morphological marker on SBI-10, of which QTLs for shoot fly egg number, shoot fly damage, and trichome density on the upper and lower leaf co-located with trichome morphology. Additional QTLs for trichome density on the upper and lower surfaces of the leaf in the same mapping population were identified on SBI-01, SBI-04, SBI-06, and SBI-10 [102] (Table 1). Two candidate genes for trichome morphology were a fibroin heavy chain and a tubulin beta-2/beta-3 chain which have been found in other species to be specifically expressed in leaf trichomes of tomato (*Solanum lycopersicum* L.) and *Arabidopsis* [102].

6.1.4 Root Systems

Many efforts to phenotype drought tolerance traits in plants have focused on the above-ground portion of the plant, because it is much easier to observe than the roots. However, as the major point of entry for water into the plant, the importance of roots should not be overlooked. Because roots are so difficult to phenotype in the field, root-related traits are ideal candidates for improvement through marker-assisted selection. To date, relatively few studies have identified QTLs associated with root traits in sorghum, but methods to more accurately and quickly phenotype roots are being developed, which should lead to better selection for root-related traits.

In sorghum, root system development begins as the radicle emerges from the seed and elongates to become the primary root [103]. Around the fourth-leaf stage, nodal roots begin to emerge from the first node of the coleoptile, and eventually these will form the bulk of the root system [103]. These nodal roots are called crown roots if they form below ground and are called brace roots if they arise from above-ground nodes [104]. The angle at which the nodal roots grow has been proposed as a trait on which to select for drought avoidance [69, 105]. In theory, a narrower (more vertical) root angle should indicate deeper root growth, and thus better ability to access deeper soil layers where more water is available [69]. In sorghum, the plant needs to grow to about the sixth-leaf stage before the root system is developed enough to observe differences in nodal root angle [103]. This contrasts with other grasses like maize which develop nodal roots much sooner [103]. Hydroponic [106] or germination paper-based methods [107] have been developed for observation of maize roots, but they may not be ideal for sorghum, because these media cannot support the plant long enough [103]. To observe sorghum root angle in-situ, a soil-based system is preferable. Custom root growth chambers were constructed, essentially miniature rhizotrons [108]. They consist of two sheets of clear acrylic 45 × 50 cm, separated on three sides by 3 mm-thick rubber strips, and held together with clamps. The space between the sheets is filled with soil. Seeds can be planted directly into the chambers, or pre-germinated and then transplanted. The chambers are placed vertically into metal tubs in a greenhouse. A black polycarbonate sheet with holes cut into it to allow seedlings to emerge is placed over the tub to keep light out. Initially a flatbed scanner was used to take images of the chambers, but for higher throughput of imaging, a box was constructed which holds a root chamber vertically between two digital cameras, which take photos of both sides of the chamber simultaneously [109]. Root angle can then be determined from the photos using software. Using these custom root chambers, four QTLs for nodal root angle in sorghum were identified from a RIL population of 141 lines, derived from BR23296 × SC170-6-8 [108] (Table 1). The percentage of variance explained by the QTLs (R^2) ranged between 6.72 and 29.78%, and the heritability for the trait was high (73.7%). Heritability for root angle was much greater than for the other traits measured such as root dry weight, shoot dry weight, and total leaf area (13.0, 21.6, and 20.4%, respectively). Some QTL were also identified for these traits, despite the relatively low heritabilities.

Using larger rhizotrons (240 cm wide × 10 cm thick × 120 cm deep) with a clear panel on the front, the nodal root angle of sorghum seedlings was shown to indeed correlate to root architecture in mature plants [105]. The narrow-angle genotype BR23296 had more roots visible at 100 cm depth directly below the plant, whereas the wide-angle genotype SC170-6-8 had more roots visible at 80 cm depth at a distance of 120 cm away from the plant. Soil cores were taken from the rhizotrons at harvest, and slight (though not statistically significant) differences were noted in the distribution of water extraction between the narrow and wide-angle genotypes.

An above-ground rhizotron constructed using cement blocks (25 m × 4 m × 1.5 m high) was filled with soil [110]. Into this were planted 184 RI lines, and the two parental lines (E36-1 and SPV570), which had contrasting phenotypes for root system size. After 110 days, one wall of the rhizotron was removed, and each plant was carefully removed from the soil with the root system intact. Roots were washed, and then various measurements were taken, including root length, number of roots per plant, root volume, fresh and dry weight, and root to shoot ratio. A genetic map containing 104 markers was constructed, and QTLs were identified on chromosomes SBI-01, SBI-04, and SBI-10 (Table 1) using composite interval mapping.

It is virtually impossible to excavate the entire root system of every plant in a field experiment, particularly when the plants are mature. However, the angle of the crown roots has been studied as a proxy for rooting depth. A population from a cross between drought-tolerant Early Hegari-Sart and drought-sensitive BK7 was developed [69]. The drought-tolerant parent has a steeper crown root angle and exhibits less wilting during dry periods. The mapping population was planted at two locations in Florida, and the roots were dug up at maturity using a technique similar to the "shovelomics" methods developed for maize [104]. After washing away the soil, the crown root angles were measured with a digital protractor. A high-density linkage map was constructed using data from over 6000 GBS-based SNP markers [63]. One QTL was identified for crown root angle on SBI-03 (Table 1). The sorghum genome sequence [56] combined with a very dense genetic map allows for the identification of candidate genes underlying QTLs. The QTL region on SBI-03 for crown root angle contains 85 genes. Among them is a putative tryptophan aminotransferase, which may function in auxin metabolism, and is homologous to a rice gene involved in root gravitropism [69].

The primary function of brace roots is to help stabilize the plant [111]. The relationship between brace roots and below-ground nodal roots has not been studied in sorghum, but in maize, a moderate correlation ($r = 0.53$) between brace root angle and crown root angle was found [104]. The underlying genetics of brace roots have been investigated in sorghum [111]. Sansui, a tall Chinese landrace with numerous brace roots on the lower 6–8 nodes, was crossed with Jiliang 2, an elite variety with brace roots only on the lowest node. The F_1 hybrid had a similar number of nodes with brace roots as Sansui. Two hundred nineteen F_2 plants were analyzed in the field for number of nodes with brace roots, along with other agronomic traits. A genetic map was constructed using 109 polymorphic SSR markers. Using inclusive composite interval mapping, two QTLs were identified on SBI-07 and SBI-06 which explained, respectively, 52.5 and 7.6% of the phenotypic variation in the number of nodes with brace roots (Table 1). These QTLs do not overlap with other root architecture QTLs identified in sorghum, but they were also identified in completely different genetic backgrounds.

Silicon aggregates have been found in the endodermis of sorghum roots, and it has been suggested that this element contributes to the crop's ability to withstand water stress [73, 112, 113]. Greater silicon content in roots of a drought-tolerant cultivar (Gadambalia, 3.5% Si) was observed as compared to a drought-sensitive cultivar (Tabat, 2.2% Si) [114]; however, it is not clear if their difference in drought tolerance is due to the difference in root silicon content. No other studies on genetic variation for silicon content in sorghum roots have been reported, so it is not certain if this trait could be a target for breeding. Based on homology with maize, three silicon transporter genes (*SbLsi1, SbLsi2,* and *SbLsi6*) have been identified in sorghum. Two of these genes (*SbLsi1* and *SbLsi2*) were shown to be highly expressed in the growing root tip in the presence or absence of silicon. *SbLsi6* showed a lower level of expression, but was slightly upregulated in the absence of silicon [113].

6.2 Drought Tolerance Traits

6.2.1 Compatible Solutes

One means by which a plant cell can cope with osmotic stress brought about by drought or high salinity is to increase the concentration of osmotically active solutes within the cell, thus lowering its osmotic potential [115]. This causes water to flow in, thus maintaining cell turgor and normal functions [116]. Compatible solutes are osmotically active small molecules that can accumulate to high concentrations within the cell without interfering with normal cellular functions [115]. They may also have protective or stabilizing effects on enzymes and membranes [116, 117]. Proline [118–120] and glycinebetaine [122] have been identified in sorghum as compatible solutes that may confer some level of tolerance to drought. Variation in the ability to accumulate glycinebetaine has been identified in both maize and sorghum [121, 122]. In sorghum, IS2319 has been identified as glycinebetaine-deficient, while P932296 is a glycinebetaine accumulator. A RIL population from a cross between these two lines was developed [117]. The concentration of leaf glycinebetaine showed a segregation pattern which suggested one major gene determines whether a line produces glycinebetaine, as well as one or more genes of smaller effect which control the amount of glycinebetaine in lines that do accumulate this solute. In the $F_{7:8}$ generation, 10 individual plants from 28 RILs were screened for differences in glycinebetaine accumulation, to identify those that were still segregating for the trait. From two segregating families, two pairs of NILs were derived for further study. Biochemical analysis suggested that the deficiency in glycinebetaine was caused by the inability to convert choline to betaine aldehyde, the immediate precursor to glycinebetaine. This step is catalyzed by choline monooxygenase (CMO); however, no difference in the expressed CMO sequences or level of gene expression between pairs of accumulating and non-accumulating NILs was found [123]. No QTLs for compatible solute accumulation have yet been reported in sorghum.

6.2.2 Pre-flowering Drought Tolerance

Pre-flowering drought stress in grain sorghum occurs when plants are experiencing significant water stress prior to anthesis, especially from panicle differentiation to anthesis. Plants that are under pre-flowering drought stress will exhibit symptoms of leaf rolling, leaf erectness, reduced height, delayed flowering, panicle blasting, saddle effect, and reduced panicle size [124]. Figure 1 demonstrates pre-flowering drought stress tolerance and susceptibility in Lubbock, TX in June 2017. The evaluation of pre-flowering drought tolerance is also confounded by plant maturity; early maturing genotypes often escape water stress and consistently express a tolerant phenotype. Sources of pre-flowering drought

Fig. 1 Pre-flowering drought tolerance (left) versus pre-flowering drought susceptibility (right). Picture taken in Lubbock, TX in June 2017

tolerance have been identified, and have been used in sorghum breeding programs for many years [124, 125].

Compared to post-flowering drought tolerance, genetic mapping studies identifying QTLs for pre-flowering stress tolerance have been somewhat limited. One of the most important RIL populations used in the genetic study of pre-flowering stress tolerance to date is "Tx7078 × B35." Six QTLs were identified that were specifically associated with pre-flowering drought tolerance [125] (Table 1). The alleles derived from Tx7078 were associated with enhanced drought tolerance at all but one QTL. Tolerance measured as grain yield per se was 42.1% of the total phenotypic variation explained, indicating that pre-flowering drought tolerance plays an important role in overall grain yield in water stress environments. Plant phenology also strongly interacts with drought tolerance and grain yield [126]. In the pre-flowering drought tolerance study, QTLs for maturity and height were often co-localized with QTL for drought tolerance [125].

Kebede et al. [127] utilized an F_7 RIL population derived from the cross SC56 × Tx7000 to identify QTL for both pre-flowering drought tolerance and post-flowering drought tolerance. Three

QTLs for pre-flowering drought were identified (Table 1). *Pfr G* was identified as a major QTL, contributing 15% and 37.7% of the phenotypic variance in two environments [127]. Most of the pre-flowering QTLs were unlinked to post-flowering drought tolerance and other agronomic QTLs; therefore, incorporation and pyramiding of both types of drought tolerance should be possible.

6.2.3 Post-flowering Drought Tolerance (Stay-Green)

Post-flowering drought tolerance, also known as stay-green, is an agronomically important drought tolerance trait that produces greater retention of green leaf area from anthesis to grain-fill maturity [128–130]. Stay-green has been important in sorghum production regions throughout the world that experience consistent water-limited conditions during the grain fill period. For example, Australia and other Mediterranean climates have utilized the stay-green trait extensively because producers routinely rely on early season rains and stored soil moisture to sustain a crop to maturity [131]. The phenotype is associated with greater green leaf area, reduced lodging, increased basal stem sugars, reduced susceptibility to charcoal rot (*Macrophomina phaseolina*), increased nitrogen content, and increased seed size at maturity [129, 132]. Figure 2 demonstrates post-flowering drought tolerance versus susceptibility in a breeding nursery in Corpus Christi, TX in 2015. Prior studies also identified a link between crop nitrogen content, extent of N-mobilization post anthesis, and the onset of leaf senescence associated with differences in stay-green [133–135]. The value of the stay-green trait in terms of overall grain yield is often debated

Fig. 2 Post-flowering drought tolerance (left) versus post-flowering drought susceptibility (right). Picture taken in Corpus Christi, TX in August 2015

and in many cases, is the result of trade-offs during the growing season between increased production in water-limited seasons and decreased overall production in high-yielding seasons. Jordan et al. [131] identified that stay-green on average was positively associated with grain yield in environments that yielded 6 metric tons per hectare or less. In high yielding environments, the positive attributes of the stay-green trait declined and in some environments, were associated with a reduction in grain yield. Nevertheless, the stay-green trait appears to be an important drought tolerance trait that often produces higher, and more stable yields, especially since sorghum is primarily produced on marginal lands where optimal growing conditions are seldom realized.

Many RIL populations have been utilized to map stay-green QTLs in sorghum [127, 136–142]. The four parental genotypes from these studies that confer stay-green are BTx642, QL41, SC56, and E36-1. A list of 39 QTLs (LOD > 2) that were identified from five studies were compiled onto a single genetic map and 12 regions of the sorghum genome that contain stay-green QTLs were identified [143]. Favorable alleles for stay-green from more than one genotype were identified in many of the regions from multiple studies which gives further support of the importance of these QTLs. Yet, to date, no genes have been identified from any of these 12 regions.

One of the most widely utilized stay-green RIL populations for many research groups is BTx642 × Tx7000, which was originally developed by Texas A&M University in the 1990s [141]. BTx642 (also known as B35) is a widely used stay-green line for both stay-green mapping and breeding. BTx642 was derived from a BC_1 selection of BTx406 and an Ethiopian durra line, IS12555, has excellent stay-green ratings, and has been used extensively for drought tolerance breeding [130, 140]. Studies have shown that the genes conferring stay-green act with varied levels of dominance and partial dominance [144], which is an important trait for the production of stay-green hybrids. Four original stay-green QTLs (*Stg1, Stg2, Stg3, Stg4*) were mapped using the BTx642 × Tx7000 population [141] and their relative ranking based on contribution to the stay-green phenotype are *Stg2* > *Stg1* > *Stg3* > *Stg4*. *Stg1* and *Stg2* are located on SBI-03, *Stg3* on SBI-02, and *Stg4* on SBI-05 [145, 146] (Table 1).

NILs of the non-stay-green line Tx7000 carrying each individual *Stg* allele have also been developed and results demonstrate that each *Stg* allele (*Stg1-Stg4*) reduced leaf senescence in the Tx7000 genetic background. Among the four *Stg* alleles, NILs containing alleles of *Stg2* had the highest green leaf area at maturity (GLAM) among all NILs analyzed, further supporting that *Stg2* has the greatest influence on the stay-green phenotype [147]. Borrell et al. [148] also investigated the mechanisms that control the stay-green trait in sorghum using the Tx7000 NILs which contain individual *Stg* QTL and concluded that *Stg* loci reduced the overall canopy size at flowering, reduced tillering, and promoted overall root growth. A reduction in total canopy size at anthesis reduced overall water demands of the plant, and allowed for greater water availability during grain filling. A positive correlation between canopy size and water use at anthesis was observed resulting in less water use in three of the four Tx7000 *Stg* NILs compared to Tx7000. Therefore, increased grain yields associated with *Stg* loci are at least partially explained by increased water use post-anthesis in response to an overall reduced canopy size [148].

Recently, a novel stay-green QTL, designated *Stg5*, has been identified on SBI-01 that co-localizes with known dhurrin biosynthetic genes [149]. Dhurrin is a cyanogenic glucoside present in sorghum that once catabolized by the enzyme dhurrinase, produces hydrogen cyanide (HCN) gas as a defense mechanism against herbivory [150]. Hayes et al. [149] identified *Stg5* on the distal arm of SBI-01 using a new digital genotyping (DG) genetic map of BTx642 × Tx7000 with a total of 897 unique markers spanning all ten chromosomes (Table 1). In addition, *Stg1-Stg4* were also reidentified in the study, confirming that stay-green is a complex trait with many potential mechanisms. Although it still remains unclear how exactly dhurrin is associated with the stay-green phenotype, it is hypothesized that dhurrin could be providing an N source for grain filling under water limiting environments, therefore delaying plant senescence. The increased production of the N rich compound dhurrin could also be altering leaf growth and development by affecting N-status or signaling in developing leaves, thus suggesting that elevated levels of leaf dhurrin could cause a reduction in leaf area, a key trait associated with the function of *Stg1-Stg4*.

Furthermore, in a greenhouse study, drought stress was imposed beginning at the boot stage in a diverse set of 107 sorghum accessions [151]. After 12 days, the percentage of dry leaves was recorded in the drought-stressed plants. A QTL for percentage of dry leaves was identified by association analysis on SBI-01 (Table 1), which may correspond to the previously identified stay-green QTL *StgF* [136], or *StgA* [142]. However, only 98 SSR markers were used in this study, so the power to detect QTLs in this study may have been limited.

7 Future Perspectives

The use of aquifers to irrigate crops began in the second half of the twentieth century, but in Asia, Africa, Australia, and North America the withdrawal of water has exceeded the recharge causing a depletion of water in the aquifers [45]. Four areas of the U.S experience groundwater depletion of which the Ogallala Aquifer, which exists under the central portion of the U.S., is depleted in Kansas and Texas [45]. As water becomes a limiting factor due to depletion of aquifers, a growing world population, water pollution, and climate change, more drought-resistant crops must be grown that can exist without irrigation. Sorghum is well adapted to hot and dry environments, is frequently grown without irrigation [3], and has a large number of genetic resources for continual drought tolerance improvement [152]. Utilization of genes in sorghum hybrids that confer pre-flowering and post flowering drought tolerance, enhance root systems, promote thick cuticles and trichome development, and enhance water use efficiency may help meet the need of growing crops in warmer and water restricted environments. Yet the effect of these genes must be studied for their impact on sorghum yield and/or growth under normal and drought stress conditions as selection for drought resistance traits can come at a cost to yield. Using this information, sorghum breeders must combine the most effective QTLs or genes controlling drought response traits into hybrids without a negative effect on yield under non-drought conditions.

Acknowledgments

The authors would like to thank Dr. Somashekhar Punnuri (Fort Valley State University, Fort Valley, GA) and Dr. Corley Holbook (USDA-ARS, Tifton, GA) for their critical comments in the development of this book chapter.

References

1. Boyer JS (1982) Plant productivity and environment. Science 218:444–448
2. Chaves MM, Maroco JP, Pereira JS (2003) Understanding plant responses to drought-from genes to the whole plant. Funct Plant Biol 30:239–264
3. Sanchez AC, Subudhi PK, Rosenow DT, Nguyen HT (2002) Mapping QTLs associated with drought resistance in sorghum (*Sorghum bicolor* L. Moench). Plant Mol Biol 48:713–726
4. Kramer PJ, Boyer JS (1995) Water relations of plants and soils. Academic Press, San Diego, CA
5. Hu H, Xiong L (2014) Genetic engineering and breeding of drought-resistant crops. Annu Rev Plant Biol 65:715–741
6. Blum A (1996) Crop responses to drought and the interpretation of adaptation. Plant Growth Regul 20:135–148
7. Saini HS, Westgate ME (1999) Reproductive development in grain crops during drought. Adv Agron 68:59–96
8. Lascano HR, Antonicelli GE, Luna CM, Melchiorre MN, Gómez LD, Racca RW, Trippi VS, Casano LM (2001) Antioxidant system response of different wheat cultivars under drought: field and in vitro studies. Funct Plant Biol 28:1095–1102
9. Levitt J (1980) Responses of plants to environmental stresses, Physiological ecology series. Academic, Michigan
10. Fang Y, Xiong L (2015) General mechanisms of drought response and their application in drought resistance improvement in plants. Cell Mol Life Sci 72:673–689
11. Mooney HA, Ehleringer J, Berry JA (1976) High photosynthetic capacity of a winter annual in Death Valley. Science 194:322–324

12. Cattivelli L, Rizza F, Badeck F-W, Mazzucotelli E, Mastrangelo AM, Francia E, Marè C, Tondelli A, Michele Stanca A (2008) Drought tolerance improvement in crop plants: an integrated view from breeding to genomics. Field Crops Res 105:1–14
13. Ebercon A, Blum A, Jordan WR (1977) A rapid colorimetric method for epicuticular wax contest of sorghum leaves. Crop Sci 17:179–180
14. Grammatikopoulos G, Manetas Y (1994) Direct absorption of water by hairy leaves of *Phlomis fruticosa* and its contribution to drought avoidance. Can J Bot 11:1805–1811
15. Munné-Bosch S, Alegre L (2004) Die and let live: leaf senescence contributes to plant survival under drought stress. Funct Plant Biol 31:203–216
16. Rohde A, Bhalerao RP (2007) Plant dormancy in the perennial context. Trends Plant Sci 12:217–223
17. O'Toole JC, Cruz RT, Singh TN (1979) Leaf rolling and transpiration. Plant Sci Lett 16:111–114
18. Pearcy RW, Ehleringer J (1984) Comparative ecophysiology of C_3 and C_4 plants. Plant Cell Envir 7(1):11
19. Malamy JE (2005) Intrinsic and environmental response pathways that regulate root system architecture. Plant Cell Environ 28:67–77
20. Sawidis T, Sofia K, Delivopoulos S (2005) The root-tuber anatomy of *Asphodelus aestivus*. Flora 200:332–338
21. Basu S, Ramegowda V, Kumar A, Pereira A (2016) Plant adaptation to drought stress. F1000Research 5:1554
22. Gill SS, Tuteja N (2010) Reactive oxygen species and antioxidant machinery in abiotic stress tolerance in crop plants. Plant Physiol Biochem 48:909–930
23. Singh M, Kumar J, Singh S, Singh VP, Prasad SM (2015) Roles of osmoprotectants in improving salinity and drought tolerance in plants: a review. Rev Environ Sci Biotechnol 14:407–426
24. Mundree SG, Baker B, Mowla S, Peters S, Marais S, Vander Willigen C, Govender K, Maredza A, Farrant JM, Thomson JA (2002) Physiological and molecular insights into drought tolerance. Afr J Biotech 1:28–38
25. Price AH, Taylor A, Ripley SJ, Griffiths A, Trewavas AJ, Knight MR (1994) Oxidative signals in tobacco increase cytosolic calcium. Plant Cell 6:1301–1310
26. Fujita M, Fujita Y, Noutoshi Y, Takahashi F, Narusaka Y, Yamaguchi-Shinozaki K et al (2006) Crosstalk between abiotic and biotic stress responses: a current view from the points of convergence in the stress signaling networks. Curr Opin Plant Biol 9:436–442
27. Atkinson NJ, Urwin PE (2012) The interaction of plant biotic and abiotic stresses: from genes to the field. J Exp Bot 63:3523–3544
28. Asselbergh B, Achuo AE, Hofte M, Van Gijsegem F (2008) Abscisic acid deficiency leads to rapid activation of tomato defence responses upon infection with *Erwinia chrysanthemi*. Mol Plant Pathol 9:11–24
29. Ton J, Flors V, Mauch-Mani B (2009) The multifaceted role of ABA in disease resistance. Trends Plant Sci 14:310–317
30. Camejo D, Guzman-Cedeno A, Moreno A (2016) Reactive oxygen species, essential molecules, during plant-pathogen interactions. Plant Physiol Biochem 103:10–23
31. Mittler R, Vanderauwera S, Gollery M, Van Breusegem F (2004) Reactive oxygen gene network of plants. Trends Plant Sci 9:490–498
32. Rodriguez MCS, Petersen M, Mundy J (2010) Mitogen-activated protein kinase signaling in plants. Annu Rev Plant Biol 61:621–649
33. Chinchilla D, Zipfel C, Robatzek S, Kemmerling B, Nurnberger T, Jones JDG, Felix G, Boller T (2007) A flagellin-induced complex of the receptor *FLS2* and *BAK1* initiates plant defence. Nature 448:497–500
34. Andreasson E, Ellis B (2010) Convergence and specificity in the *Arabidopsis* MAPK nexus. Trends Plant Sci 15:106–113
35. Miller G, Mittler R (2006) Could heat shock transcription factors function as hydrogen peroxide sensors in plants? Ann Bot 98:279–288
36. Jedmowski C, Ashoub A, Beckhaus T, Berberich T, Karas M, Brüggemann W (2014) Comparative analysis of *Sorghum bicolor* proteome in response to drought stress and following recovery. Int J Proteomics 2014:Article ID 395905. https://doi.org/10.1155/2014/395905
37. Kimber C (2000) Origins of domesticated sorghum and its early diffusion to India and China. In: Smith CW, Frederiksen RA (eds) Sorghum: origin, history, technology, and production. John Wiley, New York, pp 3–98
38. Rooney WL, Blumenthal J, Bean B, Mullet JE (2007) Designing sorghum as a dedicated bioenergy feedstock. Biofpr 1:147–157
39. Rao PS, Vinutha KS, Kumar GSA, Chiranjeevi T, Uma A, Lal P et al (2016) Sorghum: a multipurpose bioenergy crop.

In: Ciampitti I, Prasad V (eds) Sorghum: state of the art and future perspectives, agronomy monographs, vol 58. American Society of Agronomy and Crop Science Society of America, Madison, WI

40. Williams RJ, Rao KN (1981) A review of sorghum grain moulds. Trop Pest Manage 27:200–211
41. Hilley JL, Weers BD, Truong SK, McCormick RF, Mattison AJ, McKinley BA, Morishige DT, Mullet JE (2017) Sorghum *Dw2* encodes a protein kinase regulator of stem internode length. Sci Rep 7:4616
42. Blum A, Arkin GF (1984) Sorghum root growth and water-use as affected by water supply and growth duration. Field Crops Res 9:131–142
43. Stout DG, Simpson GM (1978) Drought resistance of *Sorghum bicolor*. 1. Drought avoidance mechanisms related to leaf water status. Can J Plant Sci 58:213–224
44. Tari I, Laskay G, Takács Z, Poór P (2013) Response of sorghum to abiotic stress: a review. J Agro Crop Sci 199:264–274
45. Brauer D, Baumhardt RL (2016) Future prospects for sorghum as a water-saving crop. In: Ciampitti I, Prasad V (eds) Sorghum: state of the art and future perspectives, Agron Monogr 58. ASA and CSSA, Madison, WI
46. International Crops Research Institute for the Semi-Arid Tropics (ICRISAT)/Food and Agriculture Organization (FAO) (1996) The world sorghum and millet economies. ICRISAT/FAO, Patancheru/Rome
47. O'Brien D (2017) Domestic and international sorghum marketing. In: Ciampitti I, Prasad V (eds) Sorghum: state of the art and future perspectives. Agron Monogr 58. ASA and CSSA, Madison, WI
48. Thompson LM (1963) Evaluation of weather factors in the production of grain sorghums. Agron J 55:182–185
49. Gale F, Hansen J, Jewison M (2014) China's growing demand for agricultural imports, EIB-136, U.S. Department of Agriculture, Economic Research Service, February 2014
50. Jewison M, Gale F (2012) China's market for distillers dried grains and the key influences on its longer run potential, FDS-12g-01, U.S. Department of Agriculture, Economic Research Service
51. Pratt LH, Liang C, Shah M, Sun F, Wang H, Reid SP, Gingle AR, Paterson AH, Wing R, Dean R, Klein R, Nguyen HT, Ma H-M, Zhao X, Morishige DT, Mullet JE, Cordonnier-Pratt MM (2005) Sorghum expressed sequence tags identify signature genes for drought, pathogenesis, and skotomorphogenesis from a milestone set of 16,801 unique transcripts. Plant Physiol 139:869–884
52. Soares MB, Bonaldo MF (1998) Constructing and screening normalized cDNA libraries. In: Birren B, Green ED, Klapholz S, Myers RM, Roskams J (eds) Genome analysis: a laboratory manual, vol 2. Cold Spring Harbor Laboratory Press, New York
53. Buchanan CD, Lim S, Salzman RA, Kagiampakis I, Morishige DT, Weers BD, Klein RR, Pratt LH, Cordonnier-Pratt MM, Klein PE, Mullet JE (2005) *Sorghum bicolor*'s transcriptome response to dehydration, high salinity and ABA. Plant Mol Biol 58:699–720
54. Temnykh S, DeClerck G, Lukashova A, Lipovich L, Cartinhour S, McCouch S (2001) Computational and experimental analysis of microsatellites in rice (*Oryza sativa* L.): frequency, length variation, transposon associations, and genetic marker potential. Genome Res 11:1441–1452
55. Srinivas G, Satish K, Madhusudhana R, Seetharama N (2009) Exploration and mapping of microsatellite markers from subtracted drought stress ESTs in *Sorghum bicolor* (L.) Moench. Theor Appl Genet 118:703–717
56. Paterson AH, Bowers JE, Bruggmann R, Dubchak I, Grimwood J, Gundlach H, Haberer G, Hellsten U, Mitros T, Poliakov A, Schmutz J, Spannagl M, Tang H, Wang X, Wicker T, Bharti AK, Chapman J, Feltus FA, Gowik U, Grigoriev IV, Lyons E, Maher CA, Martis M, Narechania A, Otillar RP, Penning BW, Salamov AA, Wang Y, Zhang L, Carpita NC, Freeling M, Gingle AR, Hash CT, Keller B, Klein P, Kresovich S, McCann MC, Ming R, Peterson DG, Mehboob-ur-Rahman WD, Westhoff P, Mayer KFX, Messing J, Rokhsar DS (2009) The *Sorghum bicolor* genome and the diversification of grasses. Nature 457:551–556
57. Dugas DV, Monaco MK, Olsen A, Klein RR, Kumari S, Ware D, Klein PE (2011) Functional annotation of the transcriptome of *Sorghum bicolor* in response to osmotic stress and abscisic acid. BMC Genomics 12:514–510
58. Johnson SM, Lim F-L, Finkler A, Fromm H, Slabas AR, Knight MR (2014) Transcriptomic analysis of *Sorghum bicolor* responding to combined heat and drought stress. BMC Genomics 15:456
59. Groppa MD, Tomaro ML, Benavides MP (2001) Polyamines as protectors against

cadmium or copper-induced oxidative damage in sunflower leaf discs. Plant Sci 161:481–488
60. Kaspar S, Peukert M, Svatos A, Matros A, Mock H-P (2011) MALDI-imaging mass spectrometry–an emerging technique in plant biology. Proteomics 11:1840–1850
61. Collard BCY, Jahufer MZZ, Brouwer JB et al (2005) An introduction to markers, quantitative trait loci (QTL) mapping and marker-assisted selection for crop improvement: the basic concepts. Euphytica 142:169
62. Myles S, Peiffer J, Brown PJ, Ersoz ES, Zhang Z, Costich DE, Buckler ES (2009) Association mapping: critical considerations shift from genotyping to experimental design. Plant Cell 21:2194–2202
63. Elshire RJ, Glaubitz JC, Sun Q, Poland JA, Kawamoto K, Buckler ES et al (2011) A robust, simple genotyping-by-sequencing (GBS) approach for high diversity species. PLoS One 6:e19379
64. Jiao Y, Burke JJ, Chopra R, Burow G, Chen J, Wang B, Hayes C, Emendack Y, Ware D, Xin Z (2016) A sorghum mutant resource as an efficient platform for gene discovery in grasses. Plant Cell 28:1551–1562
65. Blum A (2005) Drought resistance, water-use efficiency, and yield potential-are they compatible, dissonant, or mutually exclusive? Aust J Agric Res 56:1159–1168
66. Balota M, Payne WA, Rooney W, Rosenow D (2008) Gas exchange and transpiration ratio in sorghum. Crop Sci 48:2361–2371
67. Kapanigowda MH, Payne WA, Rooney WL, Mullet JE, Balota M (2014) Quantitative trait locus mapping of the transpiration ratio related to preflowering drought tolerance in sorghum (*Sorghum bicolor*). Func Plant Biol 41:1049–1065
68. Morishige DT, Klein PE, Hilley JL, Sahraeian SME, Sharma A, Mullet JE (2013) Digital genotyping of sorghum–a diverse plant species with a large repeat-rich genome. BMC Genomics 14:448
69. Lopez JR, Erickson JE, Munoz P, Saballos A, Felderhoff TJ, Vermerris W (2017) QTLs associated with crown root angle, stomatal conductance, and maturity in sorghum. Plant Genome 10(2). https://doi.org/10.3835/plantgenome2016.04.0038
70. Maurel C, Verdoucq L, Luu D-T, Santoni V (2008) Plant aquaporins: membrane channels with multiple integrated functions. An Rev Plant Biol 59:595–624
71. Lu ZJ, Neumann PM (1999) Water stress inhibits hydraulic conductance and leaf growth in rice seedlings but not the transport of water via mercury-sensitive water channels in the root. Plant Physiol 120:143–151
72. Hasan SA, Rabei SH, Nada RM, Abogadallah G (2016) Water use efficiency in the drought-stressed sorghum and maize in relation to expression of aquaporin genes. Biol Plantarum 61:127–137
73. Liu P, Yin L, Deng X, Wang S, Tanaka K, Zhang S (2014) Aquaporin-mediated increase in root hydraulic conductance is involved in silicon-induced improved root water uptake under osmotic stress in *Sorghum bicolor* L. J Exp Bot 65:4747–4756
74. Koch K, Ensikat HJ (2008) The hydrophobic coatings of plant surfaces: epicuticular wax crystals and their morphologies, crystallinity and molecular self-assembly. Micron 39:759–772
75. Yeats TH, Rose JKC (2013) The formation and function of plant cuticles. Plant Physiol 163:5–20
76. Becker M, Kerstiens G, Schönherr J (1986) Water permeability of plant cuticles: permeance, diffusion and partition coefficients. Trees 1:54
77. Kerier F, Schönherr J (1988) Permeation of lipophilic chemicals across plant cuticles: prediction from partition coefficients and molar volumes. Arch Environ Contam Toxicol 17:7–12
78. Lendzian KJ (1984) Permeability of plant cuticles to gaseous air pollutants. In: Koziot MJ, Whatley FR (eds) Gaseous air pollutants and plant metabolism, vol 36. Butterworths, London, pp 77–81
79. Lendzian KJ, Kerstiens G (1991) Sorption and transport of gases and vapors in plant cuticles. Rev Environ Contam Toxicol 121:65–128
80. Peters PJ, Jenks MA, Rich PJ, Axtell JD, Ejeta G (2009) Mutagenesis, selection, and allelic analysis of epicuticular wax mutants in sorghum. Crop Sci 49:1250–1258
81. Uttam GA, Praveen M, Rao YV, Tonapi VA, Madhusudhana R (2017) Molecular mapping and candidate gene analysis of a new epicuticular wax locus in sorghum (*Sorghum bicolor* L. Moench). Theor Appl Genet 130:2109–2125
82. Jenks MA, Rich PJ, Rhodes D, Ashworth EN, Axtell JD, Ding C-K (2000) Leaf sheath cuticular waxes on bloomless and sparse-bloom mutants of *Sorghum bicolor*. Phytochemistry 54:577–584
83. Agrawal VP, Lessiere R, Stumpf PK (1984) Biosynthesis of very-long-chain fatty acids in

microsomes of epidermal cells of *Allium porrum* L. Arch Biochem Biophys 230:580–589

84. Jenks MA, Joly RJ, Peters PJ, Rich PJ, Axtell JD, Ashworth EN (1994) Chemically-induced cuticle mutation affecting epidermal conductance to water vapor and disease susceptibility in *Sorghum bicolor* (L.). Moench. Plant Physiol 105:1239–1245
85. Jenks MA, Rich PJ, Peters PJ, Axtell JD, Ashworth EN (1992) Epicuticular wax morphology of bloomless (bm) mutants in *Sorghum bicolor*. Int J Plant Sci 153:311–319
86. Jordan WR, Shouse PJ, Blum A, Miller FR, Monk RL (1984) Environmental physiology of sorghum: II. Epicuticular wax load and cuticular transpiration. Crop Sci 24:1168–1173
87. Peterson GC, Suksayretrup K, Weibel DE (1982) Inheritance of some bloomless and sparse-bloom mutants in sorghum. Crop Sci 22:63–67
88. Burow GB, Franks CD, Xin Z (2008) Genetic and physiological analysis of an irradiated bloomless mutant (epicuticular wax mutant) of sorghum. Crop Sci 48:41–48
89. Burow GB, Franks CD, Acosta-Martinez V, Xin ZG (2009) Molecular mapping and characterization of BLMC, a locus for profuse wax (bloom) and enhanced cuticular features of sorghum (*Sorghum bicolor* (L.) Moench.). Theor Appl Genet 118:423–431
90. Punnuri S, Harris-Shultz K, Knoll J, Ni X, Wang H (2017) The genes *Bm2* and *Blmc* that affect epicuticular wax deposition in sorghum are allelic. Crop Sci 57:1552–1556
91. Reina JJ, Guerrero C, Heredia A (2007) Isolation, characterization, and localization of *AgaSGNH* cDNA: a new SGNH-motif plant hydrolase specific to *Agave americana* L. Leaf epidermis. J Exp Bot 58:2717–2731
92. Mizuno H, Kawahigashi H, Ogata J et al (2013) Genomic inversion caused by gamma irradiation contributes to downregulation of a WBC11 homolog in bloomless sorghum. Theor Appl Genet 126:1513–1520
93. Hülskamp M, Miséra S, Jürgens G (1994) Genetic dissection of trichome cell development in Arabidopsis. Cell 76:555–566
94. Inamdar JA, Gangadhara M (1977) Studies on the trichomes of some Euphorbiaceae. Feddes Repert 88:103–111
95. Ehleringer JR (1984) Ecology and ecophysiology of leaf pubescence in North American desert plants. In: Rodriguez E, Healy PL, Mehta I (eds) Biology and chemistry of plant trichomes. Plenum Press, New York, pp 113–132
96. Levin DA (1973) The role of trichomes in plant defense. Q Rev Biol 48:3–15
97. Buta JG, Lusby WR, Neal JW, Waters RM, Pittarelli GW (1993) Sucrose esters from *Nicotianan gossei* active against the greenhouse whitefly *Trialeuroides vaporariorum*. Phytochemistry 32:859–864
98. Ehleringer JR, Björkman O, Mooney HA (1976) Leaf pubescence: effects on absorbance and photosynthesis in a desert shrub. Science 192:376–377
99. Brewer CA, Smith WK, Vogelmann TC (1991) Functional interaction between leaf trichomes, leaf wettability and the optical properties of water droplets. Plant Cell Environ 14:955–962
100. Gonzáles WL, Negritto MA, Suárez LH, Gianoli E (2008) Induction of glandular and non-glandular trichomes by damage in leaves of *Madia sativa* under contrasting water regimes. Acta Oecol 33:128–132
101. Schreuder MDJ, Brewer CA, Heine C (2001) Modelled influences of non-exchanging trichomes on leaf boundary layers and gas exchange. J Theor Biol 210:23–32
102. Satish K, Srinivas G, Madhusudhana R, Padmaja PG, Reddy RN, Mohan SM et al (2009) Identification of quantitative trait loci for resistance to shoot fly in sorghum [*Sorghum bicolor* (L.) Moench]. Theor Appl Genet 119:1425–1439
103. Singh V, van Oosterom EJ, Jordan DR, Messina CD, Cooper M, Hammer GL (2010) Morphological and architectural development of root systems in sorghum and maize. Plant Soil 333:287–299
104. Trachsel S, Kaeppler SM, Brown KM, Lynch JP (2011) Shovelomics: high throughput phenotyping of maize (*Zea mays* L.) root architecture in the field. Plant Soil 341:75–87
105. Singh V, van Oosterom EJ, Jordan DR, Hammer GL (2012) Genetic control of nodal root angle in sorghum and its implications on water extraction. Eur J Agron 42:3–10
106. Clark RT, Famoso AN, Zhao K, Shaff JE, Craft EJ, Bustamante CD, McCouch SR, Aneshansley DJ, Kochian LV (2013) High-throughput two-dimensional root system phenotyping platform facilitates genetic analysis of root growth and development. Plant Cell Environ 36:454–466
107. Le Marié C, Kirchgessner N, Marschall D, Walter A, Hund A (2014) Rhizoslides: paper-based growth system for non-destructive, high throughput phenotyping of root development by means of image analysis. Plant Meth 10:13

108. Mace ES, Singh V, Van Oosterom EJ, Hammer GL, Hunt CH, Jordan DR (2012) QTL for nodal root angle in sorghum (*Sorghum bicolor* L. Moench) co-locate with QTL for traits associated with drought adaptation. Theor Appl Genet 124:97–109
109. Joshi DC, Singh V, Hunt C, Mace E, van Oosterom E, Sulman R, Jordan D, Hammer G (2017) Development of a phenotyping platform for high throughput screening of nodal root angle in sorghum. Plant Meth 13:56
110. Rajkumar FB, Kavil SP, Girma Y, Arun SS, Dadakhalandar D, Gurusiddesh BH, Patil AM, Thudi M, Bhairappanavar SB, Narayana YD, Krishnaraj PU, Khadi BM, Kamatar MY (2013) Molecular mapping of genomic regions harbouring QTLs for root and yield traits in sorghum (*Sorghum bicolor* L. Moench). Physiol Mol Biol Plants 19:409–419
111. Li R, Han Y, Lv P, Du R, Liu G (2014) Molecular mapping of the brace root traits in sorghum (*Sorghum bicolor* L. Moench). Breeding Sci 64:193–198
112. Coskun D, Britto DT, Huynh WQ, Kronzucker HJ (2016) The role of silicon in higher plants under salinity and drought stress. Front Plant Sci 7:1072
113. Soukup M, Martinka M, Bosnic D, Caplovicova M, Elbaum R, Lux A (2017) Formation of silica aggregates in sorghum root endodermis is predetermined by cell wall architecture and development. Ann Bot 120:739–753
114. Lux A, Luxova M, Hattori T, Inanaga S, Sugimoto Y (2002) Silicification in sorghum (*Sorghum bicolor*) cultivars with different drought tolerance. Physiol Plantarum 115:87–92
115. McNeil SD, Nuccio ML, Hanson AD (1999) Betaines and related osmoprotectants. Targets for metabolic engineering of stress resistance. Plant Physiol 120:945–949
116. Nuccio ML, Rhodes D, McNeil SD, Hanson AD (1999) Metabolic engineering of plants for osmotic stress resistance. Curr Opin Plant Biol 2:128–134
117. Mickelbart MV, Peel G, Joly RJ, Rhodes D, Ejeta G, Goldsbrough PB (2003) Development and characterization of near-isogenic lines of sorghum segregating for glycinebetaine accumulation. Physiol Plantarum 118:253–261
118. Castro NS, Ortiz-Cereceres J, Mendoza-Castillo MD, Huerta AJ, Trejo-Lopez C, Zavala-Garcia F, Martinez-Garza A (2003) Accumulation of proline in leaves of sorghum genotypes (*Sorghum bicolor* L. Moench) which differ in their response to drought. Phyton Int J Exp Bot 72:49–57
119. Burke JJ, Payton P, Chen J, Xin Z, Burow G, Hayes C (2015) Metabolic responses of two contrasting sorghums to water-deficit stress. Crop Sci 55:344–353
120. Nxele X, Klein A, Ndimba BK (2017) Drought and salinity stress alters ROS accumulation, water retention, and osmolyte content in sorghum plants. S Afr J Bot 108:261–266
121. Yang W-J, Nadolska-Orczyk A, Wood KV, Hahn DT, Rich PJ, Wood AJ, Saneoka H, Premachandra GS, Bonham CC, Rhodes JC, Joly RJ, Samaras Y, Goldsbrough PB, Rhodes D (1995) Near-isogenic lines of maize differing for glycinebetaine. Plant Physiol 107:621–630
122. Yang W-J, Rich PJ, Axtell JD, Wood KV, Bonham CC, Ejeta G, Mickelbart MV, Rhodes D (2003) Genotypic variation for glycinebetaine in *Sorghum bicolor*. Crop Sci 43:162–169
123. Peel GJ, Mickelbart MV, Rhodes D (2010) Choline metabolism in glycinebetaine accumulating and non-accumulating near-isogenic lines of *Zea mays* and *Sorghum bicolor*. Phytochemistry 71:404–414
124. Rosenow DT, Quisenberry JE, Wendt CW, Clark LE (1983) Drought tolerant sorghum and cotton germplasm. Dev Agric Managed-For Ecol 12:207–222
125. Tuinstra MR, Grote EM, Goldsbrough PB, Ejeta G (1996) Identification of quantitative trait loci associated with pre-flowering drought tolerance in sorghum. Crop Sci 36:1337–1344
126. Ludlow MM, Muchow RC (1990) A critical evaluation of traits for improving crop yields in water-limited environments. Adv Agron 43:107–153
127. Kebede H, Subudhi PK, Rosenow DT, Nguyen HT (2001) Quantitative trait loci influencing drought tolerance in grain sorghum (*Sorghum bicolor* L. Moench). Theor Appl Genet 103:266–276
128. Borrell AK, Hammer GL, Douglas ACL (2000) Does maintaining green leaf area in sorghum improve yield under drought? I. leaf growth and senescence. Crop Sci 40:1026–1037
129. Borrell AK, Hammer GL, Henzell RG (2000) Does maintaining green leaf area in sorghum improve yield under drought? II. Dry matter production and yield. Crop Sci 40:1037–1048

130. Rosenow DT, Ejeta G, Clark LE, Gilbert ML, Henzell RG, Borrell AK, Muchow RC (1996) Breeding for pre- and post-flowering drought stress resistance in sorghum. In: Rosenow D (ed) The international conference on genetic improvement of sorghum and millet, Lubbock, TX. 23–27 Sept 1996. Publication No. 97–5. International Sorghum and Millet, Lincoln, NE. p 400–424
131. Jordan DR, Hunt CH, Cruickshank AW, Borrell AK, Henzell RG (2012) The relationship between the stay-green trait and grain yield in elite sorghum hybrids grown in a range of environments. Crop Sci 52:1153–1161
132. Rosenow DT, Clark LE (1981) Drought tolerance in sorghum. In: Loden HD, Wilkinson D (eds) Proceedings of the Annual Corn and Sorghum Industry Research Conference–American Seed Trade Association, Corn and Sorghum Division, Corn and Sorghum Research Conference, Washington, DC. 9–11 Dec. 1981. American Seed Trade Association, Alexandria, VA. p 18–30
133. Borrell AK, Hammer GL (2000) Nitrogen dynamics and the physiological basis of stay-green in sorghum. Crop Sci 40:129501307
134. Borell AK, Hammer G, Van Oosterom E (2001) Stay-green: a consequence of the balance between supply and demand for nitrogen during grain filling? Ann Appl Biol 138:91–95
135. Thomas H, Ougham H, Canter P, Donnison I (2002) What stay-green mutants tell us about the nitrogen remobilization in leaf senescence. J Exp Bot 53(370)
136. Tuinstra MR, Grote EM, Goldsbrough PB, Ejeta G (1997) Genetic analysis of post-flowering drought tolerance and components of grain development in *Sorghum bicolor* (L.) Moench. Mol Breed 3:439–448
137. Crasta OR, Xu WW, Rosenow DT, Mullet J, Nguyen HT (1999) Mapping of post-flowering drought resistance traits in grain sorghum: association between QTLs influencing premature senescence and maturity. Mol Gen Genet 262(3):579–588
138. Subudhi PK, Rosenow DT, Nguyen HT (2000) Quantitative trait loci for the stay green trait in sorghum (*Sorghum bicolor* L. Moench): consistency across genetic backgrounds and environments. Theor Appl Genet 101:733–741
139. Tao YZ, Henzell RG, Jordan DR, Butler DG, Kelly AM, McIntyre CL (2000) Identification of genomic regions associated with stay green in sorghum by testing RILs in multiple environments. Theor Appl Genet 100:1225–1232
140. Xu W, Rosenow DT, Nguyen HT (2000a) Stay green trait in grain sorghum: relationship between visual rating and leaf chlorophyll concentration. Plant Breed 119:365–367
141. Xu W, Subudhi PK, Crasta OR, Rosenow DT, Mullet JE, Nguyen HT (2000b) Molecular mapping of QTLs conferring stay-green in grain sorghum (*Sorghum bicolor* L. Moench). Genome 43:461–469
142. Haussmann BIG, Mahalakshmi V, Reddy BVS, Seetharama N, Hash CT, Geiger HH (2002) QTL mapping of stay-green in two sorghum recombinant inbred populations. Theor Appl Genet 106:133–142
143. Harris K (2007) Genetic analysis of the sorghum bicolor stay-green drought tolerance trait. Dissertation, Texas A&M University, College Station, TX
144. Walula RS, Rosenow DT, Wester DB, Nguyen H (1994) Inheritance of the stay-green trait in sorghum. Crop Sci 34:970–972
145. Mace ES, Jordan DR (2010) Location of major effect genes in sorghum (*Sorghum bicolor* (L.) Moench). Theor Appl Genet 121:1339–1356
146. Weers B (2011) Integrated analysis of phenology, traits, and QTL in the drought tolerant sorghum genotypes BTx642 and RTx7000. Dissertation, Texas A&M University, College Station, TX
147. Harris K, Subudhi P, Borrell A, Jordan D, Rosenow D, Nguyen H, Klein P, Klein R, Mullet J (2006) Sorghum stay-green QTL individually reduce post-flowering drought induced leaf senescence. J Exp Bot 58:327–338
148. Borrell AK, Mullet JE, George-Jaeggli B, Oosterom EJV, Hammer GL, Klein PE, Jordan DR (2014) Drought adaptation of stay-green sorghum is associated with canopy development, leaf anatomy, root growth, and water uptake. J Exp Bot 65:6251–6263
149. Hayes CM, Weers B, Thakran M, Burow G, Xin Z, Emendack Y, Burke JJ, Rooney WL, Mullet JE (2015) Discovery of a dhurrin QTL in *Sorghum bicolor*: co-localization of dhurrin biosynthesis and a novel stay-green QTL. Crop Sci 56:104–112
150. Gleadow RM, Møller BL (2014) Cyanogenic glycosides: synthesis, physiology, and phenotypic plasticity. Annu Rev Plant Biol 65:155–195

151. Sakhi S, Shehzad T, Rehman S, Okuno K (2013) Mapping the QTLs underlying drought stress at developmental stage of sorghum (*Sorghum bicolor* (L.) Moench) by association analysis. Euphytica 193:433–450

152. Upadhyaya HD, Dwivedi SL, Wang Y, Vetriventhan M (2016) Sorghum genetic resources. In: Ciampitti I, Prasad V (eds) Sorghum: state of the art and future perspectives. Agron Monogr, vol 58. ASA and CSSA, Madison, WI

Chapter 3

Field Studies to Develop Weed Management Programs for Grain Sorghum

J. Anita Dille

Abstract

Weed management programs to be used in grain sorghum production are best investigated in field studies with naturally occurring weed populations in their relevant growing environments. Weed control tactics to be evaluated include use of crop production practices such as row spacing and seeding rates, mechanical tools, and herbicide programs with soil- and foliar-applied products.

Key words Chemical, Cultural, Efficacy, Herbicide, Tillage, Weed control

1 Introduction

The presence of weeds in grain sorghum fields results in reductions in crop yield and quality, harvest inefficiencies, harbors other disease and insect pests, and overall, loss of economic return. Numerous weed control tactics are available to control weeds in grain sorghum, including cultural, mechanical, and chemical tools that need to be combined into an integrated weed management system for greatest benefit. Recent reviews have described the impacts of weeds on grain sorghum production and options for weed management [1, 2].

Cultural weed control tactics include developing crop production systems that would increase the competitiveness of grain sorghum against the weeds, including cultivars that are more competitive, narrowing row spacing, increasing seeding rates, and modifying fertility rates and application methods [3–8]. Mechanical weed control tactics include overall tillage systems (conventional, minimum, or no-tillage) [9, 10], as well as choices within the growing crop, such as rotary hoes, inter-row crop cultivation, between-row mowing [11] or flaming.

Chemical weed control tactics include various herbicides, different application timings such as preemergence and postemergence

Zuo-Yu Zhao and Jeff Dahlberg (eds.), *Sorghum: Methods and Protocols*, Methods in Molecular Biology, vol. 1931,
https://doi.org/10.1007/978-1-4939-9039-9_3,

Fig. 1 Example of a four-row grain sorghum plot with stake placed in center between rows 2 and 3. Preemergence treatment of Dual II Magnum (s-metolachlor) was applied on the same day as planting on June 7, 2013 followed by a postemergence treatment of Huskie (bromoxynil and pyrasolfutole) and atrazine applied on June 27 to grain sorghum at the V7 growth stage. Photo taken July 3, 2013 by Dr. Curtis Thompson, Kansas State University

(Fig. 1), and residual, systemic, and contact herbicides [12–15]. This category also includes the use of herbicide-tolerant grain sorghum hybrids (e.g., ALS-inhibitor and ACCase-resistant) with its corresponding herbicide tool [16–18]. On occasion, understanding the crop tolerance to the different herbicide products is evaluated and may be in the absence of weeds [14, 15, 19].

Well-designed field experiments can be used to evaluate and compare these weed control tactics that can be recommended to grain sorghum producers to improve their productivity and profitability. The first step is to have a clear research question and/or objective in order to know whether the expected response (hypothesis) has occurred and can be accepted.

2 Materials

As these are field-based experiments, relevant grain sorghum growing environments need to be identified, field locations with uniform (or minimum level of variability in) soil properties such as soil type, texture, pH, and organic matter, and somewhat uniform but not very dense community of weed species to be controlled (*see* **Note 1**).

2.1 Factors to Consider as Materials for Cultural Weed Control Studies

1. Prior to selecting a field site, determine what has occurred in that area in previous 2 to 3 years if possible. Any carryover of previous experimental treatments should be identified and avoided. Conduct a soil test to determine fertility levels and to characterize other soil properties, to be able to fertilize adequately or to be consistent in generating similar growing conditions across multiple locations and across years.
2. The same grain sorghum hybrid should be used across sites within a year, and across years, as best as possible so this is not a confounding factor. If the same hybrid is not possible, select those with same maturity.
3. Ensure that the planter or drill is calibrated to seed at correct seeding rate. Know the crop seed viability to over seed as necessary to achieve correct plant density.

2.2 Materials for Mechanical Control Studies

1. Adjust size of plots (width) based on size of mechanical tools, such as disk, chisel plow, moldboard plow, field cultivators, inter-row cultivators, etc. and match spacing to planter row spacings.
2. Adjust size of plots (length) based on minimizing border impacts at front and back of plots.

2.3 Materials for Chemical Control Studies

1. A sprayer (backpack, bicycle, or tractor-mounted, Fig. 2) that is calibrated to deliver correct amount of spray solution that contains the required herbicide, additives, and water or fertilizer as carrier solutions.
2. Use clean tap or distilled water for the spray solution.

Fig. 2 Examples of backpack sprayer (**a**) and of tractor-mounted sprayer (**b**) to apply chemical treatments. Photos by Dr. Anita Dille, Kansas State University

If projects are mainly herbicidal evaluations, a computer-based program, ARM (Agriculture Research Management, developed by Gylling Data Management, Inc.), saves time creating trials, collecting and analyzing data, and producing the many reports required for successful and timely trial management. ARM is software to help plan, create, manage, analyze, and report planned efficacy experiments. While ARM is typically used for crop production and crop protection experiments by field or greenhouse researchers, it also works well with many non-cropland plant experiments. ARM is designed to manage and summarize on a single-trial basis.

3 Methods

3.1 Experimental Design

1. Experimental design and plot size will depend on choices of cultural, mechanical, and/or chemical control tactics as treatments. Often split-plot designs are selected to have whole plots for crop row spacing and/or tillage treatments, and to have subplots that are individual seeding rates, inter-row cultivation, or herbicide treatments.
2. A weedy all season control plot (nontreated) should be included so that the level of weed pressure can be observed and used in comparisons against imposed weed control treatments (Fig. 3).

Fig. 3 Example of a four-row nontreated plot used as a weedy check against weed control treatment plots. Grain sorghum planted but no weed control imposed. Dominant weed species is Palmer amaranth (*Amaranthus palmeri*). Photo taken June 30, 2015 by Dr. Curtis Thompson, Kansas State University

Fig. 4 Example of four-row grain sorghum plot with stake placed between rows 2 and 3. Postemergence treatment of Huskie (bromoxynil and pyrasulfotole) applied to V3 grain sorghum. "Running" weedy check strips are located just outside rows 1 (on left) and 4 (on right) where postemergence spray application did not reach. Plots were end-trimmed for consistent length. Photo taken June 30, 2015 by Dr. Curtis Thompson, Kansas State University

3. For many chemical control studies, edges of treated plots are not treated (spray swath narrower than plot width) and can be used as "running" weed checks next to the treatment (Fig. 4). This is important if weed communities are not distributed evenly across the experimental location.
4. Often a weed-free all season control plot (could be hand-weeded or safe herbicide treatments) should be included so that the crop potential can be observed that has not been affected by imposed weed control treatments.
5. There should be at least four replications of all treatments, and typically arranged in a Randomized Complete Block design.
6. A single field experiment should be repeated in space (multiple geographic locations) and in time (over two or more years) in order to account for environmental variability and to generalize results.

3.2 Plots

1. Typical dimension of a single plot is ~3 m wide, equivalent to 4-crop rows at 0.76 m spacing, and ~10 m (25 to 30 ft) long. Plots may need to be laid out prior to planting in order to apply any early pre-plant herbicidal treatments.

2. If the grain sorghum experiment is isolated in a field, at least 4 to 8 rows should be borders surrounding the entire experiment to minimize edge effects. Most data should be collected from the center two crop rows and their corresponding inter-row areas (0.38 m on either side of the crop row) for a total width of 1.5 m.

3.3 Application of Treatments

1. Pre-emergence treatments are typically applied to plots the day of or within 24 h of planting. Post-emergence treatments are applied according to experimental protocol.
2. For herbicidal treatments, a typical handheld boom sprays a swath that is 1.5 to 1.67 m wide, which would cover the center two crop rows. This leaves "running" checks of nontreated areas to compare to control provided by the treatment (Fig. 4).

3.4 Evaluation of Treatments/Data to Be Collected

1. Data to be collected include observations of weed control efficacy for each individual weed species in the experiment, typically on a scale of 0 to 100% where 0 is no control and 100% is complete weed control. If visual crop injury is to be evaluated, this is also on a scale of 0 to 100% where 0 is no crop injury and 100% is complete injury and death of the crop.
2. Other data to be collected could include weed species composition, density, and biomass. Often one to two small quadrats (0.5 by 0.5 m square) could be randomly placed if only a one time observation is taken, or could be permanently located within the plot for data collection. Grain sorghum crop growth stages are valuable data to collect at each observation time according to Vanderlip [20] and Ciampitti [21].
3. If the researcher desires information on impact of weeds by mid-bloom growth stage, 1 m row of grain sorghum plants could be clipped at the soil surface, counted and fresh weight taken. A subsample of two or three plants from the sample could be selected and fresh weight taken. Plants are brought back to the lab in order to separate into main stem leaves and sheath/stems, tiller leaves and stems, and any reproductive structures. Leaf area could be measured using a leaf area machine (LI-3100C Area Meter, LI-COR Biosciences, Lincoln, NE). All plant parts are bagged separately, dried in an oven, and weighed. Total dry weight and total component parts from 1 m row of grain sorghum can be calculated based on the ratio of dry and fresh weights of subsample.
4. Final grain sorghum yield and yield components can be determined at the end of the growing season (*see* **Note 2**). Yield components include final plant stand (# m^{-1} row), number of sorghum heads (# m^{-1} row), seeds/head, and weight of seeds (g $head^{-1}$). Final grain yields should be summarized in kg ha^{-1} at 14.5% moisture content.

3.5 Statistical Analysis

1. Depending on the experimental design and the type of data collected, appropriate analyses should be selected.
2. Recent reviews were published in Weed Science that summarize current approaches to statistical analyses [22].

4 Notes

1. If insufficient weed species occur, crop mimics could be seeded across the study. For example, Rox orange [*Sorghum bicolor* (L.) Moench] can substitute for annual grasses (or shattercane specifically), while sunflower (*Helianthus annuus*) can be a substitute for a large-seeded broadleaf weed species. Also, seed of the weed species could be over seeded in the experimental area, preferably at the same biological time as natural seed rain (fall for summer annual weed species, late spring for winter annual weed species). Seed viability should be tested before seeding to know what density level might be expected.
2. Ensure that each plot is trimmed to the exact length desired for combine harvesting (Fig. 4). If sorghum heads or grain is to be collected by hand, or if there appears to be edge effect, do not collect from the first or last meter of row. If a biomass sample was removed during the season, do not forget to adjust area that grain sorghum was harvested from.

References

1. Stahlman PW, Wicks GA (2000) Weeds and their control in grain sorghum. Chapter 3.5. In: Smith CW (ed) Sorghum: origin, history, technology and production. John Wiley & Sons, Inc., New York, pp 535–590
2. Thompson CR, Dille JA, Peterson DE (2017) Weed competition and management in sorghum. In: Ciampitti I, Prasad V (eds) Sorghum: state of the art and future perspectives, vol 58. American Society of Agronomy, Madison. https://doi.org/10.2134/agronmonogr58.2014.0071
3. Hewitt CA (2015) Effect of row spacing and seeding rate on grain sorghum tolerance of weeds. MSc Thesis, Department of Agronomy, Kansas State University. p. 94
4. Shaffer GE (2016) Critical duration of grass weed interference in grain sorghum. MSc Thesis, Department of Agronomy, Kansas State University. p. 83
5. Staggenborg SA, Fjell DL, Gordon WB, Marsh BH, Devlin DL (1997) Reducing weed competition in grain sorghum with cultural practices. Proc NCWSS 52:68–69
6. Staggenborg S, Gordon B, Taylor R, Duncan S, Fjell D (1999) Narrow-row grain sorghum production in Kansas. Kansas State University Research and Extension Publication, MF2388. Manhattan, KS, p 4
7. Peerzada AM, Ali HH, Chauhan BS (2017) Weed management in sorghum [*Sorghum bicolor* (L.) Moench] using crop competition: a review. Crop Prot 95:74–80
8. Unruh BJ (2013) Influence of nitrogen on weed growth and competition with grain sorghum. MSc Thesis, Department of Agronomy, Kansas State University. p. 66
9. Vencill WK, Banks PA (1994) Effects of tillage systems and weed management on weed populations in grain sorghum (*Sorghum bicolor*). Weed Sci 42:541–547
10. Walker RH, Norris BE Jr, McGuire JA (1991) Grass weed management systems in grain sorghum (*Sorghum bicolor*). Weed Technol 5:54–60
11. Donald WW (2007) Between-row mowing systems control summer annual weeds in no-till grain sorghum. Weed Technol 21:511–517

12. Geier PW, Stahlman PW, Regehr DL, Olson BL (2009) Preemergence herbicide efficacy and phytotoxicity in grain sorghum. Weed Technol 23:197–201
13. Reddy SS, Stahlman PW, Geier PW, Thompson CR, Currie RS, Schlegel AJ, Olson BL, Lally NG (2013) Weed control and crop safety with premixed pyrasulfotole and bromoxynil in grain sorghum. Weed Technol 27:664–670
14. Fromme DD, Dotray PA, Grichar WJ, Fernandez CJ (2012) Weed control and grain sorghum (*Sorghum bicolor*) tolerance to pyrasulfotole plus bromoxynil. Int J Agron. https://doi.org/10.1155/2012/951454
15. Fromme DD, Grichar WJ, Dotray PA, Fernandez CJ (2014) Grain sorghum tolerance and weed control with pyrasulfotole plus bromoxynil combinations. Crop Manage. https://doi.org/10.1094/CM-2013-0010-RS
16. Hennigh DS, Al-Khatib K, Currie RS, Tuinstra MR, Geier PW, Stahlman PW, Claassen MM (2010) Weed control with selected herbicides in acetolactate synthase-resistant sorghum. Crop Prot 29:879–883
17. Hennigh DS, Al-Khatib K, Tuinstra MR (2010) Postemergence weed control in acetolactate synthase-resistant grain sorghum. Weed Technol 24:219–225
18. VanLoenen EA (2018) Evaluation of herbicide programs in acetolactate synthase-resistant grain sorghum. MSc Thesis, Department of Agronomy, Kansas State University. p. 64
19. Keeling JW, Brown BA, Reed JD, Dotray PA (2013) Grain sorghum response to saflufenacil applied preemergence. Crop Prot 46:1–6
20. Vanderlip RL (1993) How a sorghum plant develops. Kansas State University, Publication, Manhattan S3, p 20
21. Ciampitti IA (2015) Sorghum growth and development. Kansas State University, Publication, Manhattan MF3234 poster
22. Ritz C, Kniss AR, Streibig JC (2015) Research methods in weed science: statistics. Weed Sci 63(Special Issue):166–187

Chapter 4

Discovery of Sorghum Haploid Induction System

Tanveer Hussain and Cleve Franks

Abstract

Sorghum is the fifth most important cereal grain crop after corn, wheat, rice, and pearl millet in the world. Conventional sorghum breeding relies on multiple generations of self-pollination to achieve the adequate levels of homozygosity for hybrid evaluation, which adds several years and great cost to the breeding process. As in maize, doubled haploid (DH) is the key technology to speed up the breeding process in sorghum. Through 3 years of efforts, two haploid inducer lines, SMHI01 and SMHI02, were discovered by screening 4000 germplasms worldwide. These two inducers have been evaluated in different growth environments and have shown to generate haploids at frequency of 1–2%. The putative haploids produced with these two inducers were evaluated and ploidy was confirmed cytologically and biochemically. The discovery of these inducer lines is the first step toward a revolutionary change in sorghum breeding.

Key words Haploid, Doubled haploid, Haploid inducer, Sorghum, Microspores, Pollen, Plant regeneration, Chromosome doubling

1 Introduction

1.1 Haploid and Doubled Haploid

Haploid plants are sporophytes that contain a single set of chromosomes (n), in contrast to diploid plants, which contain two sets of chromosomes (2n), or polyploid plants, which contain three or more sets of chromosomes (i.e., 3n, 4n, etc.). Haploids originate spontaneously at a very low frequency in almost every plant species (e.g., maize, wheat, rice, soybean, barley, rye, tomato, potato). The first report of a spontaneous haploid plant, *Datura stramonium*, was published by Blakeslee et al. in 1922 [1] and was followed by reports of spontaneous hybrids in many other species. In maize (*Zea mays* L.), the occurrence of haploids was reported by Chase [2, 3] when he reported a spontaneous haploid induction of 1/1000 seeds from breeding material used in the central corn belt of the United States. Haploid plants are valuable because they can be used to develop homozygous doubled haploid (DH) lines for genetic studies and plant breeding without requiring generations of self-pollination that is typically required to produce homozygous plants. Because spontaneous haploid production is

Zuo-Yu Zhao and Jeff Dahlberg (eds.), *Sorghum: Methods and Protocols*, Methods in Molecular Biology, vol. 1931,
https://doi.org/10.1007/978-1-4939-9039-9_4,

not frequent enough to take advantage of for plant breeding, laboratory-based methods are developed to create haploids. For example, haploid production by chromosome elimination through wide crosses has been reported in barley (*Hordeum vulgare* L.) [4]. Similarly, Guha and Maheshwari [5] developed an anther culture technique for the production of haploids in the laboratory.

A DH plant or cell is one that is developed by the doubling of a haploid set of chromosomes. As spontaneous doubling occurs infrequently and inconsistently, artificial chromosome doubling of haploids, with the use of antimitotic agents, is required for plant breeding applications. Colchicine is the chemical agent most commonly used to induce chromosome doubling. During the late 1970s, the process used to create DH plants was slow, and DHs were not widely utilized because of the high labor and time demands [6]. With technological innovation and increased demand for end-use applications of DH lines, such as for marker identification and gene mapping, the development of DHs in higher plants has received increased attention [6]. DH methodologies have now been applied to over 250 species, and DHs are used in breeding programs in a number of crop species [7–9], including barley, maize, rapeseed (*Brassica napus* L.), and wheat (*Triticum aestivum* L.) [10, 11].

Two main systems have been widely used to generate haploid plants for breeding purposes, maternal and paternal haploid induction systems. A maternal haploid system is exclusively derived from the seed parent plant [12] and induction is caused by the pollinating parent [13]. This induction system can produce haploid plants from a given female line with an appropriate haploid inducer line or through wide hybridization. Another system that has been reported in maize is the *ig*-induced haploid, where the haploid produced in this case represents the genome of pollen parent [14]. A paternal haploid system achieves haploidy through the process of androgenesis from male reproductive tissue-induced through anther or isolated microspore culture. Anther culture is the preferred technique for DH production in many crops because this approach allows the large-scale anther culture formation and application to a broad range of genotypes [15]. In contrast, an isolated microspore culture is performed by removing somatic anther tissue, and requires advanced equipment and skills compared to anther culture. Although microspore culture provides an enhanced approach for exploring cellular, molecular, and biochemical processes during pollen embryogenesis, it is more complex than anther culture, and therefore it is developed and applied in fewer species [16].

In maize, maternal haploids occur spontaneously at a very low frequency. With such a low haploid induction frequency, breeding with spontaneous haploids was thought to be unrealistic for commercial breeding [17]. However, the rate at which maternal haploids are induced can be increased by pollination with haploid-inducing pollen of the same species (called a haploid inducer line). When an inducer line is used to pollinate a diploid plant, the majority of embryos produced are regular hybrid embryos, but a smaller proportion of haploid maternal embryos with normal triploid endosperms are also produced. The first genuine haploid inducer line was "Stock 6" in maize, with a haploid induction frequency of up to 2–3% [13]. Through modification of this line, enhanced haploid frequency of 8–12% is achievable, making it feasible to generate adequate numbers for breeding in maize.

In recent years, several crops have seen successfully implemented DH technology for commercial breeding. However, development of this technology has lagged in sorghum, where the lack of DH breeding methods has slowed development of new hybrids. Conventional sorghum breeding relies on multiple generations of self-pollination to achieve the adequate levels of homozygosity for hybrid evaluation, which adds several years and great cost to the breeding process. Corteva Agriscience™, Agriculture Division of DowDuPont™ has discovered two sorghum haploid inducer lines, which enable the creation of DHs in sorghum, and represent the first step to significantly accelerating the sorghum breeding process. The development of haploid and DH sorghum capabilities represents a major leap forward in sorghum breeding technology and will provide sorghum breeders with new capacity to developing higher-yielding sorghum hybrids.

1.2 Advantage of DH for Plant Breeding

Homozygous plants are the fundamental requirement for hybrid development and commercialization. Haploids plants are the critical component for the process of developing DH system. Prior to DH technology, achieving complete homozygosity in plants required several generations of self-pollination, and required intensive labor and time resources. DH production is an important advancement in plant breeding because it greatly reduces the number of cycles of self-pollination and time needed to obtain homozygous plants, thereby reducing the duration of the breeding cycle for a given crop species.

DH production systems allow a completely homozygous line to develop from heterozygous parents in a single generation (Fig. 1) [18, 19]. In open-pollinating crops, such as maize, DH systems offer a promising alternative to recurrent self-pollination for rapid inbred line development [9], whereas, in self-pollinating species (e.g., wheat), DH production systems can be used to increase the efficiency of cultivar development in breeding programs [10, 18, 20]. In addition to saving resources and time, DH production systems also allow for increased selection efficiency over conventional practices because of increased additive genetic variance, absence of dominance variation and within-family segregation [19]. DH lines increase the efficacy of selection, reduce the number of breeding cycles, and reduce the effort for line maintenance

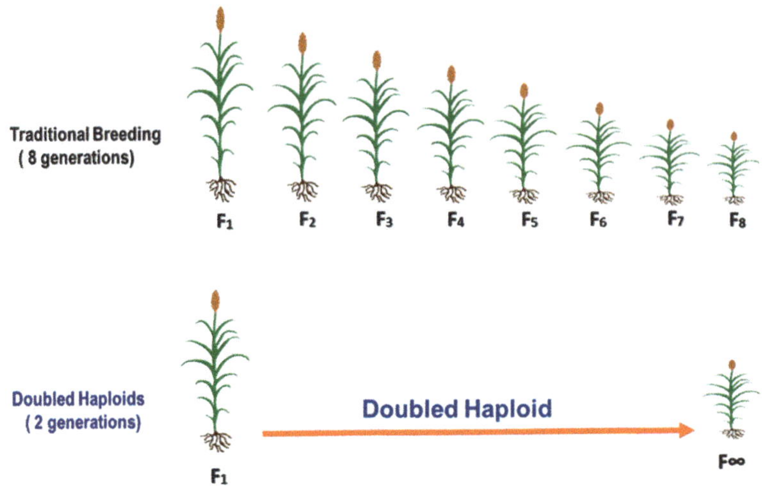

Fig. 1 Conventional vs. doubled haploid breeding

[12, 21, 22]. Therefore, an application of a DH production system could speed up the cultivar and population development for crop improvement.

As an example, the use of haploid technology has completely transformed the process of commercial maize improvement, and has allowed maize breeders to better leverage other supporting technologies, including marker-assisted selection and whole genome predictions. In conventional breeding, the progeny of breeding crosses must be self-pollinated for five or more generations in order to attain a level of homozygosity suitable for the creation of experimental hybrids, and this process is then usually continued for several more generations thereafter to completely eliminate any visual segregation. In the creation of DHs, this entire process is condensed to a single step, so that completely homozygous breeding lines are generated in 1 year. With DH technology, the process that normally takes at least seven generations to complete can be accomplished in two. This precludes the need for the breeder to carry forward all of the generations required for conventional inbreeding, and allows a much larger set of finished germplasm to be evaluated each year.

1.3 Chromosome Doubling of Haploid and DH Production

Haploid plants generated through any of the haploid production methods described above are sterile due to abnormalities in chromosome segregation during meiosis. To be utilized in a breeding program, chromosomes in the haploid plants should be doubled to restore fertility and to achieve homozygous fertile plants. Therefore, DH are haploids that have undergone the process of chromosome doubling (Fig. 2). It is one of the most critical steps in the DH production system. Although chromosome doubling can occur spontaneously and sporadically at a very low rate in some

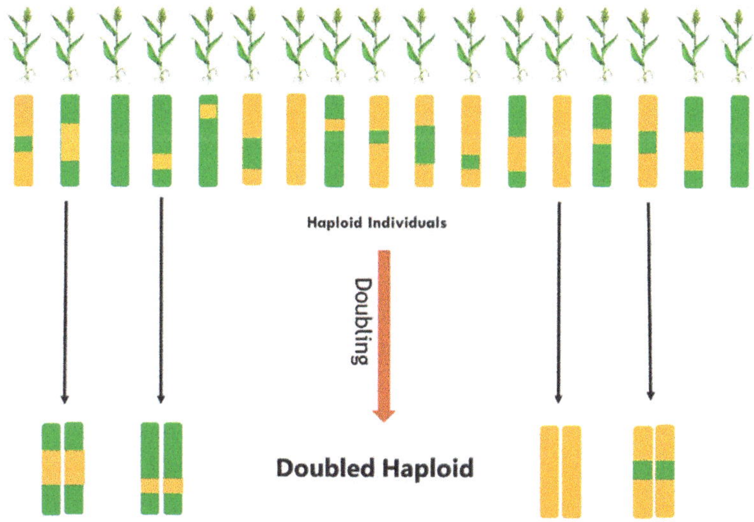

Fig. 2 Diagram showing process of chromosome doubling of haploids

species, for a DH production system, it is accomplished with the use of antimitotic agents. While there are several types of chemicals used to induce chromosome doubling in various plant species, colchicine is the most commonly used chromosome doubling agent. Other chemicals used for chromosome doubling include oryzalin (3,5-dinitro-N4, N4-dipropylsulfanilamide), trifluralin, amiprophosmethyl, pronamide, and nitrous oxide (N_2O). All of these chemicals are known to disrupt mitosis by binding to tubulin and disrupting the formation of microtubules and spindle fibers during cell division [23]. This prevents the polar migration of chromosomes to opposite poles, resulting in a cell with a doubled chromosome number.

The timing and tissue of application are the most important factors to consider when chromosome doubling is being achieved using chemical agents. Treatment of roots with doubling agents remains the foremost technique in practice. However, this technique typically results in chimeric doubling of plant gametophytes. In a number of species, like fruits and vegetables, apical meristem, secondary buds, and tillers can also be exposed to colchicine [24]. Colchicine treatment of maize seedlings is reported to double the chromosome number in the tassel or ear but not in both, which makes the self-pollination unattainable [25]. While chromosome doubling of haploid plants has been routinely and successfully performed using colchicine, the success and efficiency of haploid induction varies in different crop species. High mortality and abnormal plant development can be attributed to colchicine treatment. This could explain the low frequency of DH production in maize.

In this study, we highlight the recent advancements in the field of the haploid induction discovery in sorghum. The goal of this research is to identify sorghum genotypes that would serve as an inducer of haploidy, which represents the first step in the creation of DH lines in sorghum.

2 Materials and Methods

1. At the beginning of this study, 1500 sorghum hybrids were planted and evaluated in Puerto Vallarta, Mexico. In subsequent cycles, using the nurseries in alternating environments, 1000 to 1500 hybrids were concurrently created and evaluated every 6 months. More than 4000 elite and exotic sorghum breeding lines and accessions were evaluated for their capacity to induce haploidy in F1 progeny.
2. Hybrid crosses were made by crossing potential inducers to cytoplasmic male sterile (CMS) females line (A1 cytoplasm), and the progenies of these crosses were visually evaluated for the appearance of putative haploids. This general cycle of crossing and evaluation was repeated every growing season, in both summer and winter nurseries, continuously for 3 years.
3. Once putative inducers were identified based on the appearance of haploids, a wide range of sorghum germplasm was crossed with these putative inducer lines in order to validate the haploid induction capability of these inducers. When putative haploids were identified, remnant seed of hybrids with possible haploids were regrown for a second field evaluation and tissue samples were taken for flow cytometry assays to confirm ploidy levels. After the confirmation of the haploid chromosome in these lines, the putative haploid inducers were affirmed.
4. Analyses used to confirm haploidy included flow cytometry and chromosome counts under light microscope. Leaf punches from putative haploid and diploid plants were assayed with a flow cytometer to determine their ploidy levels. Flow cytometry measures the amount of nuclear DNA present, reflected as R-value. Dividing somatic cells isolated from root tips of putative haploid and diploid plants were observed under light microscope to count the chromosome numbers and to determine either 10 (1n) or 20 (2n) in their genome.

3 Results and Discussion

3.1 Discovery of Two Haploid Inducer Lines: SMHI01 and SMHI02

Through continuous evaluation of the haploid induction capability of those 4000 sorghum germplasms over six planting seasons, two sorghum inducer lines were identified that exhibited consistent

Fig. 3 A typical haploid sorghum plant

induction of haploid plants (Fig. 3) in different sorghum genetic backgrounds. There two inducer lines were identified as *S*orghum *M*aternal *H*aploid *I*nducer 01 and 02 (SMHI01 and SMHI02, Fig. 4). SMHI01 was adapted grain sorghum variety with red-seeded and pigmented plants. This line can completely restore the fertility of A1 CMS cytoplasm. SMHI02 was also adapted grain sorghum variety, with white-seeds, dark glumes, and pigmented plants. This line also can completely restore fertility of A1 CMS cytoplasm. The relative maturity (the days from germination to mid-anthesis) of these two lines was 68 days. The haploid induction frequency of these two inducers ranged from 1.5 to 2.0% for SMHI01 and from 1.0 to 1.5% for SMHI02, depending on the different sorghum genetic backgrounds used in haploid induction. This induction capacity of these two inducers is in the similar range of maize "Stock 6" (1–2%) [13].

3.2 Confirmation of Haploids Induced by These Two Inducers

The haploid plants derived from the F1 seeds produced from the female plants pollinated with these inducers showed unique plant features, similar to haploid plants of other crops, such as maize. These unique features include short plant status, narrow leaves, and thin stocks (Fig. 3). If no chromosome doubling treatment is applied, most or all the florets will not be fertile, and no fertile pollen will be produced, and very few or no seeds will be produced. Besides visual observations of phenotype, we also conducted flow

Fig. 4 Sorghum haploid inducer lines, SMHI01 (left) and SMHI02 (right)

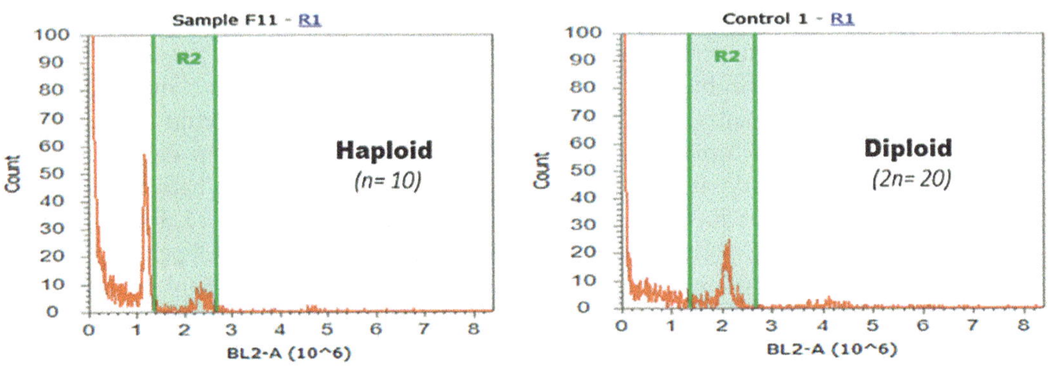

Fig. 5 Example of flow cytometry results. Left graph shows the result from a haploid plant, and the right graph is from a diploid plant as control

cytometry assays with leaf punches for all haploids. The flow cytometry results further confirmed haploid status (Fig. 5). In addition, chromosome numbers in the root tip cells of some haploid plants and diploid controls were counted under light microscope (Fig. 6). Figure 6 shows two cells, the left haploid cell having

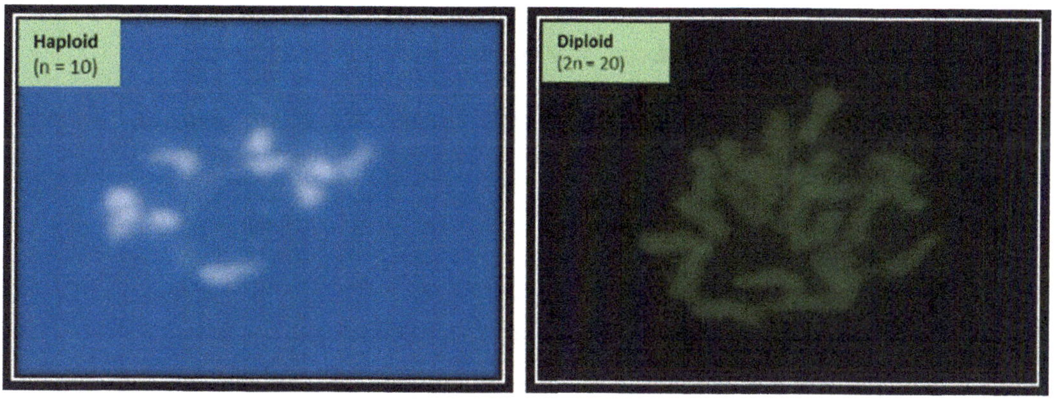

Fig. 6 Cytological count of sorghum chromosome number of haploid and diploid plants

Table 1
Results of field testing of haploid inducers at Johnston, Iowa State

Pedigree	Cross type	Total plants	Diploid plants	Diploid plants (%)	Haploid plants	Haploid induction rate (%)
SMHI02/SMHI01	F1 hybrid[a]	183	180	98.4%	3	1.6%
PHPA-1/SMHI01	F1 hybrid	74	73	98.5%	1	1.4%
PHPA-2/SMHI01	F1 hybrid	71	70	96.6%	1	1.4%
PHPB-1/SMHI01	F1 hybrid[a]	35	32	91.4%	3	8.6%
Sum		363	355	97.8%	8	2.2%

[a]Emasculation

10 (1n) chromosomes and the right diploid cell having 20 chromosomes (2n).

3.3 Verification of the Stability of Haploid Inducers

To validate the stability of these inducers in a wider range of crosses, we crossed these two inducers to both sorghum sterile (A) and fertile (B) lines and evaluated their haploid production. When crossing to the fertile sorghum lines, the anthers from the female plants were emasculated. The F1 plants derived from these F1 seeds were screened for haploids, and the results showed the haploid plants were recovered from both crosses (Tables 1 and 2). These experiments were conducted in three field locations with different growth environments, Iowa (Table 1), Kansas (Table 2), and Texas (Table 3). In one of these experiments conducted in Iowa, 3 haploid plants were identified among 35 plants screened, it gave 8.6% of haploid induction rate (Table 1), much higher than all other crosses. This cross was PHPB-1 as the female parent and pollinated by inducer SMHI01. Since the populations screened were relatively small, it is hard to draw any final conclusions. In addition, we had

Table 2
Results of field testing of haploid inducers at Manhattan, Kansas State

Pedigree	Cross type	Total plants	Diploid plants	Diploid plants (%)	Haploid plant	Haploid induction rate (%)
PHPA-2/SMHI01	F1 hybrid	300	298	99.2%	2	0.7%
PHPB-2/SMHI01	F1 hybrid[a]	180	179	99.4%	1	0.6%
PHPB-3/SMHI01	F1 hybrid[a]	195	194	99.0%	1	0.5%
Sum		675	671	99.4%	4	0.6%

[a]Emasculation

Table 3
Results of field testing of haploid inducers at Plainview, Texas State

Pedigree	Cross type	Total plants	Diploid plants	Diploid plants (%)	Haploid plants	Haploid induction rate (%)
PHPA-1/SMHI01	F1 hybrid[a]	170	167	98.2%	3	1.76%
PHPA-1/SMHI02	F1 hybrid	137	135	98.5%	2	1.46%
Sum		307	302	98.4%	4	1.62%

[a]Emasculation

made an intercrossing of these inducers themselves to test overall induction rates (Table 1). The haploid induction frequency was varied in these three locations; however, the overall frequency of haploid induction by these two inducers was still in the range of 1–2%, as mentioned above. Based on these observations, it can be concluded that these two inducers, SMHI01 and SMHI02, have stable capability to induce 1–2% haploid production across different sorghum genotypes and in different growing environments.

Similar to "Stock 6," which was discovered in maize in the 1950s [13], this discovery is the first step toward the creation of DHs in sorghum and holds the potential to create a powerful tool for a revolutionary change in sorghum breeding in near future. DH is expected to dramatically improve the speed and efficiency of sorghum breeding.

Acknowledgments

This study is a collaborative research project between DowDuPont™ and the United Sorghum Checkoff Program. All research was conducted by DowDuPont™ scientists and the research funding was provided by the United Sorghum Checkoff Program.

References

1. Blakeslee AF, Belling J, Farnham ME, Bergner AD (1922) A haploid mutant in the jimson weed, *Datura stramonium*. Science 55:646–647
2. Chase SS (1949) Monoploid frequencies in a commercial double cross hybrid main and in its component single hybrids and inbred lines. Genetics 34:328–332
3. Chase SS (1952) Production of homozygous diploids of maize from monoploids. Agron J 44:263–267
4. Kasha KJ, Kao KN (1970) High frequency haploid production in barley (*Hordeum vulgare* L.). Nature 225:874–876
5. Guha S, Maheswari SC (1964) In vitro production of embryos from anthers of Datura. Nature 204:497
6. Forster BP, Haberle-Bors E, Kasha KJ, Touraev A (2007) The resurgence of haploids in higher plants. Trends Plant Sci 12:368–375
7. Snape J (1989) Doubled haploid breeding: theoretical basis and practical applications. In: Mujeeb-Kazi A, Sitch LA (eds) Review of advances in plant biotechnology. IRRI. International Maize and Wheat Improvement Center, Edo Mex, pp 19–30
8. Raina SK (2010) Doubled haploid breeding in cereals. In: Janick J (ed) Plant breeding reviews, vol 15. John Wiley & Sons, Oxford
9. Murovec J, Bohanec B (2011) Haploids and doubled haploids in plant breeding. In: Abdurakhmonov I (ed) Plant breeding, INTECH pp 87–106. WWW.intechopen.com
10. Liu W, Zheng MY, Polle EA, Konzak CF (2002) Highly efficient doubled-haploid production in wheat (*Triticum aestivum* L.) via induced microspore embryogenesis. Crop Sci 42:686–692
11. Melchinger AE, Longin CF, Utz HF, Reif JC (2005) Hybrid maize breeding with doubled haploid lines: quantitative genetic and selection theory for optimum allocation of resources. In: Proceedings of the forty first annual Illinois corn breeders' school, Illinois, United States pp 8–21
12. Röber FK, Gordillo GA, Geiger HH (2005) In vivo haploid induction in maize- performance of new inducers and significance of doubled haploid lines in hybrid breeding. Maydica 50:275–283
13. Coe EH (1959) A line of maize with high haploid frequency. Am Nat 93:381–382
14. Pollacsek M (1992) Management of the ig gene for haploid induction in maize. Agronomie 12(3):247–251
15. Sopory S, Munshi M (1996) Anther culture. In: Mohan JM et al (eds) In vitro haploid production in higher plants, vol 1. Kluwer Academic Publishers, Dordrecht, pp 145–176
16. Nitsch C (1977) Culture of isolated microspores. In: Reinert J, YPS B (eds) Plant cell, tissue and organ culture. Springer, Berlin Heidelberg New York, pp 268–278
17. Geiger HH (2009) Doubled haploids. In: Bennetzen JL, Hake S (eds) Handbook of maize. Springer, New York, pp 641–657
18. Zheng MY, Weng Y, Liu W, Konzak CF (2002) The effect of ovary-conditioned medium on microspore embryogenesis in common wheat (*Triticum aestivum* L.). Plant Cell Rep 20:802–807
19. Forster BP, Thomas WTB (2010) Doubled haploids in genetics and plant breeding. In: Janick J (ed) Plant breeding reviews, vol 25. John Wiley & Sons, Oxford
20. Kunz C, Islam SMS, Berberat J, Peter SO, Buter B, Stamp P, Schmid JE (1999) Assessment and improvement of wheat microspore derived embryo induction and regeneration. J Plant Physiol 156:190–196
21. Chang MT, Coe EH (2009) Doubled haploids. In: Kriz AL, Larkins A (eds) Biotechnology in agriculture and forestry, vol 63. Springer, New York, pp 127–142
22. Geiger HH, Gordillo GA (2009) Doubled haploids in hybrid maize breeding. Maydica 54:485–499
23. Hansen NJP, Andersen SB (1996) *In vitro* chromosome doubling potential of colchicine, oryzalin, trifluralin and APM in *Brassica napus* microspore culture. Euphytica 88:159–164
24. Mathias R, Robbelen G (1991) Effective Diploidization of microspore-derived haploids of rape (*Brassica napus* L.) by in vitro colchicine treatment. Plant Breed 106(1):82–84
25. Wan Y, Petilino JF, Widholm JM (1989) Efficient production of doubled haploid plants through colchicine treatment of anther-derived maize callus. Theor Appl Genet 77 (6):889–892

Chapter 5

Development of a Pedigreed Sorghum Mutant Library

Junping Chen, Guihua Zou, and Zhanguo Xin

Abstract

Induced mutagenesis is a powerful approach to generate variations for elucidation of gene function and to create new traits for breeding. Here, we described a procedure to develop a pedigreed mutant library through chemical mutagenesis with ethylmethane sulfonate (EMS) treated seeds in sorghum and discussed its potential to generate new traits for sorghum improvement. Unlike random mutagenesis, a pedigreed mutant library, once properly established, can serve as a powerful resource to isolate and recover mutations of both agronomical and biological importance. With the development of affordable and high-throughput next-generation sequencing technologies, identification of causal mutations from a mutant library with a uniform genetic background becomes increasingly efficient and cost-effective. Fast causal gene discovery from mutant libraries combined with precise genome editing techniques will accelerate incorporation of new traits and revolutionize crop breeding.

Key words *Sorghum bicolor*, Sorghum, Ethylmethane sulfonate, EMS, Mutagenesis, Mutant library.

1 Introduction

Random mutagenesis has long been applied to sorghum to isolate novel phenotypes that may have potential application in breeding [1, 2]. Various mutagens, such as X-ray and γ-irradiation, EMS, methyl methane sulfonate (MMS), diethyl sulfate (DES), *N*-Nitroso methyl urea (NMU), *N*-Nitroso ethyl urea (NEU), or combinations of chemical and irradiation mutagens, have been used [1, 3–5]. Beneficial mutations, such as high protein digestibility, high lysine, brown midrib, and others, have been isolated [6–10]. In random mutagenesis, the mutagen treated seeds (M_1) are planted to produce "self-fertilized" M_2 seeds, which are harvested in pools. Mutants with desired phenotypes are selected from M_2 plants and preserved. The rest of plants are usually discarded.

In contrast with random mutagenesis, a pedigreed mutant library consists of M_4 seed pools derived from individual mutagenized M_1 seeds by pedigree or single seed descent regardless of the phenotype displayed by the mutant plants [11]. The advantage of the pedigreed mutant library is its ability to preserve most

Zuo-Yu Zhao and Jeff Dahlberg (eds.), *Sorghum: Methods and Protocols*, Methods in Molecular Biology, vol. 1931,
https://doi.org/10.1007/978-1-4939-9039-9_5, © Springer Science+Business Media, LLC, part of Springer Nature 2019

mutations, including recessive lethal mutations, allows repeatedly screen of variety of traits, and recovers mutants that cannot produce seeds in the primary screens. Furthermore, a mutant library derived from a uniform genetic background provides a unique resource to identify the causal mutations efficiently through bioinformatic analysis of next-generation sequencing data from bulked mutants isolated from segregating F_2 population [12, 13].

The generation of a high-quality sorghum reference genome sequence and the bioinformatic annotation of the genes in the sorghum genome have made it possible to study gene function on a genome-wide scale, and to compare gene function with other plants [14, 15]. A systematic mutant library that contains multiple mutations for all genes in the sorghum genome is a valuable resource to deduce the functions of sorghum genes. Several mutant libraries have been constructed [11, 16–18]. Jiao et al. (2016) sequenced 256 mutant lines selected from a large pedigreed mutant library via next-generation sequencing platform of Illumina High-Seq2000 [19]. This sequencing effort identified ~1.8 million canonical EMS-induced G/C to A/T mutations, covering 94% of the genes. Comparison with the natural variations derived from sequencing of the natural accessions and genotyping by sequencing of the sorghum association panels, 97.5% of the induced mutations in the pedigreed mutant library are shown to be novel [19–22]. Based on this estimation, the entire 6400 pedigreed mutant lines may possess over 40 million novel mutations. Thus, the mutant library may provide a valuable genetic resource to isolate novel mutants to understand the adaptability of sorghum to a wide range of hostile environment and to discover novel traits for sorghum improvement.

Consistent with the number of novel mutations observed, the mutant library also displays many novel phenotypes [11, 23, 24]. One of the novel phenotypes is the multiple-seeded (*msd*) mutant (Fig. 1). Sorghum panicles bear two types of spikelets, which are named according to their mode of attachment to inflorescence branches. The sessile spikelets are attached to a flower branch directly, while the pedicellate spikelets are attached to a flower branch through a short petiole called pedicel. The terminal spikes on a flower branch consist of one sessile spikelet and two pedicellate spikelets, while the spikes below the terminal spike consist of one sessile and one pedicellate spikelet. In sorghum and many other grasses, only the sessile spikelets are capable of producing both male and female floral organs and set seeds. The pedicellate spikelets, which either produce no floral organ or only anthers, eventually abort and produce no seed. In contrast, all spikelets in the *msd* mutants develop both male and female floral organs and produce normal seeds [25]. The causal gene mutations for the *msd1* mutants have been identified by MutMap [13]. It encodes a Teosinte branched/Cycloidea/PCF (TCP) transcription factor.

Fig. 1 Morphological comparison of BTx623 (WT) and *msd1* panicles. (**A**) The *msd1-1* mutant has increased number of primary branches. (**B**) The *msd1-1* mutant has larger primary branches. (**C**) All spikelets in *msd1-1* are fertile, while only the sessile spikelet is fertile in BTx623 (WT). (**D**) The *msd1-1* mutant has enlarged panicle (Copyright permission from Crop Sciences 501382943)

During panicle development, MSD1 activates jasmonic acid-induced programed cell death pathway that leads to the arrest of the pedicellate spikelets. Mutations in the *Msd1* gene block the cell death pathway and restore the normal development of the pedicellate spikelets into seeds [26]. Six independent point mutations in the *Msd1* gene are identified from the mutant library. Consistent with the fact that the *msd* phenotype has not been observed from the current available sorghum germplasm collections, none of these causal mutations can be found in the sequenced natural accessions.

Another novel phenotype that may have potential to improve sorghum yield is the distinctive erect leaf (*erl*) mutants (Fig. 2). Erect leaf is an architectural trait that leads the eightfold yield gain in maize hybrids released during the post-Green Revolution period [27, 28]. Compared with modern maize hybrids, sorghum exhibits an open canopy with wide leaf angles that are almost parallel to the ground. Sorghum architecture has not been improved in the last 50 years due to lack of genetic resources [29]. We have isolated a series of *erl* mutants [23, 30]. Among the 6400 M_3 field plots in the

Fig. 2 Leaf architecture of BTx623 and two mutants: *erl1-1* and *erl1-2*. The pictures of BTx623 and two mutants: *erl1-1* and *erl1-2* are taken at physiological maturity (Copyright permission from Journal of Plant Registrations 501382962)

field, over 50 segregated for leaf angles [30]. Eleven of these mutants were confirmed in M_4 generation. Several mutants have similar or slightly bigger panicles than the wild type (WT). These *erl* mutants may prove to be useful for improving sorghum biomass and grain yield based on the yield gains brought about by the improved leaf angle in maize hybrids [31]. The causal mutations for the erect leaf traits have not been identified. Given the novelty of the phenotype, we expect the mutations that lead to erect leaf architecture may not exist in the natural sorghum collections.

The above two examples illustrate that mutant library can be a valuable resource to isolate novel trait for sorghum improvement. It can be reasonably assumed that other novel traits also exist in the library. It will take a community effort and collective expertise to identify all novel traits to accelerate sorghum improvement.

The rapid development of genome editing technologies will revolutionize trait development and breeding strategies in crops [32, 33]. If a target gene to be edited is known, genome editing can be very useful to alter the target gene without affecting other parts of the genome. However, if the target gene is not known, it is impractical to edit every gene in a genome to identify the target genes underlying a superior trait. On the other hand, chemical mutagenesis, especially EMS, can induce hundreds and thousands of mutations in single mutated plant [19, 34]. It is relatively easy to identify useful traits by high through screens of a manageable number of mutant populations. However, the induced mutants suffer from the presence of high number of unrelated mutations, which will require multiple generations of backcrosses to remove. A combined approach of induced mutation and genome editing may prove to be a powerful approach to transform agricultural through modern breeding. For example, new traits can be discovered through screening the induced mutant library and the causal mutations can be identified by fast forward genetic approach, such as MutMap [12, 13, 35]. Once the target genes are known, beneficial mutations can be introduced into elite lines through genome editing, eliminating the need for extensive backcrosses. At present, a new transformation platform that is cultivar-independent has been developed to apply genome editing technology in sorghum [36]. It is exciting to combine the mutant library for fast discovery of new lead genes and efficiently introduce the novel traits to elite lines by editing the target genes, avoiding the lengthy recurrent backcrosses required for reducing the background mutations.

This chapter detailed a protocol for constructing pedigreed mutant library in sorghum, which is also applicable to other crops. With the rapid development in next-generation sequencing technology, dramatic reduction in sequencing cost, and increase in sequencing quality and data output, high-quality mutant library has become an ever-powerful resource for discovery of genes underlying agronomy traits.

2 Materials

2.1 Sorghum Seeds

2.1.1 Obtain Sorghum Inbred Seeds

Sorghum seeds can be requested from Germplasm Resources Information Network (GRIN) (https://www.ars-grin.gov). We selected BTx623 for constructing the mutant library because it is used as the inbred line to generate the reference genome sequence for sorghum and as a parent of many mapping populations [15, 37–40].

2.1.2 Generate Single Seed Descent with Phenotype Uniformity

1. Grow 12 sorghum plants in greenhouse and select one plant that is most typical of the phenotypes of the original line to produce self-fertilized seeds.
2. Harvest the self-fertilized panicle from this plant.
3. Grow 12 new plants from the single panicle harvested above and select one most typical plant again to produce self-fertilized seeds.
4. Repeat this process for six generations to achieve phenotype uniformity (*see* **Note 1**).

2.2 Chemicals and Solutions

1. Obtain EMS and sodium thiosulfate from Sigma.
2. Prepare 0.1–0.3% (v/v) EMS suspension: add 100, 200, or 300 μl EMS to 100 ml of deionized water containing 0.05%

tween-20 as a surfactant (EMS is a none-polar chemical with slight solubility in water).

3. Prepare sodium thiosulfate solution: Dissolve 200 g of sodium thiosulfate in 1 l of tap water and add 4 g of sodium hydroxide slowly (pellet by pellet) to avoid splash.

2.3 Growing Sorghum in Greenhouse/Field

Sorghum plants can be grown in greenhouse under natural light. It is best to control the temperature between 25 and 30 °C. The optimum temperature is 28 °C. Sorghum is a short-day plant. Day length should be controlled around 12 h for normal flowering time of about 60 days. In field planting, sorghum can be planted when soil temperature is above 15 °C. Although sorghum seeds can germinate at 12 °C, poor germination and seedling growth can reduce plant health and grain yield. Regardless in greenhouse or field, to ensure high quality of the mutant library, sorghum panicles must be bagged before anthesis. Pollination bags can be purchased from Midco (https://www.midcoglobal.com).

2.4 Chemical Hood and Other Protection/Safety Items

1. Require a chemical hood with good ventilation during EMS-treatment to avoid EMS evaporation into environment.
2. Wear double layers of latex gloves during working with EMS to avoid skin-contact with EMS.

2.5 Bioinformatic Software or Computer Program to Record and Analysis Data

1. Record plant phenotypes and pedigrees in Excel, Access, or other database programs.
2. Conduct sequencing analysis using online software (Refer to Jiao et al. for detail [19]).

2.6 Sorghum Seed Storage

1. Store sorghum seeds at room temperature for less than 5 years.
2. Store sorghum seeds at −18 °C with low humidity for longer-term (more than 30 years).

3 Methods

3.1 Generate the Mutagenized Seeds

1. Soak ~100 g dry seeds in 200 ml EMS suspension in 1.5 l flask for 16 h in a chemical hood (*see* **Note 2**).
2. Decant EMS solution into a 2.5 l flask containing 1 l 0.1 M NaOH, 20% $Na_2S_2O_3$ and incubate overnight to deactivate EMS in a chemical hood. The deactivated EMS solution can be disposed into drains.
3. Wash the treated seeds 10 times with tap water, changing washing water every 30 min (*see* **Note 3**).
4. Air-dry the washed seeds. From this point, the seeds can be handled as normal seeds. The air-dried EMS-treated seeds (M_1 seeds) can be stored at room temperature for a few months without negative effect on seed germination or mutation rate.

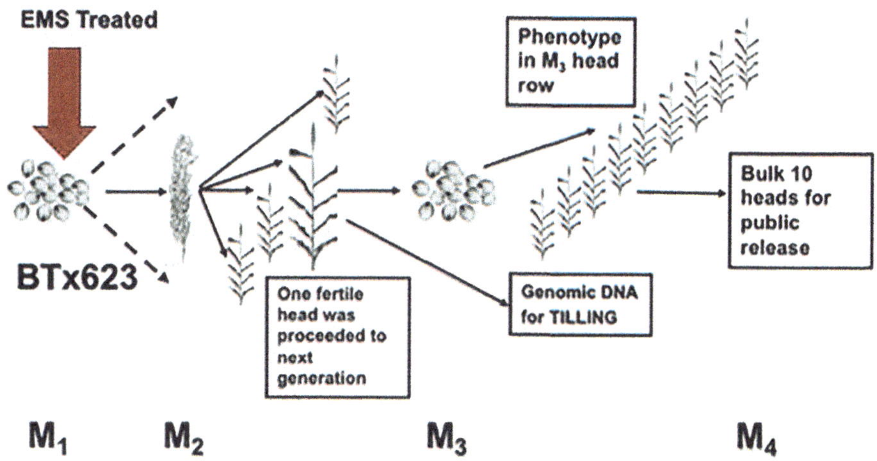

Fig. 3 Scheme to develop the pedigreed mutant library in sorghum. The M_1 is propagated by single seed descent until M_3 plants. The graph showed the process from a single M_1 seed to the buked M_4 seeds deposited in the mutant library. A single M_2 plant derived from each M_1 head row is sampled for DNA analysis or genotyping and the same plant is propagated to produce M_3 seeds. Phenotype is usually conducted with M_3 plants

3.2 Establish the Pedigreed Mutant Library

Figure 3 showed the steps generating pedigreed seed pools (*see* Note 4).

1. Plant M_1 seeds in the field at 100,000 seeds per hectare and grow M_1 plants.
2. Bag every panicle on these M_1 plants before anthesis with rain-proof paper pollination bags. Self-pollinate these M_1 plants and produce M_2 seeds.
3. Select the M_1 panicles with seed setting of 30% or less and discard the panicles with seed setting similar to WT plants.
4. Plant single head row plots with approximately 40 plants for each M_2 pool and grow these M_2 plots.
5. Select one M_2 plant in each plot that is fertile and produce self-pollinated M_3 seeds. Starting at this generation, each selected M_2 plant forms an individual mutant pool.
6. Plant one row plots with the M_3 seeds from each of the M_3 seed pools and grow M_3 plants.
7. Evaluate and record mutant phenotypes in each M_3 row plot. Collect a few of seeds from those M_3 plants displaying desired mutant phenotypes.
8. Bulk at least 10 panicles per M_3 row plot as one pedigreed M_4 seed pool and combine all these M_4 seed pools to establish a mutant library.

3.3 Confirm the Nature of Mutants

1. Plant one row plot with the M_4 seeds from each of M_4 seed pools and observe if all these M_4 plants have the similar mutant phenotype to the M_3 plant.
2. Backcross one typical M_4 mutant plant to the WT sorghum line that is used to generate the mutant library and generate F_1 (M_4BC_1) seeds.
3. Evaluate the F_1 plant phenotype and determine the nature of the mutation in each pool. If the F_1 plants display the phenotype similar to the WT, the mutation is recessive. If the F_1 plants display the phenotype similar to the mutant, the mutation is dominant.
4. Self-pollinate some of these F_1 plants and produce F_2 ($M_4BC_1F_2$) seeds.
5. Plant about 100 F_2 seeds in each pool to determine the segregation ratio of the mutant phenotype vs. WT in F_2 plants. A single recessive mutation shows approximately 1:3 segregation of the mutant to WT phenotype. While a single dominant mutation shows 3:1 segregation of the mutant to WT phenotype. Only those mutants following Mendelian single gene segregations are considered true mutants and will be subjected to further studies (*see* **Note 5**).
6. If the mutant does not segregate as a single gene mutation, discard the mutant and the F_2 population. Most mutant phenotypes we observed so far were single gene recessive. Some mutants displayed multiple phenotypes, which would segregate independently as single gene recessive mutations.

3.4 Eliminate Background Mutations

1. Select one mutant from the F_2 populations that showed a segregation ratio consistent with a single gene mutation.
2. Backcross the F_2 plant to the WT and generate $M_4BC_2F_1$.
3. Grow at least 8 $M_4BC_2F_1$ plants and cross these plants to WT again to generate $M_4BC_3F_1$ plants.
4. Self-pollinate the $M_4BC_3F_1$ plants to produce $M_4BC_3F_2$ seeds.
5. Grow at 12 $M_4BC_3F_2$ plants from each of these $M_4BC_3F_2$ seeds.
6. Identify the backcrossed mutants from these $M_4BC_3F_2$ populations.
7. After these backcrosses, 87.5% of the background mutations can be removed. The mutants should be good for most phenotypical and physiological analyses (*see* **Note 6**).

4 Notes

1. To construct a high-quality mutant library, it is necessary to start with a highly pure line. Initial observation showed that the BTx623 inbred line obtained from GRIN displayed large variations in plant height and panicle size. To ensure uniformity of the genetic background, the BTx623 seeds were purified through single seed descent for six generations with selection of one plant with typical BTx623 phenotype among 25 plants planted at each generation. We believe this process can be applied to construct mutant library in other sorghum lines or other crops even when variation in visual phenotype is not apparent. If space is limited, the single seed descent can be generated with 12 plants at each generation.
2. In our experience, direct treatment of dry seeds with EMS is more advantageous in generating efficient and uniform mutations than the seeds imbibed first and then treated with EMS. Many EMS mutagenesis works are conducted with imbibed seeds, which often lead to low or inconsistent mutation rates among mutated plants. The container size and EMS solution volume can be different from what we used. It is better to keep the weight ratio of dry seeds to EMS suspension at 1 to 2 to ensure efficient mutagenesis and utilization of EMS.
3. Thoroughly washing and soaking the EMS-treated seeds in tap water is necessary to remove as much as unbound EMS from the seed coat and inside of the seeds to reduce the chemical toxicity of the EMS and increase mutation efficiency. After 10 plus times of washing and soaking, the seeds can be handled as regular seeds.
4. To make pedigreed mutant library, each seed pool in the library must be derived from an individual mutagenized cell through single seed descent (Fig. 3). The notation in describing the generations of mutagenesis is different from transformation [41, 42]. The M_1 seeds are referred to the seeds that are treated with EMS. The M_1 plants are referred to the plants developed from the M_1 seeds. M_2 seeds are produced on an M_1 plant. M_2 plants are developed from M_2 seeds. M_3 seeds are produced on a M_2 plant. M_3 plants are developed from M_3 seeds. M_4 seeds are produced from an M_3 plant. In Mutant library, the M_4 seeds are bulked from multiple M_3 plants.

   A single plant embryo is made of multiple germ cells, each of which can be independently mutated, thus a M_1 plant can be genetically mosaic. The M_2 seeds are usually a mixture of seeds derived from different independently mutagenized germ cells. A sorghum embryo has been estimated to have 2 to 13 germ cells [43]. Because it is not known how many germ cells are mutated in each of the M_1 seeds, it is safe to select only one M_2 plant to produce M_3 seeds. Because the M_3 seeds are produced from a single M_2 plant, likely, they are derived from a single mutated germ cell, it is not necessary to conduct single seed descent from M_3 generation on. Thus, 10 M_3 panicles can be bulked to make the M_4 seed pools to be deposited in the mutant library. To preserve most mutations and ensure all mutations in each line of the mutant library are derived from a single germ cell, M_4 seeds, derived from pooled M_3 plants, should be kept in the library. Further advancement through self-fertilization may lead to homozygosity of lethal or highly damaging mutations, which may depress the plant vigor and lead to loss of the mutations. When the M_4 seed pools run low for particular lines, the M_4 seeds can be re-planted. At least 20 fertile plants should be bulked to replace the seeds. The bulked seeds are factually M_4 seeds because most of the lethal mutations are still at heterozygous state in bulked seeds.

5. EMS can induce mutations in both germ and somatic cells. Many mutations can be lethal at varied seedling/plant stages or/and non-fertile (varied stages of gamete development of either male or female side) etc. Most mutation phenotypes observed on M_1 plants are from EMS-induced mutations in somatic cells and, therefore, are not heritable. The identification and isolation of mutants should be started in M_2 plants or later generations regardless of the mutant phenotypes even for dominant mutants. Furthermore, any mutant phenotypes observed in M_2 or M_3 plants must be confirmed in two aspects: (1) heritability and (2) segregation with Mendelian ratio in the following generation. A confirmed mutant through the above steps is backcrossed to WT BTx623. The $M_4BC_1F_1$ plants are evaluated for the nature of mutation, i.e., dominant or recessive. If the mutant is dominant, the phenotypes of the $M_4BC_1F_1$ plants will be the same or similar to the phenotype observed in the mutants; otherwise, it is WT phenotype if the mutant is recessive. The nature of the mutants will be further confirmed by the phenotype segregation in the following generation generated by $M_4BC_1F_1$ self-pollination. A single gene dominant mutant will display statistically 3:1 segregation ratio of mutant phenotype to WT phenotype; while a single gene recessive mutant will show 1:3 of mutant phenotype to WT phenotype. Because of the high frequency of induced mutations, combinations of multiple background mutations can lead to some complicated phenotypes. Thus, only mutants confirmed by progeny testing and segregation analysis can be considered as true mutants. Only after these stringent validations, can the mutants be used for genetic and breeding studies.

6. In accessions from natural germplasm collections, landraces, or breeding lines, most mutations that reduce the fitness or adaptability of the plants are sorted out through natural or artificial selections from the thousands of years of breeding practice or millions of years natural selection. Unlike natural accessions, pedigreed mutant library is recently generated without any applied selection pressure. This feature allows the selection of unique mutations that may be of biologically or even of agronomically interest. But due to the presence of a large number of undesirable background mutations, it may hinder the direct use of mutants in breeding. To eliminate these undesirable background mutations for biological studies, at least three recurrent backcrosses to the WT parent should be conducted and it can remove majority (~87.5%) of the unlinked background mutations. For breeding applications, at least 6 backcrosses, which can remove 98.5% mutations unlinked to the causal gene, are recommended before testing for potential in breeding. Due to the presence of a large number of background mutations, any mutant, no matter how promising the phenotype is, should not be used directly for breeding observation without sufficient backcrosses.

   With the development of variety-independent transformation and genome editing techniques, the lengthy procedures of elimination of background mutations can be replaced by re-creating the mutant from the original parent by editing the causal gene, provided the causal gene for the mutant phenotype is known [36]. This is particularly useful for the introduction of the mutation into other elite lines suitable for breeding.

Acknowledgment

We thank the United Sorghum Checkoff for funding and Lan Liu-Gitz for technical support.

Disclaimer: Mention of trade names or commercial products in this article is solely for the purpose of providing specific information and does not imply recommendation or endorsement by the U.S. Department of Agriculture.

USDA is an equal opportunity provider and employer.

References

1. Quinby JR, Karper RE (1942) Inheritance of mature plant characters in sorghum. Induced by Radiation J Hered 33:323–327
2. Gaul H (1964) Mutations in plant breeding. Radiat Bot 4:155–232
3. Sree Ramulu K (1970) Induced systematic mutations in sorghum. Mutat Res Fundam Mol Mech Mutagen 10:77–80
4. Sree Ramulu K (1970) Sensitivity and induction of mutations in sorghum. Mutat Res Fundam Mol Mech Mutagen 10:197–206

5. Sree Ramulu K, Sree Rangasamy SR (1972) An estimation of the number of initials in grain sorghum using mutagenic treatments. Radiat Bot 12:37–43
6. Quinby JR (1975) The genetics of sorghum improvement. J Hered 66:56–62
7. Ejeta G, Axtell J (1985) Mutant gene in sorghum causing leaf "reddening" and increased protein concentration in the grain. J Hered 76:301–302
8. Oria MP, Hamaker BR, Axtell JD, Huang CP (2000) A highly digestible sorghum mutant cultivar exhibits a unique folded structure of endosperm protein bodies. Proc Natl Acad Sci U S A 97:5065–5070
9. Singh R, Axtell JD (1973) High lysine mutant gene (hl) that improves protein quality and biological value of grain sorghum. Crop Sci 13:535–539
10. Porter KS, Anxtell JD, Lechtenberg VL, Colenbrander VF (1978) Phenotype, fiber composition, and in vitro dry matter disappearance of chemically induced brown midrib (bmr) mutants of sorghum. Crop Sci 18:205–208
11. Xin Z, Wang ML, Barkley NA, Burow G, Franks C, Pederson G et al (2008) Applying genotyping (TILLING) and phenotyping analyses to elucidate gene function in a chemically induced sorghum mutant population. BMC Plant Biol 8:103
12. Abe A, Kosugi S, Yoshida K, Natsume S, Takagi H, Kanzaki H et al (2012) Genome sequencing reveals agronomically important loci in rice using MutMap. Nat Biotechnol 30:174–178
13. Jiao Y, Burow G, Gladman N, Acosta-Martinez V, Chen J, Burke J et al (2018) Efficient identification of causal mutations through sequencing of bulked f2 from two allelic bloomless mutants. Front Plant Sci 8:2267
14. Paterson AH (2008) Genomics of sorghum. Int J Plant Genomics 2008:362451
15. Paterson AH, Bowers JE, Bruggmann R, Dubchak I, Grimwood J, Gundlach H et al (2009) The *Sorghum bicolor* genome and the diversification of grasses. Nature 457:551–556
16. Blomstedt CK, Gleadow RM, O'Donnell N, Naur P, Jensen K, Laursen T et al (2012) A combined biochemical screen and TILLING approach identifies mutations in Sorghum bicolor L. Moench resulting in acyanogenic forage production. Plant Biotechnol J 10:54–66
17. Krothapalli K, Buescher EM, Li X, Brown E, Chapple C, Dilkes BP, Tuinstra MR (2013) Forward genetics by genome sequencing reveals that rapid cyanide release deters insect herbivory of *Sorghum bicolor*. Genetics 195:309–318
18. Rizal G, Karki S, Alcasid M, Montecillo F, Acebron K, Larazo N et al (2014) Shortening the breeding cycle of sorghum, a model crop for research. Crop Sci 54:520
19. Jiao Y, Burke JJ, Chopra R, Burow G, Chen J, Wang B et al (2016) A sorghum mutant resource as an efficient platform for gene discovery in grasses. Plant Cell 28:1551–1562
20. Mace ES, Hunt CH, Jordan DR (2013) Supermodels: sorghum and maize provide mutual insight into the genetics of flowering time. Theor Appl Genet 126:1377–1395
21. Lasky JR, Upadhyaya HD, Ramu P, Deshpande S, Hash CT, Bonnette J et al (2015) Genome-environment associations in sorghum landraces predict adaptive traits. Sci Adv 1(6):e1400218
22. Morris GP, Ramu P, Deshpande SP, Hash CT, Shah T, Upadhyaya HD et al (2013) Population genomic and genome-wide association studies of agroclimatic traits in sorghum. Proc Natl Acad Sci U S A 110:453–458
23. Xin Z, Gitz D, Burow G, Hayes C, Burke JJ (2015) Registration of two allelic erect leaf mutants of sorghum. J Plant Regist 9:254–257
24. Xin Z, Burow GB, Burke JJ 2014 Multi-seed mutant of sorghum for increasing grain yield. US Patent US 2014/0068798 Al, 6 Mar 2014
25. Burow G, Xin Z, Hayes C, Burke J (2014) Characterization of a multiseeded mutant of sorghum for increasing grain yield. Crop Sci 54:2030–2037
26. Jiao Y, Lee YK, Gladman N, Chopra R, Christensen SA, Regulski M et al (2018) MSD1 regulates pedicellate spikelet fertility in sorghum through the jasmonic acid pathway. Nat Commun 9:822. https://doi.org/10.1038/s41467-018-03238-4
27. Duvick DN, Cassman KG (1999) Post-green revolution trends in yield potential of temperate maize in the north-Central United States. Crop Sci 39:1622–1630
28. Tian F, Bradbury PJ, Brown PJ, Hung H, Sun Q et al (2011) Genome-wide association study of leaf architecture in the maize nested association mapping population. Nat Genet 43:159–162
29. Assefa Y, Staggenborg SA (2011) Phenotypic changes in grain sorghum over the last five decades. J Agron Crop Sci 197:249–257
30. Xin Z, Wang M, Burow G, Burke J (2009) An induced sorghum mutant population suitable for bioenergy research. Bioenergy Res 2:10–16

31. Dhugga KS (2007) Maize biomass yield and composition for biofuels. Crop Sci 47:2211–2227
32. Miao J, Guo D, Zhang J, Huang Q, Qin G, Zhang X et al (2013) Targeted mutagenesis in rice using CRISPR-Cas system. Cell Res 23:1233–1236
33. Zhou B, Yue Y, Sun X, Ding Z, Ma W, Zhao M (2017) Maize kernel weight responses to sowing date-associated variation in weather conditions. Crop J 5:43–51
34. Greene EA, Codomo CA, Taylor NE, Henikoff JG, Till BJ, Reynolds SH et al (2003) Spectrum of chemically induced mutations from a large-scale reverse-genetic screen in Arabidopsis. Genetics 164:731–740
35. Schneeberger K, Ossowski S, Lanz C, Juul T, Petersen AH, Nielsen KL et al (2009) SHOREmap: simultaneous mapping and mutation identification by deep sequencing. Nat Methods 6:550–551
36. Che P, Anand A, Wu E, Sander JD, Simon MK, Zhu W et al (2018) Developing a flexible, high-efficiency agrobacterium-mediated sorghum transformation system with broad application. Plant Biotechnol J 16(7):1388–1395. https://doi.org/10.1111/pbi.12879
37. Burow GB, Klein RR, Franks CD, Klein PE, Schertz KF, Pederson GA et al (2011) Registration of the BTx623/IS3620C recombinant inbred mapping population of sorghum. J Plant Regist 5:141–145
38. Feltus FA, Hart GE, Schertz KF, Casa AM, Kresovich S, Abraham S et al (2006) Alignment of genetic maps and QTLs between inter- and intra-specific sorghum populations. Theor Appl Genet 112:1295–1305
39. Menz MA, Klein RR, Mullet JE, Obert JA, Unruh NC, Klein PE (2002) A high-density genetic map of *Sorghum bicolor* (L.) Moench based on 2926 AFLP, RFLP and SSR markers. Plant Mol Biol 48:483–499
40. Brown PJ, Klein PE, Bortiri E, Acharya CB, Rooney WL, Kresovich S (2006) Inheritance of inflorescence architecture in sorghum. Theor Appl Genet 113:931–942
41. Henikoff S, Till BJ, Comai L (2004) TILLING. Traditional mutagenesis meets functional genomics. Plant Physiol 135:630–636
42. Howe A, Sato S, Dweikat I, Fromm M, Clemente T (2006) Rapid and reproducible agrobacterium-mediated transformation of sorghum. Plant Cell Rep 25:784–791
43. Peters PJ, Jenks MA, Rich PJ, Axtell JD, Ejeta G (2009) Mutagenesis, selection, and allelic analysis of epicuticular wax mutants in sorghum. Crop Sci 49:1250–1258

Chapter 6

Marker Assisted Selection in Sorghum Using KASP Assay for the Detection of Single Nucleotide Polymorphism/ Insertion Deletion

Gloria Burow, Ratan Chopra, Halee Hughes, Zhanguo Xin, and John Burke

Abstract

Marker assisted selection (MAS), an advance tool in plant breeding that allows accurate and efficient introgression of important agronomic trait(s) from a germplasm source to desired elite lines, has been applied to sorghum recently. Here, we report the methods for the deployment of MAS for trait introgression using endpoint genotyping technology for single nucleotide polymorphism (SNP)/insertion deletion (InDel) coupled with an application of KASP (Kompetitive Allele Specific Polymerase Chain Reaction [PCR]) chemistry allowing for the selection of parents for generational advancement without going through the laborious and time consuming phenotypic selection and additional generations for selection of desired individuals. This MAS-SNP marker assisted backcrossing scheme can be applied to accurately select heterozygotes for use as an allele donor parent in each backcross generation, thus expediting the backcrossing scheme and resulting in time savings of 3 years compared to conventional methods of introgression practiced in sorghum breeding and improvement.

Key words Marker assisted selection (MAS), Single nucleotide polymorphism (SNP), Endpoint genotyping, Kompetitive Allele Specific PCR(KASP), Cluster analysis

1 Introduction

Marker assisted selection (MAS) refers to the selection of target trait(s) based on DNA markers that specify the variation from the actual gene that underpins the trait or from closely linked gene (s) [1, 2]. MAS is useful as a breeding tool for the maintenance of recessive alleles during backcrossing and for expediting backcross breeding in general [3]. In sorghum, marker assisted selection for a number of traits is an important tool that can significantly increase efficiency and genetic gain in recurrent selection and backcross (introgression) breeding programs. The availability of the sorghum genome sequence [4] (https://phytozome.jgi.doe.gov, www.gramene.org), and of vital genomic resources such as mutant populations [5, 6] and recombinant inbred mapping populations [7]

Zuo-Yu Zhao and Jeff Dahlberg (eds.), *Sorghum: Methods and Protocols*, Methods in Molecular Biology, vol. 1931,
https://doi.org/10.1007/978-1-4939-9039-9_6,

have made possible the discovery of genes that control a number important agronomic traits which are controlled by recessive genes, such as brown midrib (*bmr*), absence of epicuticular wax (bloomless, *bm*), and maturity (*ma*) genes [6–11]. The specific nucleotide polymorphism that resulted in the bloomless phenotype has been validated using segregating breeding populations and during introgression into different genetic backgrounds [6].

The translation of causal SNPs into perfect DNA markers for rapid assay and accurate prediction of a phenotype is an important goal and application of marker assisted selection during introgression and for a majority of desirable traits needed in continued crop improvement [12]. Here, we focus on the description of methods pertaining to development and assay of SNP/InDel markers found in the causal gene using the SNPs for bloomless sorghum as a case study to demonstrate rapid genotyping and efficient selection for the trait in a marker assisted backcrossing project [6]. The application of the KASP assay is described here as the genotyping technology, because of its ease of use, consistency, and cost effectiveness [13–15]. KASP chemistry has been documented to be an effective method for medium throughput genotyping projects previously in maize [15] and recently in other economically important cereal crops. In sorghum, marker assisted backcrossing using KASP can expedite the development of near isogenic lines to 3 years compared to 5–6 years using conventional method of backcrossing and improved overall efficiency both in cost and accuracy of introgression (Fig. 1).

2 Materials

2.1 Segregating Population and Tissue Samples

1. Germplasm materials; specific crosses and populations used for backcross introgression populations for epicuticular wax (EW) mutants or bloomless (underpinned by *Sbcer5* and *Sbcer6*) sorghum using source such as: ARS 20 [6] to another B (or maintainer line), serving as the recurrent parent.
2. Tissue samples from a small section of young leaves at seedling stage (5–7 day old plants; 3–4 leaf stage) grown in small seedling pots in greenhouses or under controlled conditions.

2.2 Endpoint Genotyping Coupled with KASP Assay

1. Genomic DNA

   High-quality genomic DNA samples are necessary for clear and consistent allele calls using KASP assay. Genomic DNA in the amount of 10–20 ng from each individual and germplasm/line of interest to be genotyped is required (*see* **Note 1**).

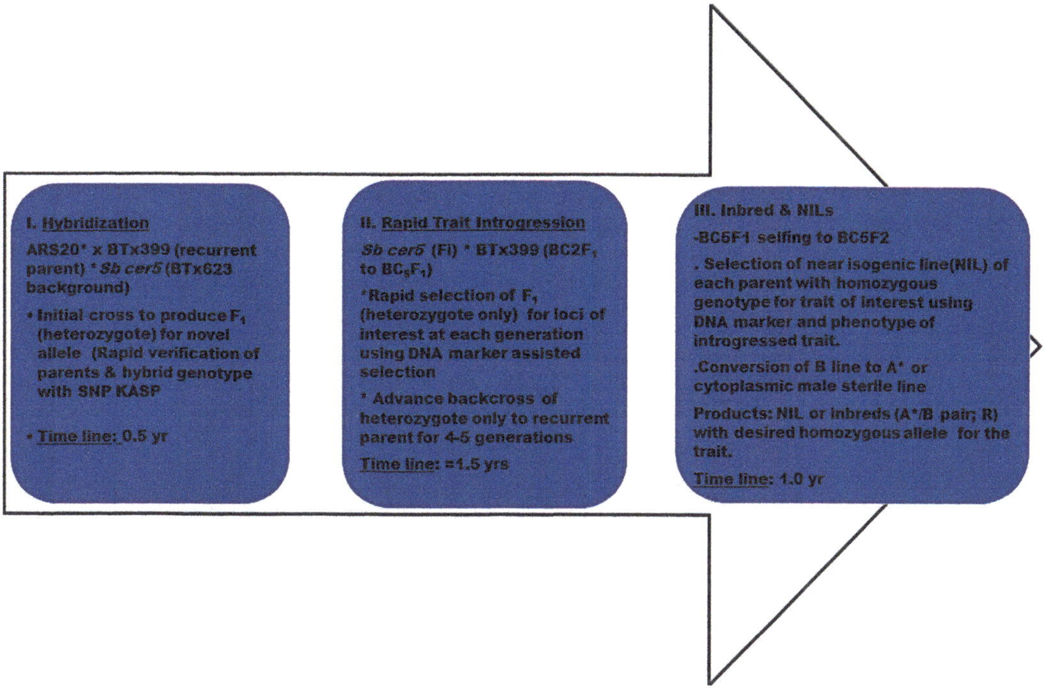

Fig. 1 DNA marker assisted backcrossing scheme for sorghum using Sb cer5 and Sb cer6 as examples for introgression into a recurrent parent

2. SNP Primers

   Three primers for each genotyping reaction are needed; two competitive forward primers; a wild type allele primer (commonly referred to as Allele X); mutant allele primer (referred to as Allele Y) and a common reverse primer.

3. KASP assay components

   KASP master mix in 2× format from LGC® genomics.

4. Equipment

   (a) Tissue grinder such as Qiagen® Tissue Lyser II for use in grinding small quantities of leaf tissue to fine powder with good consistency between samples.

   (b) An ultraviolet/visible spectrophotometer for the quantification of genomic DNA and normalization of DNA content to 10–20 ng/μl (*see* **Note 2**).

   (c) Polypropylene white or black 96-well PCR plates compatible with fluorescence energy transfer (FRET) analysis.

   (d) Multichannel micropipette.

   (e) For the endpoint genotyping analysis, an integrated machine capable of polymerase chain reaction (PCR), fluorescent detection and equipped with data analysis software such as Roche Light Cycler 480® (software v1.5)(= LC480) is one of the good options for performing genotyping with KASP chemistry (*see* **Note 3**).

3 Methods

3.1 Design and Development of KASP SNP Primers

In house design of SNP primers for KASP is a great time saving measure and protocol, which adds value in terms of flexibility, cost effectiveness, and faster turn-around time for obtaining genotype results (*see* **Note 4**).

1. Download the DNA sequence of the gene of interest from https://phytozome.jgi.doe.gov by clicking the sorghum image with caption *Sorghum bicolor* v3.1.1 in the home page of the website or through the pulldown menu for *Sorghum bicolor* v3.1.1 under the species heading.
2. Using the adjacent key word search option, copy or type in the specific gene ID for the gene of interest; for example the search for EW genes utilizes *Sb cer5*, gene ID = Sobic.009G083300 and *Sb cer6*, gene ID = Sobic.001G453200 as the search parameter. Once they are entered click on GO.
3. This will open a new window with the search result for the gene of interest, then click on G = gene view; which will lead to the Gene Info page; in the Gene Info page, click on genomic sequence, then copy and paste the genomic sequence to a word processor.
4. Using the now downloaded DNA sequence of interest, locate the SNP or InDel in the sequence, and use this nucleotide as the focal point to obtain 200 nucleotides each from both the 5′ and 3′ directions. It is important to make sure to replace the original SNP/InDel nucleotide variant to its new designated IUPAC CODE (*see* **Note 5**) in the sequence. This 400 nucleotides region of the gene with SNP/InDel replaced with its IUPAC designation will be used as sequence for designing the allele-specific KASP primers.
5. In a desktop or laptop, open the website of BatchPrimer3 (https://probes.pw.usda.gov/cgi-bin/batchprimer3/batchprimer3.cgi) and within the program choose the primer type: *allele-specific primers and allele-flanking primers* from the pulldown options. Then, input sequence of interest by pasting the DNA sequence from (3.1d) in the space provided.
6. Design the KASP primer to span 50–60 nucleotides on each side of the specified SNP by modifying a number of critical parameters within the *SNP or allele-specific primers settings* section of the Batch Primer 3 program from the default settings. Recommended values are: Primer size:16–25 nucleotides; Primer Tm: default of 57 is acceptable; Primer GC%: Min = 40; Max= 70. Scroll down to the section on *General settings for generic primers*, modify product size as; Min = 80 Max = 120; followed by modifying Max 3′ stability to: 4.

All other default parameters can be maintained in the program. Then scroll back up near the heading of the webpage, click on pick primers.

7. This will lead to a report in a separate page which specifies two forward primers under the allele-specific primer heading and a reverse primer under the allele flanking primer heading. Choose the two forward primers from allele-specific primer results as KASP forward primers and the reverse primer as a common primer. The forward primers need to be tailed with KASP compatible oligo sequence; Tail A = GAAGGTGACCAAGTT-CATGCT; Tail B = GAAGGTCGGAGTCAACGGATT [13].
8. The newly designed primers can now be outsourced for oligo synthesis and validated.
9. An example of in-house designed KASP primers validated for *Sb cer5* and *Sb cer6* is presented in Table 1.

3.2 Endpoint Genotyping and KASP Assay

Endpoint genotyping refers to a system where fluorescent reads are taken post PCR and relies on differential incorporation of fluorescence between the two PCR products to produce results. KASP is an end-point genotyping system and fluorescent reads are taken after amplification. All the procedures can be performed at room temperature unless otherwise indicated. Master mixes are prepared using ultrapure water (prepared to a sensitivity of 18 MΩ-cm using reverse osmosis column).

1. Turn ON and allow Roche LC 480 or compatible FRET machine to run for at least 15 min before proceeding to the preparation of the reaction mix. Obtain white 96-well Roche plates that are for FRET and Roche LC 480® compatible machines.
2. The basic reaction mix (~10 μl total) per sample or well for KASP genotyping consisting of the following reagents per reaction:

   1 μl of g DNA (conc of 10–20 ng/μl).

   5 μl of KASP 2× Mastermix (KASP MM).

   0.14 μl of KASP SNP primer.

   4 μl of PCR water.
3. Because of the minute amount of volumes, it is best to prepare a master mix (MMix) cocktail for KASP MM, SNP Primer, and H_2O without the genomic DNA. The volumes needed to prepare a MMix for 50 and 100 reactions for ½ or full 96-well plate assay are shown in Table 2 (*see* **Note 6**).
4. Dispense 9 μl each of MMix into wells of the white Roche 96-well plate.
5. Add 1 μl of genomic DNA using multichannel pipette.

Table 1
List of allele-specific forward and common reverse KASP primers for sorghum bloomless trait [6]

Given gene name (trait)	Sorghum gene	Mutation position	Forward primers	Common primer
Sb cer5 (bloomless)	Sobic.009G083300	Chr9:12577924	AlleleX = ACTCTTCGATGACCTTTGCC	GCGTCACCGAAGTAGACACA
			AlleleY=CACTCTTCGATGACCTTTGCT	
Sb cer6 (bloomless)	Sobic.001G453200	Chr1:65789406	AlleleX = TGACGAGCTGCGCCTC	CACTACATCCCGCCCAAC
			AlleleY = TGACGAGCTGCGCCTT	

Table 2
Reagents and volume for preparing MMix for 50 and 100 reactions of KASP assay for sorghum genotyping

Reagents	Volume (μl) needed to prepare 50 KASP assays	Volume (μl) needed to prepare 50 KASP assays
2× KASP master mix	250	500
KASP SNP primer assay mix	7	14
Water	200	400
Genomic DNA	Added separately to each well	Added separately to each well

6. Mix each assay mix in the wells by pipetting up and down carefully 3–5×; making sure to avoid cross contamination between wells.
7. Seal with LC480 plastic seal tightly, making sure to prevent spills and cross contamination. Cover or wrap in aluminum foil to prevent light exposure and quenching.
8. Centrifuge at 1606 × *g* for 5 min, cover plates with aluminum foil when transporting.
9. Go to the LC480 machine, place assay plate in the plate holder making sure to line up the plate according to the notch in the plate.
10. Proceed with specific KASP assay genotyping protocol validated to work for sorghum in the Roche 480 Light Cycler with the following linked programs: Program 1- enzyme activation at 94 °C for 15 min, 1 cycle; then Program 2- touchdown PCR consisting of a denaturation step at 94 °C followed by actual touch down steps from 65 °C to 57 °C, dropping at 0.6 °C per cycle for 10 cycles; then Program 3-thermocycling, consisting of a denaturation step at 94 °C for 20 s followed by rapid decline to a target temperature of 55 °C for 1 min for annealing and elongation for 30 cycles.
11. Complete the KASP PCR assay with a quantification step (Program 4) consisting of 37 °C for 3 min, followed by 37 °C single data acquisition for 1 s (*see* **Note 7**).

3.3 Data Acquisition and Analysis

1. The data to be gathered are fluorescence values for 523–558 nm and 483–533 nm, presented in a table format that can be exported as a MS Excel file for easier handling and back up of data.
2. Allele calls based on fluorescence data can be obtained using allele calling software embedded within the endpoint genotyping suite of Roche LC480 and will be presented as graphical clusters with differential colors.

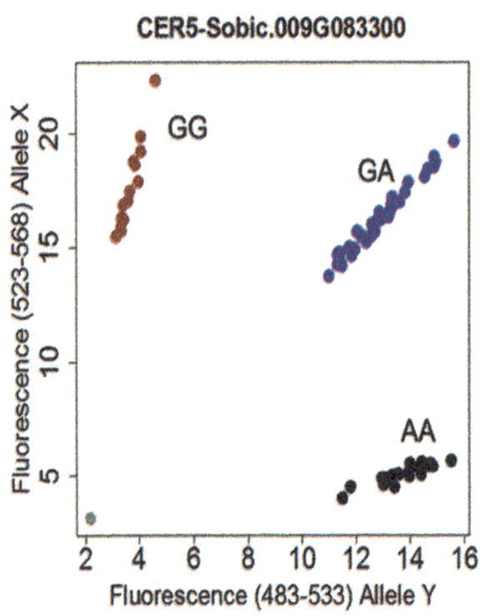

Fig. 2 Genotype plot for F_2 populations of *Sb cer5* (Sobic.009G083300) and *Sb cer6* (Sobic.001G453200) using KASP assay targeting the specific SNP variant that control the bloomless trait in sorghum. In this case, 48 offspring from an F_2 population was genotyped and three distinct clusters were obtained from the endpoint-KASP genotyping assay

3. Further analysis of allele calls for genotyping can be performed using the cluster analysis function within the MS Excel program or using K-means cluster analysis in R.
4. Distinct separation of fluorescence data in the graph allows the separation and identification of homozygotes (wild type versus homozygous mutants as examples). Heterozygotes will form distinct cluster between the two homozygote clusters. A graphical representation of the results obtained for *Sbcer5* and *Sbcer6* is presented in Fig. 2.
5. In cases where the clusters are not distinct, recycling of the plate can be performed for 5–10 more cycles using Program 3, followed by a quantification step using Program 4 (*see* **Note 8**).

4 Notes

1. There are a number of commercial genomic DNA extraction kits and technology that work well with sorghum including DNeasy plant mini kit (Qiagen) and Gen Elute Plant DNA miniprep kit to mention some.
2. A compact uv spectrophotometer that requires very tiny volume (i.e., 1 μl) of g DNA sample for analysis (such as Nanodrop) works well for the determination of sorghum genomic DNA concentration.
3. PCR and fluorescence detection can be separated from each other such that a regular Peltier PCR machine such as MJ200

can be used for the PCR step in the KASP assay. The FRET capable fluorescence machine provided with correct filters for 483–533 and 523–568 can be used for measuring the fluorescence data.

4. Alternatively, combined KASP primers design and synthesis can be outsourced to LGC Genomics through submission of sequence data highlighting the specific SNP/InDel to be targeted. LGC genomics offer two formats for design of SNP primers compatible with KASP master mix, such as KASP by design (KBD), where the end user will submit the DNA sequence highlighting SNP of focus and will deliver unvalidated SNP primers in assay mix format. The KASP on demand (KOD) format also requires the end user to submit DNA sequences highlighting the SNP of focus, but will deliver guaranteed optimized SNP marker and assay mix.
5. The SNP will vary between the parents and lines to be analyzed. The standard designations for IUPAC Code for nucleotide change that BatchPrimer3 accepts are as follows: R = A or G; Y = C or T; S = G or C; W = A or T; K = G or T; and M = A or C.
6. During the preparation of MMix remember to: (1) Turn off light when handling KASP 2× MM, reagents are light sensitive; and, (2) It is suggested to dispense MM first to the wells then add appropriate amount of genomic DNA (usually 1 μl/well).
7. It is important to allow for more than 3 min to cool down to 37 °C before reading the fluorescence value to obtain consistent results.
8. In a number of cases, there are no distinct clusters that are obtained from the assay and most of the time this is resolved by recycling the plate and adding additional PCR cycles to the run.

Acknowledgment

The authors gratefully acknowledge the support of US Sorghum Checkoff Program for funding the research described in this chapter.

Disclaimer: Mention of trade names or commercial products in this article is solely for the purpose of providing specific information and does not imply recommendation or endorsement by the U.S. Department of Agriculture. USDA is an equal opportunity provider and employer.

References

1. Ribaut JM, Hoisington D (1998) Marker assisted selection: new tools and strategies. Trends Plant Sci 3:236–239
2. Xu Y, Crouch JH (2008) Marker assisted selection in plant breeding: from publication to practice. Crop Sci 48:391–407
3. Bernardo R, Charcosset A (2006) Usefulness of gene information in marker assisted recurrent selection: a simulation appraisal. Crop Sci 46:614–621
4. Paterson AH, Bowers JE, Bruggmann R, Dubchak I, Grimwood J, Gundlach H et al (2009) The *Sorghum bicolor* genome and the diversification of grasses. Nature 457:551–556
5. Xin Z, Wang M, Burow G, Burke J (2009) An induced sorghum mutant population suitable for bioenergy research. Bioenergy Res 2:10–16
6. Jiao Y, Burke JJ, Chopra R, Burow G, Chen J, Wang B et al (2016) A sorghum mutant resource as an efficient platform for gene discovery in grasses. Plant Cell 28:1551–1562. https://doi.org/10.1105/tpc.16.00373
7. Murphy RL, Klein RR, Morishige DT, Brady JA, Rooney WL, Miller FR et al (2011) Coincident light and clock regulation of *pseudoresponse regulator protein 37*(*PRR37*) controls photperiod flowering in sorghum. PNAS 108:16469–16474
8. Klein R, Miller FR, Dugas DV, Brown PJ, Burrell AM, Klein PE (2015) Allelic variants in the *PRR37* gene and the human-mediated dispersal and diversificaiton of sorghum. Theor Appl Genet 128:1669–1683
9. Bout S, Vermerris W (2003) A candidate-gene approach to clone the sorghum brown midrib gene encoding caffeic acid O-methyl transferase. Mol Gen Genomics 269:205–214
10. Sattler SE, Saathoff AJ, Haas EJ, Palmer NA, Funnell-Harris DL, Sarath G et al (2009) A nonsense mutation in a cinnamyl alcohol dehydrogenasegene is responsible for the sorghum *brown midrib6* phenotype. Plant Physiol 150:584–595
11. Sattler SE, Palmer NA, Saballos A, Greene AM, Xin Z, Sarath G et al (2012) Identification and characterizaiton of four missense mutations in *brown midrib12* (*Bmr12*), the caffeic o-methyltransferase (COMT) of sorghum. Bioenergy Res 5:855–865
12. Franks CD, Mayor L (2014) Molecular breeding. In: Wang Y-H, Uphadyaya HD, Kole C (eds) Genetics, genomics and breeding of sorghum. CRC Press, Boca Raton, pp 169–181
13. Rosas JE, Bonnecarrère V, Pérez de Vida F (2014) One-step, codominant detection of imidazolinone resistance mutations in weedy rice (*Oryza sativa* L.). Electron J Biotechnol 17:95–101
14. He C, Holme J, Anthony J (2014) SNP genotyping: the KASP assay. In: Fleury D, Whitford R (eds) Crop breeding. Methods in molecular biology (methods and protocols), vol 1145. Humana Press, New York, NY
15. Semagn K, Babu R, Hearne S, Olsen M (2013) Single nucleotide polymorphism genotyping using Kompetitive Allele Specific PCR (KASP): overview of the technology and its application in crop improvement. Mol Breed 33:1–14

Part II

Grain Compounds, Nutrition, and Digestibility

Chapter 7

Assaying Sorghum Nutritional Quality

Kwaku G. Duodu

Abstract

Sorghum (*Sorghum bicolor* L. Moench) is a major drought-tolerant cereal crop grown mainly in the semi-arid regions of the world. It is an important basic food cereal in these regions which encompasses most of the developing world in many parts of Africa and Asia. Therefore, sorghum is an important source of nutrients for millions of inhabitants in these regions. In light of this, the nutritional quality of sorghum and how this is assessed is of major research interest. Various assays have been used to determine the contents of macronutrients and micronutrients in sorghum including how digestible and bioaccessible these are. A wide range of indices of sorghum nutritional quality has been generated. Advances in analytical instrumentation have contributed significantly to enhancing the capacity of analysts and researchers to broaden the scope of assays for sorghum nutritional quality and also to improve their accuracy.

Key words Sorghum, Nutritional quality, Starch, Protein, Digestibility, Dietary fiber, Vitamins, Minerals, Bioaccessibility

1 Introduction

Sorghum (*Sorghum bicolor* L. Moench) is the fifth most important cereal crop in the world in terms of production. According to FAOSTAT data, the top five sorghum producers in the world in terms of average production over the period 1994 to 2016 were the USA (11.7 million tonnes), Nigeria (7.4 million tonnes), India (7.3 million tonnes), Mexico (6.0 million tonnes), and Sudan (4.5 million tonnes) [1]. Compared to other cereals such as maize, sorghum has the distinct advantage of being drought-tolerant with low water requirements. In contrast to countries such as the USA where sorghum is primarily used as animal feed, many subsistence farmers in the semi-arid tropics in countries such as India and in sub-Saharan Africa cultivate sorghum as a staple food crop for human consumption. In this regard, sorghum is important to the world food economy and plays an important role in contributing to household food security in the most food-insecure regions of the world.

Zuo-Yu Zhao and Jeff Dahlberg (eds.), *Sorghum: Methods and Protocols*, Methods in Molecular Biology, vol. 1931, https://doi.org/10.1007/978-1-4939-9039-9_7,

Sorghum is processed into a wide range of food products such as porridges (fermented and non-fermented; soft or stiff), flatbreads, and beverages (non-alcoholic and alcoholic) for human consumption. For millions of people living in the semi-arid regions of the world, sorghum is a principal source of energy, protein, vitamins, and minerals. The nutritional quality of sorghum is therefore of increasing interest with a focus on strategies for enhancing the nutritional quality of sorghum. Some of these strategies include the use of simple and more traditional processing techniques such as fermentation and malting and more sophisticated approaches such as the use of genetic engineering.

Therefore, the assaying of sorghum nutritional quality forms an essential aspect of research on sorghum nutrition. A very wide range of assay protocols have been used to determine sorghum nutritional quality. These assays are used to determine the content of nutrients in sorghum, their digestibility and bioaccessibility. This chapter presents a review of some of the important indices of sorghum nutritional quality and how these are assayed. The focus will be on in vitro methods and will discuss the assaying of sorghum nutritional quality parameters related to macronutrients such as starch, protein, and dietary fiber and micronutrients such as vitamins and minerals.

2 In Vitro Starch Digestibility

The determination of in vitro starch digestibility continues to be one of the major indices of sorghum nutritional quality. In grain-based diets, starch is the most important source of glucose which is utilized in respiratory metabolism for production of energy. The need to determine in vitro starch digestibility of sorghum has become even more significant due to its direct implications for blood sugar levels and diabetes.

In principle, determination of sorghum in vitro starch digestibility proceeds by two main processes: (1) treatment of the sorghum grain or sorghum-based food with one or more starch-hydrolyzing enzymes which is invariably broken down to glucose, and (2) determination of the glucose concentration spectrophotometrically using some appropriate chromogenic reagent which reacts with the glucose. In an early study of sorghum starch digestibility [2], sorghum *ogi* (a soft porridge) was treated with glucoamylase and glucose concentration determined using a hexokinase assay procedure. Hexokinase catalyzes the phosphorylation of glucose by adenosine triphosphate (ATP) to glucose-6-phosphate which is then oxidized to 6-phosphogluconate in a reaction with oxidized nicotinamide adenine dinucleotide (NAD) catalyzed by glucose-6-phosphate dehydrogenase. During this oxidation, an equimolar amount of NAD is reduced to NADH. The consequent increase in absorbance at 340 nm is directly proportional to glucose concentration.

Sorghum starch digestibility analysis has also been conducted by first digestion of sorghum with glucoamylase followed by determination of reducing sugars using the Nelson-Somogyi method [3] which is based on the absorbance at 520 nm of a colored complex between a copper oxidized sugar and arsenomolybdate. Other workers [4, 5] also used the Nelson-Somogyi method to determine reducing sugar concentration after digestion of cooked sorghum flours with porcine pancreatic alpha-amylase. Tawaba et al. [6] determined in vitro digestibility of red sorghum starch by digestion with alpha- and beta-amylase followed by the determination of glucose concentration with the Nelson-Somogyi method.

The dinitrosalycylic acid (DNS) method has also been used for the determination of reducing sugar after starch hydrolysis of cooked sorghum with alpha-amylase [7, 8]. In the DNS method, an alkaline solution of 3,5-dinitrosalicylic acid reacts with reducing sugars and gets converted to 3-amino-5-nitrosalicylic acid with an orange color. The intensity of the color, read at wavelength of 510 nm is an index of reducing sugar content. Ezeogu et al. [9] determined in vitro starch digestibility by digestion of cooked sorghum porridge with porcine pancreatic α-amylase. Soluble dextrins in supernatants were digested to glucose using glucoamylase from *Aspergillus oryzae*. Glucose was determined using the DNS reagent. Starch digestibility was expressed in terms of percentage of the total amount of starch at the start of the reaction, as determined by the Megazyme Total Starch Assay Procedure (amyloglucosidase/alpha-amylase method).

A somewhat different approach to the above has been used to determine in vitro starch digestibility of sorghum whereby different starch fractions namely, rapidly digestible starch (RDS), slowly digestible starch (SDS), and resistant starch (RS) were measured [10] according to the classical procedure of Englyst et al. [11]. The food sample was incubated with invertase, pancreatic α-amylase, and amyloglucosidase. Glucose released after 20 min was measured and referred to as rapidly available glucose (RAG). Glucose released after 120 min of incubation represented slowly digestible starch (SDS). Total glucose (TG) was obtained by gelatinization of the starch in boiling water, treatment with 7 M KOH followed by complete enzymatic hydrolysis with amyloglucosidase. Resistant starch (RS) was determined as the starch left unhydrolyzed after 120 min incubation. Free glucose (FG) was determined as the amount of glucose in the sample treated with acetate buffer without enzymatic hydrolysis.

Determination of glucose was done using the glucose oxidase/peroxidase method. The principle of this method is that glucose is oxidized to gluconic acid and hydrogen peroxide by glucose oxidase. Hydrogen peroxide reacts with o-dianisidine in the presence of peroxidase to form a colored product (oxidized o-anisidine). Oxidized o-dianisidine reacts with sulfuric acid to form a more stable colored product. The intensity of the pink color measured at 540 nm is proportional to the original glucose concentration.

The values for total starch (TS), RDS, SDS, and RS were calculated as follows:

$$\begin{aligned} \mathrm{TS} &= (\mathrm{TG} - \mathrm{FG}) \times 0.9; \mathrm{RDS} = (\mathrm{RAG} - \mathrm{FG}) \times 0.9; \mathrm{SDS} \\ &= (\text{Glucose released after 120 min} - \mathrm{FG}) \times 0.9; \mathrm{RS} \\ &= \mathrm{TS} - (\mathrm{RDS} + \mathrm{SDS}) \end{aligned}$$

The factor 0.9 in the above equations is derived from (162 ÷ 180) which is used to convert from free D-glucose as determined in the assay, to anhydro-D-glucose as it occurs in starch [12]. Starch values are thus calculated as (glucose X 0.9) [13]. In the procedure above, starch digestibility was calculated by expressing the RDS as a percentage of the TS.

This strategy of determining different starch fractions as a way of estimating sorghum starch digestibility has been applied by other workers [14]. RDS, SDS, and RS were determined according to Englyst et al. [13]. After pre-incubation with pepsin–HCl solution, the various categories of starch were measured after degradation in an enzyme mixture containing pancreatin, amyloglucosidase, and invertase enzymes. Glucose released from samples after 20 min and 120 min of incubation was measured colorimetrically using the glucose oxidase method and then converted to RDS, SDS, and RS indexes with the equations proposed in the Englyst et al. method [11]. The Englyst et al. method [11] has been used for raw and processed sorghum flours [15] and extruded whole grain sorghum breakfast cereals [16]. The same method described by Englyst et al. [17] has been used for sorghum starch [18]. Food samples were digested with alpha-amylase and amyloglucosidase. Glucose released after 20 min and 120 min was determined by the reaction with glucose oxidase reagent with absorbance read at 510 nm using the Megazyme glucose assay kit. In this assay, RDS, SDS, and RS were expressed as a percentage of total starch.

Some of the enzymatic hydrolysis procedures used during the determination of in vitro starch digestibility involve the use of protease enzymes. These protease enzymes may be utilized as part of a simulated in vitro gastrointestinal digestion or simply for the purpose of reducing the protein content in order to improve accessibility of starch hydrolysing enzymes to the starch. A method consisting of a 3-step enzymatic hydrolysis mimicking digestion in the mouth, stomach, and small intestine in a closed system followed by the measurement of glucose has been described [19, 20]. The first step involved digestion with artificial porcine salivary amylase (mimic digestion in the mouth), followed by digestion with porcine pepsin (mimic digestion in the stomach) and finally, digestion with pancreatin, protease, alpha-amylase, and amyloglucosidase (mimic digestion in the small intestine). Glucose was analyzed using the glucose oxidase colorimetric method. Similarly, Giuberti et al. [14] employed a two-step enzymatic hydrolysis simulating gastric (pepsin) and pancreatic (pancreatin, amyloglucosidase, and invertase) phases. Glucose released was determined using the colorimetric glucose oxidase method. In determining the effect of fermentation on in vitro starch digestibility of sorghum, a pre-incubation with pepsin was conducted to reduce protein interference before digestion with alpha-amylase and amyloglucosidase [21]. Glucose content was determined using the glucose oxidase/peroxidase reagent and converted to starch content by multiplying by 0.9. Starch digestibility was reported as concentration of starch hydrolysed expressed as a percentage of total starch.

The in vitro starch hydrolysis method used by Goni et al. [22] to estimate glycemic index has been used extensively to determine in vitro starch digestibility of sorghum and sorghum-based food products. The procedure measures the rate of starch digestion and the hydrolytic process is governed by a first order reaction equation. Food samples are digested with pepsin, followed by alpha-amylase and then amyloglucosidase for complete hydrolysis of starch to glucose. The amount of reducing sugars released is expressed as a percentage of total available carbohydrates and plotted against digestion time. A hydrolysis index parameter is calculated as the area under the hydrolysis curve for the samples expressed as a percentage of the corresponding area under the hydrolysis curve for a reference white wheat bread sample. The Goni et al. [22] method has been used by various authors for the determination of in vitro starch digestibility of hard and soft endosperm sorghum porridges [23, 24], gluten-free sorghum pasta [25], gluten-free sorghum bread [26, 27], popped sorghum [28], Algerian sorghum cultivars [29], and sorghum starch [30, 31].

A method originally developed to measure soluble and resistant starch in plant samples [32] has been used to determine in vitro starch digestibility of grain sorghum [33]. It involved digestion of samples with α-amylase and amyloglucosidase to glucose and determination of soluble starch content by measuring of glucose concentration using the glucose oxidase/peroxidase reagent. Total starch content was then measured using the polarimetric or Ewers' method as per Mitchel [34]. In this method, the starch is released from the sample by boiling in dilute hydrochloric acid (HCl). This procedure effectively gelatinizes the starch granules and simultaneously hydrolyzes the starch to glucose in a single step. The acid also helps to break down the endosperm tissue, ensuring complete release of the starch granules from the protein matrix. Substances that may interfere with the measurement are removed by filtration/clarification and then glucose concentration is determined by measuring the angle of polarization or optical rotation.

3 Protein Quality

As a food staple, many people rely on cereal grains such as sorghum for their daily supply of various nutrients including protein. This is also particularly the case given that other sources of protein such as from animals tend to be expensive and so are inaccessible to poor rural community dwellers in these regions. The protein quality of sorghum is therefore an important parameter and has been the subject of extensive research. This has also generated an end goal of attaining improved sorghum protein quality which has become a driver for the investigation of food processing techniques, plant breeding techniques, and even genetic engineering as agents for improving sorghum protein quality. In this chapter, a few indices of sorghum protein quality will be discussed.

3.1 Amino Acid Composition

The amino acid composition is one of the most basic indicators of protein quality. In general, a protein may be considered of good quality if it is a good source of indispensable (or essential) amino acids. The indispensable amino acids (lysine, histidine, threonine, valine, isoleucine, leucine, methionine, phenylalanine, and tryptophan) are those that are not synthesized by the body and therefore must be provided in the diet. Various workers have reported on the amino acid composition of sorghum and sorghum foods.

Some useful reviews on amino acid analysis have been published [35–37]. Broadly speaking, amino acid analysis involves two steps: hydrolysis of the substrate and chromatographic separation and detection of amino acids in the hydrolysates. The most common hydrolytic procedure used is hydrolysis with 6 N HCl for 18–24 h at 110 °C. Acid hydrolysis has been used by many workers for amino acid analysis of sorghum and sorghum products [38–45]. Some problems associated with acid hydrolysis include partial or total destruction and incomplete hydrolysis of some amino acids [46]. Apart from the use of acid, other agents used for the hydrolysis process include microwave radiation, alkali (for the determination of tryptophan), and enzymes [35].

Derivatization is an important feature of the chromatographic process for separation and detection of amino acids. Post-column derivatization (which was developed first) involves separation of the amino acid residues on a cation exchange column and the eluting amino acids are reacted with either a chromophore or a fluorophore before detection [35]. Derivatization with ninhydrin is one of the most common post-derivatization techniques. For pre-column derivatization, the amino acids are derivatized first in the hydrolysate before separation using usually reversed-phase High Performance Liquid Chromatography (HPLC) [35]. A typically widely used pre-derivatization technique is the Pico-Tag method developed by Waters Associates which involves the use of phenylisothiocyanate as a derivatization agent to produce phenylthiocarbamyl derivates that are detected via UV absorbance [47].

3.2 In Vitro Protein Digestibility

In vitro protein digestibility is widely used as an index of protein quality. Essentially, it is a measure of how susceptible a protein is to proteolysis. Highly digestible proteins are of better quality since they would provide more amino acids for absorption upon proteolysis. The in vitro digestibility of sorghum proteins has been an area of extensive research for many years. The process for the determination of in vitro protein digestibility involves initially subjecting the food sample to proteolysis. The digested sample is then centrifuged and the protein digestibility is computed by expressing the protein content of the supernatant (solubilized nitrogen) as a percentage of the total protein content of the undigested sample. The protein content is normally determined as total nitrogen using either the Kjeldahl or Dumas combustion method. The proteolysis step can be conducted using a single enzyme or a combination of enzymes.

Akingbala et al. [2] used a single enzyme assay to determine the in vitro protein digestibility of *ogi* (a soft porridge) prepared with sorghum flour. In the assay, *ogi* porridge was treated with pronase (a *Streptomyces griseus* protease) enzyme. Free amino nitrogen was determined using an automated ninhydrin assay procedure. In the assay, ninhydrin (2,2-dihydroxyindane-1,3-dione) which is yellow in color and an oxidising agent, reacts with α-amino acids at pH 6–7 in an oxidative decarboxylation reaction to produce a blue-colored complex which is measured at 570 nm. This method of pronase digestion and nitrogen determination using the ninhydrin assay to determine the in vitro protein digestibility has been used for sorghum and maize tortillas [48].

The most popular and extensively used single enzyme method for in vitro protein digestibility of sorghum is the pepsin digestibility assay first reported by Axtell et al. [49]. The sorghum samples were digested with pepsin, centrifuged and the solubilized nitrogen in the supernatant was determined and expressed as a percentage of total nitrogen in the original undigested sample. Over the years, this method has been used by many workers and formed the basis for modifications in some instances for the determination of in vitro protein digestibility of sorghum or sorghum food products [7, 16, 21, 33, 48, 50–63].

In a departure from the usual determination of total nitrogen by either the Kjeldahl or Dumas combustion assays, Elkonin et al. [33] did a quantitative estimation of protein digestibility after pepsin digestion was done using an electrophoresis-based assay as reported earlier [64, 65]. Digested and control samples were subjected to SDS-PAGE under reducing conditions. Gels were scanned and the amount of protein, expressed as volume (intensity × area) of kafirin bands or total protein bands in the lane, was quantified. Digestibility value was determined as percent ratio of difference between protein volume in the control and digested sample to protein volume in the control sample.

Various combinations of enzymes have been used in multi-enzyme in vitro protein digestibility assays for sorghum grain and foods such as pepsin and pancreatin with nitrogen determination using the micro-Kjeldahl procedure [3, 25, 28, 66, 67]. A combination of trypsin, chymotrypsin, and peptidase enzymes has also been used [48].

3.3 Protein Digestibility Corrected Amino Acid Score (PDCAAS)

PDCAAS is a standard measure of protein quality based on human essential or indispensable amino acid requirements and the ability to digest it. Essentially, it provides an indication of how well the protein in a food can be used by the body. It was proposed in 1989 by a FAO/WHO Expert Consultation on Protein Quality Evaluation as an index of protein quality [68]. To calculate the PDCAAS, the content of the first limiting indispensable amino acid of the protein under study (lysine, in the case of sorghum protein) is expressed as a percentage of the content of the same amino acid in a reference pattern of indispensable amino acids. This reference pattern is based on the indispensable amino acid requirements of preschool-age children. The percentage obtained is corrected for the true fecal digestibility of the protein under study as measured in a rat assay [69].

The major advantages of the PDCAAS are the fact that it is simple and can be directly related to human protein requirements [70]. However, it has some limitations that relate to validity of basing the score on amino acid requirements of preschool-age children, correction for fecal instead of ileal digestibility, and truncation of PDCAAS values in instances where they exceed 100% [69, 70]. In light of these limitations, the FAO recently proposed the replacement of the PDCAAS with the digestible indispensable amino acid score (DIAAS) [71, 72]. The DIAAS is based on true ileal amino acid digestibility determined for each amino acid individually, and lysine availability estimates, utilizing non-truncated scores for food ingredients [72].

Notwithstanding its limitations, the PDCAAS still enjoys the status of being a standard and widely used index of protein quality and has been used extensively to evaluate protein quality of sorghum and sorghum-based foods. Sorghum has rather low PDCAAS and this has been attributed to the fact that sorghum protein is deficient in the indispensable amino acid lysine and has poor protein digestibility especially when wet cooked [73]. PDCAAS values reported for sorghum and sorghum-based food products include 0.24–0.58 for Algerian sorghum cultivars [74], 0.09–0.34 for raw sorghum flour and sorghum-based foods [75], 0.19–0.37 for raw sorghum flour, 0.20–0.36 for fermented sorghum flour, and 0.12–0.28 for sorghum *kisra* (a naturally lactic acid bacteria- and yeast-fermented sorghum pancake-like flatbread consumed in Sudan) [76], 0.34–0.36 for decorticated sorghum flour and 0.29–0.31 for extruded sorghum flour [77]. Comparatively, PDCAAS values of 0.74 for raw quality protein maize [75] and 0.89 and higher for dehulled and micronized cowpea flour [77] have been reported. As would be expected, the addition of legumes such as cowpea to sorghum in the preparation of composite foods improves the PDCAAS [75, 77].

3.4 Reactive (Chemically Available) Lysine Content

Owing to the fact that lysine is the limiting indispensable amino acid in sorghum protein, the reactive (or available) lysine content may be used as an index of sorghum protein quality. The "reactive lysine" refers to those lysine molecules that possess a chemically reactive (unblocked) ε-amino group and thus considered nutritionally available. A most commonly used method for the determination of reactive lysine content is the rapid dye-binding lysine (DBL) method as modified by Kim et al. [78]. The food sample is first shaken with Crocein Orange G dye and the amount of dye bound or dye-binding capacity (DBC) is determined by measuring the extinction of the dye at 482 nm. This provides a measure of the sum of total histidine, arginine, and reactive lysine. The food sample is then treated with propionic anhydride which masks the reactive lysine groups by neutralizing the basicity of free ε-NH_2 groups of reactive lysine in a propionylation reaction. The DBC of the food sample treated with propionic anhydride measures histidine and arginine. Therefore, the difference between the DBC of the food not treated with the propionylation agent and that of the food treated with it gives a measure of the reactive lysine content. Reactive lysine contents of 1.51–2.82 g/100 g protein for sorghum flour and sorghum-based foods [75] and 1.31–2.72 g/100 g protein for sorghum flour and sorghum *kisra* [76] have been reported.

4 Dietary Fiber

Dietary fiber is arguably the most complex nutrient in foods. The quest for a definition of dietary fiber has a long history and the evolution of such a definition is outlined to varying degrees in various articles [79–81]. In 2008 at a meeting in Cape Town, South Africa, the Codex Committee on Nutrition and Foods for Special Dietary Uses proposed the following consensus definition for dietary fiber: "Dietary fiber means carbohydrate polymers with ten or more monomeric units, which are not hydrolyzed by the endogenous enzymes in the small intestine of humans and belong to the following categories:

- Edible carbohydrate polymers naturally occurring in the food as consumed.

- Carbohydrate polymers, which have been obtained from food raw material by physical, enzymatic, or chemical means and which have been shown to have a physiological effect of benefit to health as demonstrated by generally accepted scientific evidence to competent authorities.
- Synthetic carbohydrate polymers that have been shown to have a physiological effect of benefit to health as demonstrated by generally accepted scientific evidence to competent authorities" [82].

From this definition therefore, dietary fiber consists of a wide range of carbohydrate components such as cellulose, hemicelluloses, soluble non-cellulosic polysaccharides, and RS. Lignin is also an important non-carbohydrate component of dietary fiber. Dietary fiber is regarded as an important nutrient of interest primarily due to the belief that it can exert physiological effects and contribute positively to the health, quality of life, and well-being of the consumer. Various studies have presented evidences that suggest either a preventive or therapeutic role of dietary fiber for various chronic diseases and conditions such as type 2 diabetes, coronary heart disease, obesity, constipation, diverticular disease, and colon cancer [80].

Analyses of dietary fiber content of sorghum and sorghum-based food products have been focused on the determination of total dietary fiber content which is obtained as the sum of soluble and insoluble dietary fiber. In some instances [83], dietary fiber has been determined by difference from the sum of ash, protein, fat, and starch contents. However, total dietary fiber content of sorghum and sorghum foods have been mostly determined using enzymatic-gravimetric methods [84–88]. Such enzymatic-gravimetric methods have been developed and published by bodies such as the Association of Official Analytical Chemists (AOAC) and the American Association of Cereal Chemists International (AACCI). The Megazyme Total Dietary Fiber Assay kit [89] presents an integrated enzymatic-gravimetric assay procedure for total dietary fiber which can be used with the AOAC and AACCI methods. In summary, it involves enzymatic treatment with thermostable alpha-amylase (for hydrolysis and depolymerization of starch), protease (to solubilize and depolymerize proteins), and amyloglucosidase (to hydrolyze starch fragments to glucose). After centrifugation, the filtrate is treated with ethanol to produce a precipitate which is dried and weighed. The final weight after correction for protein and ash content represents soluble dietary fiber. The residue obtained after the enzymatic treatments is similarly dried, weighed and represents insoluble dietary fiber after correction for protein and ash content. The sum of soluble and insoluble dietary fiber represents total dietary fiber.

5 Vitamins

As a major cereal staple for millions of people in countries within the semi-arid tropics where it is utilized for human consumption, sorghum is heavily relied upon for daily supply of micronutrients such as vitamins. Therefore, the analysis of vitamin content of sorghum and sorghum foods has been of keen interest to researchers. There has been interest in aspects such as the fate of vitamins during processing and the effect of initiatives such as biofortification and natural fortification on vitamin content of sorghum foods.

Broadly speaking, methods used for analysis of vitamins in sorghum and sorghum foods have been mostly chemical and on occasion, microbiological methods. Chemical methods have been usually spectrophotometric (colorimetric or fluorimetric) and there has been growing use of chromatographic methods. Assay protocols for vitamins involve the use of an appropriate extraction method for the vitamin of interest which is intended to quantitatively extract the vitamin from the food matrix in a form that can be easily measured. The extraction is often followed by a purification step, conversion of the vitamin into the appropriate analyte form before final analysis with the method of choice such as either spectrophotometry or chromatography. For assay of vitamins in sorghum and sorghum foods, extensive use has been made of official methods of analysis of AOAC and AACCI.

5.1 Vitamin A and Carotenoids

In sorghum and sorghum foods, vitamin A is normally analyzed as provitamin A carotenoids, mostly as beta-carotene. The AOAC method for determination of carotenoids (Method 970.64 or 45.1.014) [90] has been used for sorghum and sorghum food products [39, 91]. The method is spectrophotometric and bears similarities with the AACC Method 86.05 [92] for vitamin A and carotenoids in enriched cereals and mixed feeds in terms of utilization of open column chromatography. The AOAC method involves extraction of the sample with acetone-hexane, filtration, and removal of acetone. The hexane extracts are then applied to an activated MgO_2 diatomaceous earth column and the carotenoids are eluted with acetone and hexane and determined using UV-Visible spectrophotometry. In the AACC method, samples are saponified with potassium hydroxide and extracted with hexane. The hexane extracts are applied to an alumina column and carotene is eluted with 4% acetone in hexane and determined using UV-Visible spectrophotometry at 440 nm.

Although spectrophotometric methods are relatively inexpensive and do not require specialized equipment, they tend to be time-consuming and do not always result in complete separation of carotenoid species. HPLC methods are able to achieve better separation of carotenoid species. In one such analysis, yellow endosperm sorghums were extracted with hexane and carotenoid identification and quantification was done using reversed-phase HPLC with absorbance detection [93]. Quantification was done using regression curves of beta-carotene and lutein standards [93]. Lipkie et al. [94] determined beta-carotene content of transgenic provitamin A-fortified sorghum grains and porridges by reversed-phase HPLC with diode array detection at 450 nm. Carotenoids were extracted using cold acetone followed by methyl tert-butyl ether. Extractions were carried out in yellow light to prevent photo-oxidation of carotenoids. Extracts were dried under a nitrogen stream, resolubilized in 1:1 methanol: ethyl acetate and analyzed by HPLC.

Normal phase HPLC with diode array detection has been used to determine carotenoids in heat-processed sorghum grains [95, 96]. Samples were homogenized in cold acetone and carotenoids extracted from the acetone by partitioning into petroleum ether. Extracts were evaporated and redissolved in hexane: isopropanol (90:10 v/v). These were filtered before HPLC analysis using a silica column. The HPLC-DAD system was operated by scanning from 350 to 600 nm with detection at 450 nm. External standards of lutein and zeaxanthin were used.

5.2 Carotenoid Bioaccessibility

Apart from the content of carotenoids, there has also been interest in carotenoid bioaccessibility. Lipkie et al. [94] determined carotenoid bioaccessibility of transgenic provitamin A-fortified sorghum porridge using a simulated in vitro digestion technique. Samples were mixed with oil to facilitate micellarization of carotenes which is otherwise minimal in the absence of the added lipid. The simulated in vitro digestion process consisted of an oral phase (with alpha amylase), gastric phase (with pepsin), and an intestinal phase (using pancreatin, lipase, and bile extract). Part of the intestinal digest was centrifuged and filtered to isolate the aqueous micellar fraction. The intestinal digest and aqueous micellar fraction were analyzed for beta-carotene content. These were extracted with 1:3 acetone: petroleum ether containing 0.1% w/v butylated hydroxytoluene, dried under nitrogen and resolubilized in 1:1 methanol: ethyl acetate before HPLC analysis. A micellarization efficiency parameter was determined by expressing the concentration of beta-carotene in the aqueous micellar fraction as a percentage of the concentration of beta-carotene in the intestinal digests. Bioaccessible beta-carotene content was computed as the product of the beta-carotene content of the undigested food sample and the micellarization efficiency [94].

5.3 Vitamin E (Tocopherols and Tocotrienols)

The AACC Method 86.06 [92] outlines a procedure for the determination of alpha-tocopherol using reverse-phase HPLC with fluorescence detection which is applicable to all foods. Samples are saponified in an alkaline aqueous ethanol, neutralized and filtered which converts tocopherol esters to tocopherol. Alpha-tocopherol is determined using reverse-phase HPLC with fluorescence detection (excitation 290 nm, emission 330 nm). Normal phase HPLC with fluorescence detection has been used by various authors [95–97] as per Pinheiro-Sant'Ana et al. [98] for determination of tocopherols and tocotrienols in sorghum and sorghum foods. Extraction was done by homogenization of the samples in a solvent system containing isopropanol, hexane, and ethyl acetate. Extracts were dried, redissolved in hexane and filtered before HPLC analysis using a silica column with fluorescence detection (excitation 290 nm, emission 330 nm). External standards of alpha-tocopherol, alpha-tocotrienol, gamma-tocopherol, delta-tocopherol, and delta-tocotrienol were used.

5.4 Water-Soluble Vitamins

Due to the fact that sorghum is not regarded as an important source of vitamin C, the level of B-vitamins in sorghum and sorghum foods has been of interest. Methods for determination of B-vitamins in sorghum and sorghum foods fall into two main categories—microbiological and chemical. Chemical methods have been used more extensively and they may be spectrophotometric or chromatographic.

In general, microbiological methods proceed by the measurement of the growth of specific microorganisms in response to the vitamin of interest. Essentially, the growth of the microorganism is proportional to the concentration of the vitamin. AACC Methods 86–31, 86–40, 86–47, 86–51, and 86–72 [92] are microbiological methods for the determination of vitamin B6 complex, vitamin B12, total folate, niacin, and riboflavin respectively. These methods involve the use of *Saccharomyces uvarum* (ATCC No. 9080) for vitamin B6 complex, *Lactobacillus delbrueckii* ssp. *lactis* for vitamin B12, *Lactobacillus casei* subsp. *rhamnosus* (ATCC No. 7469) for total folate, *Lactobacillus plantarum* (ATCC No. 8014) for niacin, and *Lactobacillus casei* for riboflavin. Similar microorganisms have been used for determination of B-vitamins in sorghum using microbiological methods [99]. Specifically, *Lactobacillus fermenti* (ATCC No. 9338), *Lactobacillus casei* (ATCC No. 7469), and *Saccharomyces carlsbergensis* (ATCC No. 9080) were used for measurement of thiamine, riboflavin, and vitamin B6, respectively. *Lactobacillus plantarum* (ATCC No. 8014) was used for estimation of niacin and biotin. Total folate was determined using *Lactobacillus casei* (ATCC No. 7469) [99].

Chemical methods for determination of B-vitamins involve extraction of the sample with mineral acids and this may be accompanied by enzyme treatments. For example, thiamine, riboflavin, and niacin may be extracted with acids such as HCl or H_2SO_4 and then treated with phosphatase enzyme to hydrolyze phosphate ester forms of the vitamin and with protease enzymes such as trypsin or papain to release the vitamin from proteins. Thiamine is usually converted to thiochrome (its fluorescent form) by oxidation with potassium ferricyanide under alkaline conditions before analysis using fluorescence spectrophotometry or by HPLC with fluorescence detection. Riboflavin has fluorescent properties and therefore can be determined directly by fluorescence spectrophotometry or by HPLC with fluorescence detection. Riboflavin and niacin can also be determined colorimetrically using UV spectrophotometry or HPLC with UV detection.

AOAC spectrophotometric methods as outlined above have been applied for determination of B-vitamins in sorghum and sorghum products [39, 100]. Duodu et al. [101] used HPLC to determine thiamine in sorghum porridge. Thiamine extraction was done by acid treatment (0.4 M H_2SO_4) and hydrolysis of sample with diastase enzyme. Thiamine was converted to thiochrome by treatment with alkaline ferricyanide and separation and quantification was by reversed-phase HPLC with fluorescence detection.

6 Minerals

Mineral analysis of sorghum and sorghum foods has been conducted extensively. The mineral elements in foods are located in the ash component and therefore a first step in the determination of minerals involves an ashing process. The ashing process may be a dry ashing which involves total combustion of the organic material at very high temperature (as is done during determination of ash content during proximate analysis), or a wet ashing which involves treatment with concentrated mineral acids, or a combination of the two.

Ashing procedures applied to sorghum and sorghum foods that proceed by dry ashing alone would normally involve incineration of the sample at temperatures greater than 500 °C followed by dissolution of the ash in concentrated hydrochloric acid before further analysis [39, 102]. Dry ashing at high temperature is also sometimes followed by wet acid digestion using either concentrated nitric acid [103–105] or concentrated hydrochloric acid [106]. Ashing procedures by wet acid digestion alone have involved using a mixture of hydrochloric acid and nitric acid [107–109], a 5:1 mixture of nitric acid and perchloric acid [110, 111], perchloric acid alone [112], or a mixture of nitric acid and hydrogen peroxide [113]. Microwave-assisted acid digestion procedures are also being used which involve the utilization of concentrated nitric acid [114] or a mixture of concentrated nitric acid and hydrogen peroxide [115–117].

After ashing, minerals such as sodium and potassium in sorghum and sorghum products have been analyzed using flame photometry (mainly for sodium and potassium) [39, 106, 118]. By far, atomic absorption spectrophotometry has been used most extensively for the determination of a wide range of mineral and transition metal elements in sorghum and sorghum products [39, 91, 102–106, 110, 111, 116, 118–121]. There is currently an increasing trend toward the use of Inductively Coupled Plasma Atomic Emission Spectrometry (ICP-AES) [113, 115, 117, 122, 123] or Inductively Coupled Plasma-Optical Emission Spectrometry (ICP-OES) [108, 109, 112, 114, 124, 125]. Alternative methods that have been used for mineral analysis include X-ray fluorescence [100] and high-performance ion chromatography [107].

6.1 Mineral Bioaccessibility

Bioaccessibility of a nutrient may be defined as the quantity or fraction of the nutrient that is released from the food matrix in the gastrointestinal tract and becomes available for absorption [126]. Determination of bioaccessibility requires the use of in vitro methods. From a nutritional standpoint, it is more desirable to determine the bioavailability which may be defined as the fraction of an ingested nutrient or compound that reaches the systemic circulation and is utilized [127]. Bioavailability is determined using in vivo methods with animal or human subjects. Due to limitations such as ethical constraints and high cost of in vivo studies, many research studies determine the bioaccessibility of nutrients and in this regard, the bioaccessibility of minerals in sorghum and sorghum products has been of interest.

One way of determining mineral bioaccessibility of sorghum and sorghum products has been by a simple determination of the acid extractability of minerals from the food sample. Bioaccessibility of minerals from sorghum in terms of their extractability in hydrochloric acid as outlined by Chauhan and Mahjan [128] has been conducted [106, 110]. Samples were extracted with 0.03 M HCl for 3 h at 37 °C. and filtered. The filtrate was dried at 100 °C, acid-digested and mineral content determined using atomic absorption spectrophotometry. Mineral content of the filtrate was expressed as a percentage of the total mineral content as a measure of the mineral extractability. Towo et al. [107] determined iron bioaccessibility of fermented sorghum gruels in terms of iron solubility after enzyme digestion. Samples were digested with pepsin followed by pancreatin and bile solution. After centrifugation, the supernatants were analyzed for soluble iron using high-performance ion chromatography and expressed as a percentage of the total amount of iron in the sample.

The mineral dialyzability assay [129, 130] has been used by various workers [102, 108, 109, 113, 119, 120, 124]. This assay measures the amount of minerals that passes through dialysis tubing of specified molecular weight cut off from samples subjected to a simulated in vitro digestion procedure which is referred to as the mineral dialyzability and gives a reflection of the amount of minerals that are bioaccessible from the digested food sample. In the assay, food samples are digested with pepsin for 2 h after which dialysis tubing (molecular mass cut off 10 kDa) containing sodium bicarbonate solution is inserted into the gastric digest and incubated until a pH of 5 is obtained for the digest. Pancreatin-bile mixture is added to the gastric digest and the mixture is incubated until a pH of 7 is attained. Zn and Fe in the dialysate is determined using atomic absorption spectrophotometry and expressed as a percentage of the total Zn and Fe content in the food as the bioaccessibility. In essence, the dialyzable portion of the total minerals in the food sample represents the bioaccessible minerals. In a modification to this assay, Drago et al. [130] and Llopart and Drago [120] used piperazine N,N′-bis(2-ethanesulphonic acid) (PIPES) buffer in the dialysis bags instead of sodium bicarbonate solution as used by Luten et al. [129].

The determination of mineral uptake in cell tissue culture models has also been used to measure mineral bioaccessibility. Kruger et al. [124] used a Caco-2 cell mineral uptake with radioisotope labeling technique to estimate bioaccessibility of iron and zinc from sorghum porridge. Cooked sorghum was first subjected to simulated in vitro digestion with pepsin (gastric phase) and pancreatin (intestinal phase). For radioisotope labeling, the intrinsic iron and zinc in the digested sorghum was exchanged with ^{59}Fe from $^{59}FeSO_4$ and ^{65}Zn from $^{65}ZnCl_2$. Caco-2 cells were incubated with the labeled cooked sorghum digest for 6 h. The radioactivity in the labeled digested sorghum and the Caco-2 cells (after incubation) was counted using a gamma scintillation counter. The mineral uptake was then determined by expressing the radioactivity in the Caco-2 cells (after incubation with the labeled sorghum digest) as a percentage of the total radioactivity of the incubation mixture.

7 Conclusions

Assaying sorghum nutritional quality broadly consists of determination of the content, digestibility, and bioaccessibility of nutrients. A wide range of assays have been used to determine these nutritional quality parameters for macronutrients (starch, proteins, and dietary fiber) and micronutrients (vitamins and minerals) in sorghum. Most of these assay procedures are well documented in published methods of analyses for organizations such as AOAC and AACCI. In recent years, advances in analytical instrumentation have enhanced the scope of sorghum nutritional quality analyses.

References

1. FAOSTAT (2018) Production: crops. www.fao.org/faostat. Accessed Feb 2018
2. Akingbala JO, Rooney LW, Faubion JM (1981) Physical, chemical and sensory evaluation of *ogi* from sorghum of differing kernel characteristics. J Food Sci 46:1532–1536
3. Muindi PJ, Thomke S, Ekman R (1981) Effect of Magadi soda treatment on the tannin

content and in-vitro nutritive value of grain sorghums. J Sci Food Agr 32:25–34
4. Zhang G, Hamaker BR (1998) Low α-amylase starch digestibility of cooked sorghum flours and the effect of protein. Cereal Chem 75:710–713
5. Benmoussa M, Suhendra B, Aboubacar A, Hamaker BR (2006) Distinctive sorghum starch granule morphologies appear to improve raw starch digestibility. Starch/Stärke 58:92–99
6. Tawaba JCB, Tshiala H, Kibal I, Buetusiwa T, de Dieu Minengu J (2016) Effects of phenolic compounds on the hydrolysis of red sorghum starch by extracted red sorghum malt α- and β-amylases. Starch/Stärke 67:854–859
7. Elkhalifa AEO, Chandrashekar A, Mohamed BE, El Tinay AH (1999) Effect of reducing agents on the in vitro protein and starch digestibilities of cooked sorghum. Food Chem 66:323–326
8. Elkhalifa AEO, Schiffler B, Bernhard R (2004) Effect of fermentation on the starch digestibility, resistant starch and some physicochemical properties of sorghum flour. Nahrung/Food 48:91–94
9. Ezeogu LI, Duodu KG, Taylor JRN (2005) Effects of endosperm texture and cooking conditions on the in vitro starch digestibility of sorghum and maize flours. J Cereal Sci 42:33–44
10. Aarathi A, Urooj A, Puttaraj S (2003) In vitro starch digestibility and nutritionally important starch fractions in cereals and their mixtures. Starch/Stärke 55:94–99
11. Englyst HN, Kingman SM, Cummings JH (1992) Classification and measurement of nutritionally important starch fractions. Eur J Clin Nutr 46:223–250
12. McCleary BV, Gibson TS, Solah V, Mugford DC (1994) Total starch measurement in cereal products: Interlaboratory evaluation of a rapid enzymic test procedure. Cereal Chem 71(5):501–505
13. Englyst HN, Veenstra J, Hudson GJ (1996) Measurement of rapidly available glucose (RAG) in plant foods: a potential in vitro predictor of the glycemic response. Br J Nutr 75:327–337
14. Giuberti G, Gallo A, Cerioli C, Masoero F (2012) In vitro starch digestion and predicted glycemic index of cereal grains commonly utilized in pig nutrition. Anim Feed Sci Tech 174:163–173
15. Mkandawire NL, Kaufman RC, Bean SR, Weller CL, Jackson DS, Rose DJ (2013) Effects of sorghum (*Sorghum bicolor* (L.) Moench) tannins on α-amylase activity and in vitro digestibility of starch in raw and processed flours. J Agric Food Chem 61:4448–4454
16. Mkandawire NL, Weier SA, Weller CL, Jackson DS, Rose DJ (2015) Composition, in vitro digestibility, and sensory evaluation of extruded whole grain sorghum breakfast cereals. LWT–Food Sci Technol 62:662–667
17. Englyst KN, Hudson GJ, Englyst HN (2000) Starch analysis in food. In: Meyers RA (ed) Encyclopedia of analytical chemistry. Wiley, New York
18. Amoako DB, Awika JM (2016) Polymeric tannins significantly alter properties and in vitro digestibility of partially gelatinized intact starch granule. Food Chem 208:10–17
19. Al-Rabadi GJS, Gilbert RG, Gidley MJ (2009) Effect of particle size on kinetics of starch digestion in milled barley and sorghum grains by porcine alpha-amylase. J Cereal Sci 50:198–204
20. Al-Rabadi GJ, Torley PJ, Williams BA, Bryden WL, Gidley MJ (2012) Particle size heterogeneity in milled barley and sorghum grains: effects on physico-chemical properties and starch digestibility. J Cereal Sci 56:396–403
21. Pranotoa Y, Anggrahini S, Efendi Z (2013) Effect of natural and *Lactobacillus plantarum* fermentation on in-vitro protein and starch digestibilities of sorghum flour. Food Biosci 2:46–52
22. Goñi I, Garcia-Alonso A, Saura-Calixto F (1997) A starch hydrolysis procedure to estimate glycemic index. Nutr Res 17:427–437
23. Lemlioglu-Austin D, Turner ND, McDonough CM, Rooney LW (2012) Effects of brans from specialty sorghum varieties on in vitro starch digestibility of soft and hard sorghum endosperm porridges. Cereal Chem 89:190–197
24. Lemlioglu-Austin D, Turner ND, McDonough CM, Rooney LW (2012) Effects of sorghum [*Sorghum bicolor* (L.) Moench] crude extracts on starch digestibility, estimated glycemic index (EGI), and resistant starch (RS) contents of porridges. Molecules 17:11124–11138
25. Susanna S, Prabhasankar P (2013) A study on development of gluten free pasta and its biochemical and immunological validation. LWT–Food Sci Technol 50:613–621
26. Wolter A, Hager AS, Zannini E, Arendt EK (2013) In vitro starch digestibility and predicted glycaemic indexes of buckwheat, oat, quinoa, sorghum, teff and commercial gluten-free bread. J Cereal Sci 58:431–436

27. Wolter A, Hager AS, Zannini E, Arendt EK (2014) Influence of sourdough on in vitro starch digestibility and predicted glycemic indices of gluten-free breads. Food and Function 5:564–572
28. Saravanabavan SN, Shivanna MM, Bhattacharya S (2013) Effect of popping on sorghum starch digestibility and predicted glycemic index. J Food Sci Technol 50:387–392
29. Souilah R, Djabali D, Belhadi B, Mokrane H, Boudries N, Nadjemi B (2014) In vitro starch digestion in sorghum flour from Algerian cultivars. Food Sci Nutr 2(3):251–259
30. Liu H, Fan F, Cao R, Blanchard C, Wang M (2016) Physicochemical properties and in vitro digestibility of sorghum starch altered by high hydrostatic pressure. Int J Biol Macromol 92:753–760
31. Liu H, Wang L, Shen M, Guo X, lv M, Wang M (2016) Changes in physicochemical properties and in vitro digestibility of tartary buckwheat and sorghum starches induced by annealing. Starch/Stärke 68:709–718
32. McCleary BV, Monaghan DA (2002) Measurement of resistant starch. J AOAC Int 85:665–675
33. Elkonin LA, Italianskaya JV, Fadeeva IY, Bychkova VV, Kozhemyakin VV (2013) In vitro protein digestibility in grain sorghum: effect of genotype and interaction with starch digestibility. Euphytica 193:327–337
34. Mitchell GB (1999) Methods of starch analysis. Starch/Stärke 42:131–134
35. Fountoulakis M, Lahm HW (1998) Hydrolysis and amino acid composition analysis of proteins. J Chromatogr A 826:109–134
36. Rigas PG (2012) Review: liquid chromatography–post column derivatization for amino acid analysis: strategies, instrumentation and applications. Instrum Sci Technol 40:161–193
37. Rigas PG (2013) Post-column labeling techniques in amino acid analysis by liquid chromatography. Anal Bioanal Chem 405:7957–7992
38. Neucere NJ, Sumrell G (1979) Protein fractions from five varieties of grain sorghum: amino acid composition and solubility properties. J Agric Food Chem 27:809–812
39. Khalil JK, Sawaya WN, Safi WJ, Al-Mohammad HM (1984) Chemical composition and nutritional quality of sorghum flour and bread. Plant Food Hum Nutr 34:141–150
40. Ejeta G, Hassen MM, Mertz ET (1987) In vitro digestibility and amino acid composition of pearl millet (*Pennisetum typhoides*) and other cereals. Proc Natl Acad Sci U S A 84:6016–6019
41. Ahmed ZS, Abd El-Moneim GM, Yassen AAE (1996) Comparative studies on protein fractions, and amino acid composition from sorghum and pearl millet. Nahrung 40:305–309
42. Sanni AI, Asiedu M, Ayernor GS (2001) Influence of processing conditions on the nutritive value of *Ogi-baba*, a Nigerian fermented sorghum gruel. Plant Food Hum Nutr 56:217–223
43. Ebadi MR, Pourreza J, Jamalian J, Edriss MA, Samie AH, Mirhadi SA (2005) Amino acid content and availability in low, medium and high tannin sorghum grain for poultry. Int J Poult Sci 4:27–31
44. Agu HO, Jideani IA, Yusuf IZ (2007) Nutrient and sensory properties of dambu produced from different cereal grains. Nutr Food Sci 37:272–281
45. Mabelebele M, Siwela M, Gous RM, Iji PA (2015) Chemical composition and nutritive value of south African sorghum varieties as feed for broiler chickens. S Afr J Anim Sci 45:206–213
46. Kaiser FE, Gehrke CW, Zumwalt RW, Kuo KC (1974) Amino acid analysis: hydrolysis, ion-exchange cleanup, derivatization and quantitation by gas-liquid chromatography. J Chromatogr 94:113–133
47. White JA, Hart RJ, Fry JC (1986) An evaluation of the waters Pico-tag system for the amino acid analysis of food materials. J Autom Chem 8:170–177
48. Serna-Saldivar SO, Knabe DA, Rooney LW, Tanksley TD Jr, Sproule AM (1988) Nutritional value of sorghum and maize tortillas. J Cereal Sci 7:83–94
49. Axtell JD, Kirles AW, Hassen MM, Mason ND, Mertz ET, Munck L (1981) Digestibility of sorghum proteins. Proc Natl Acad Sci U S A 78:1333–1335
50. Mertz ET, Hassen MM, Cairns-Whittern C, Kirleis AW, Tu L, Axtell J (1984) Pepsin digestibility of proteins in sorghum and other major cereals. Proc Natl Acad Sci U S A 81:1–2
51. Hamaker BR, Kirleis AW, Mertz ET, Axtell JD (1986) Effect of cooking on the protein profiles and in vitro digestibility of sorghum and maize. J Agric Food Chem 34:647–649
52. Hamaker BR, Kirleis AW, Butler LG, Axtell JD, Mertz ET (1987) Improving the in vitro protein digestibility of sorghum with

reducing agents. Proc Natl Acad Sci U S A 84:626–628
53. Hassan IAG, El Tinay AH (1995) Effect of fermentation on tannin content and in-vitro protein and starch digestibilities of two sorghum cultivars. Food Chem 53:149–151
54. Oria MP, Hamaker BR, Shull JM (1995) Resistance of sorghum α-, β- and γ-kafirins to pepsin digestion. J Agric Food Chem 43:2148–2153
55. Parker ML, Grant A, Rigby AN, Belton PS, Taylor JRN (1999) Effects of popping on the endosperm cell walls of sorghum and maize. J Cereal Sci 30:209–216
56. Hicks C, Bean SR, Lookhart GL, Pedersen JF, Kofoid KD, Tuinstra MR (2001) Genetic analysis of kafirins and their phenotypic correlations with feed quality traits, in vitro digestibility, and seed weight in grain sorghum. Cereal Chem 78:412–416
57. Duodu KG, Nunes A, Delgadillo I, Parker ML, Mills ENC, Belton PS, Taylor JRN (2002) Effect of grain structure and cooking on sorghum and maize in vitro protein digestibility. J Cereal Sci 35:161–174
58. Nunes A, Correia I, Barros A, Delgadillo I (2004) Sequential in vitro pepsin digestion of uncooked and cooked sorghum and maize samples. J Agric Food Chem 52:2052–2058
59. Correia I, Nunes A, Saraiva JA, Barros AS, Delgadillo I (2011) High pressure treatments largely avoid/revert decrease of cooked sorghum protein digestibility when applied before/after cooking. LWT–Food Sci Technol 44:1245–1249
60. Da Silva LS, Taylor J, Taylor JRN (2011) Transgenic sorghum with altered kafirin synthesis: kafirin solubility, polymerization, and protein digestion. J Agric Food Chem 59:9265–9270
61. Licata R, Chu J, Wang S, Coorey R, James A, Zhao Y, Johnson S (2014) Determination of formulation and processing factors affecting slowly digestible starch, protein digestibility and antioxidant capacity of extruded sorghum–maize composite flour. Int J Food Sci Technol 49:1408–1419
62. Elhassan MSM, Emmambux MN, Hays DB, Peterson GC, Taylor JRN (2015) Novel biofortified sorghum lines with combined waxy (high amylopectin) starch and high protein digestibility traits: effects on endosperm and flour properties. J Cereal Sci 65:132–139
63. Thin T, Myat L, Ryu GH (2016) The effects of CO_2 injection and barrel temperatures on the physiochemical and antioxidant properties of extruded cereals. Prev Nutr Food Sci 21:271–280
64. Aboubacar A, Axtell JD, Huang CP, Hamaker BR (2001) A rapid protein digestibility assay for identifying highly digestible sorghum lines. Cereal Chem 78:160–165
65. Wong JH, Marx DB, Wilson JD, Buchanan BB, Lemaux PG, Pedersen JF (2010) Principal component analysis and biochemical characterization of protein and starch reveal primary targets for improving sorghum grain. Plant Sci 179:598–611
66. Jood S, Kapoor AC (1992) Effect of storage and insect infestation on protein and starch digestibility of cereal grains. Food Chem 44:209–212
67. Elmaki HB, Babiker EE, El Tinay AH (1999) Changes in chemical composition, grain malting, starch and tannin contents and protein digestibility during germination of sorghum cultivars. Food Chem 64:331–336
68. FAO/WHO Expert Consultation (1990) Protein quality evaluation. Food and Agricultural Organization of the United Nations, FAO Food and Nutrition Paper 51, Rome
69. Schaafsma G (2000) The protein digestibility-corrected amino acid score. J Nutr 130:1865S–1867S
70. Schaafsma G (2012) Advantages and limitations of the protein digestibility-corrected amino acid score (PDCAAS) as a method for evaluating protein quality in human diets. Br J Nutr 108:S333–S336
71. FAO (Food and Agriculture Organisation) (2013) Dietary protein quality evaluation in human nutrition, Report of an FAO expert consultation. FAO food and nutrition paper no. 92. FAO, Rome (Italy)
72. Rutherford SM, Fanning AC, Miller BJ, Moughan PJ (2015) Protein digestibility-corrected amino acid scores and digestible indispensable amino acid scores differentially describe protein quality in growing male rats. J Nutr 145:372–379
73. Henley EC, Taylor JRN, Obukosia SD (2010) The importance of dietary protein in human health: combating protein deficiency in sub-Saharan Africa through transgenic biofortified sorghum. Adv Food Nutr Res 60:21–52
74. Mokrane H, Amoura H, Belhaneche-Bensemra N, Courtin CM, Delcour JA, Nadjemi B (2010) Assessment of Algerian sorghum protein quality [*Sorghum bicolor* (L.) Moench] using amino acid analysis and in vitro pepsin digestibility. Food Chem 121:719–723

75. Anyango JO, de Kock HL, Taylor JRN (2011) Impact of cowpea addition on the protein digestibility corrected amino acid score and other protein quality parameters of traditional African foods made from non-tannin and tannin sorghum. Food Chem 124:775–780
76. AwadElkareem AM, Taylor JRN (2011) Protein quality and physical characteristics of *kisra* (fermented sorghum pancake-like flatbread) made from tannin and non-tannin sorghum cultivars. Cereal Chem 88:344–348
77. Vilakati N, MacIntyre U, Oelofse A, Taylor JRN (2015) Influence of micronization (infrared treatment) on the protein and functional quality of a ready-to-eat sorghum-cowpea African porridge for young child-feeding. LWT–Food Sci Technol 63:1191–1198
78. Kim JS, Kim KJ, Ma WCJ, Chung HY (2007) Development of a method to quantify lysine in small amount of rice grain. Korean J Sanitation 22:75–84
79. Lupton JR, Betteridge VA, Pijls LTJ (2009) Codex final definition of dietary fibre: issues of implementation. Qual Assur Saf Crop Food 1(4):206–212
80. Mann JI, Cummings JH (2009) Possible implications for health of the different definitions of dietary fibre. Nutr Metab Cardiovasc Dis 19:226–229
81. McCleary BV (2010) Development of an integrated total dietary fibre method consistent with the codex Alimentarius definition. Cereal Foods World 55:24–28
82. Codex Alimentarius Commission (2009) Report of the 30th Session of the Codex Committee on Nutrition and Foods for Special Dietary Uses, Cape Town, South Africa, 3–7 November 2008. http://www.fao.org/input/download/report/732/nf31_01e.pdf. Assessed 4 Feb 2018
83. Dovi KAP, Chiremba C, Taylor JRN, De Kock HL (2018) Rapid sensory profiling and hedonic rating of whole grain sorghum-cowpea composite biscuits by low income consumers. J Sci Food Agric 98:905–913
84. Bach Knudsen KE, Munck L (1988) Effect of cooking, pH and polyphenol level on carbohydrate composition and nutritional quality of a sorghum *(Sorghum bicolor* (L.) Moench) food, ugali. Br J Nutr 59:31–47
85. Barikmo I, Ouattara F, Oshaug A (2004) Protein, carbohydrate and fibre in cereals from Mali—how to fit the results in a food composition table and database. J Food Comp Anal 17:291–300
86. Amalraj A, Pius A (2015) Influence of oxalate, phytate, tannin, dietary fibre, and cooking on calcium bioavailability of commonly consumed cereals and millets in India. Cereal Chem 92:389–394
87. Moraes EA, Da Silva Marineli R, Lenquiste SA, Steel CJ, De Menezes CB, Queiroz VAV, Júnior MRM (2015) Sorghum flour fractions: correlations among polysaccharides, phenolic compounds, antioxidant activity and glycemic index. Food Chem 180:116–123
88. Rao BD, Kulkarni DB, Kavitha C (2018) Study on evaluation of starch, dietary fibre and mineral composition of cookies developed from 12 sorghum cultivars. Food Chem 238:82–86
89. Megazyme (2017) Total dietary fibre assay procedure K-TDFR-100A/K-TDFR-200A. https://secure.megazyme.com/files/booklet/k-tdfr_data.pdf. Assessed 4 Feb 2018
90. AOAC (2000) Official methods of analysis of AOAC International, 17th ed. Official method 970.64 or 45.1.014. AOAC International, Gaithersburg, MD
91. Elemo GN, Elemo BO, Okafor JNC (2011) Preparation and nutritional composition of a weaning food formulated from germinated sorghum (*Sorghum bicolor*) and steamed cooked cowpea (*Vigna unguiculata* Walp). Am J Food Technol 6:413–421
92. AACC (2004) American Association of Cereal Chemists approved methods, 10th ed. Methods 86–05, 86–06, 86–31, 86–40, 86–47, 86–51, 86–72. AACC International, St Paul, MN
93. Fernandez MGS, Kapran I, Souley S, Abdou M, Maiga IH, Acharya CB, Hamblin MT, Kresovich S (2009) Collection and characterization of yellow endosperm sorghums from West Africa for biofortification. Genet Resour Crop Ev 56:991–1000
94. Lipkie TE, De Moura FF, Zhao ZY, Albertsen MC, Che P, Glassman K, Ferruzzi MG (2013) Bioaccessibility of carotenoids from transgenic provitamin a biofortified sorghum. J Agric Food Chem 61:5764–5771
95. Cardoso LDL, Montini TA, Pinheiro SS, Pinheiro-Sant'Ana HM, Martino HSD, Moreira AVB (2014) Effects of processing with dry heat and wet heat on the antioxidant profile of sorghum. Food Chem 152:210–217
96. Cardoso LDM, Pinheiro SS, Da Silva LL, De Menezes CB, De Carvalho CWP, Tardin FD, Queiroz VAV, Martino HSD, Pinheiro-Sant'Ana HM (2015) Tocochromanols and carotenoids in sorghum (*Sorghum bicolor* L.):

diversity and stability to the heat treatment. Food Chem 172:900–908

97. Anunciação PC, Cardoso LDM, Gomes JVP, Lucia CMD, Carvalho CWP, Galdeano MC, Queiroz VAV, Alfenas RDCG, Martino HSD, Pinheiro-Sant'Ana HM (2017) Comparing sorghum and wheat whole grain breakfast cereals: sensorial acceptance and bioactive compound content. Food Chem 221:984–989
98. Pinheiro-Sant'Ana HM, Guinazi M, da Silva Oliveira D, Della Lucia CM, de Lazzari Reis B, Brandão SCC (2011) Method for simultaneous analysis of eight vitamin E isomers in various foods by high performance liquid chromatography and fluorescence detection. J Chromatogr A 1218:8496–8502
99. Hegedus M, Pedersen B, Eggum BO (1985) The influence of milling on the nutritive value of flour from cereal grains. 7. Vitamins and tryptophan. Plant Food Hum Nutr 35:175–180
100. Mahgoub SEO, Ahmed BM, Ahmed MMO, El Agib ENAA (1999) Effect of traditional Sudanese processing of *kisra* bread and *hulumur* drink on their thiamine, riboflavin and mineral contents. Food Chem 67:129–133
101. Duodu KG, Minnaar A, Taylor JRN (1999) Effect of cooking and irradiation on the labile vitamins and antinutrient content of a traditional African sorghum porridge and spinach relish. Food Chem 66:21–27
102. Gautam S, Platel K, Srinivasan K (2010) Higher bioaccessibility of iron and zinc from food grains in the presence of garlic and onion. J Agric Food Chem 58:8426–8429
103. Lestienne I, Icard-Vernière C, Mouquet C, Picq C, Trèche S (2005) Effects of soaking whole cereal and legume seeds on iron, zinc and phytate contents. Food Chem 89:421–425
104. Shegro A, Labuschagne MT, Shargie NG, Van Biljon A (2013) Multivariate analysis of nutritional diversity in sorghum landrace accessions from Western Ethiopia. J Biol Sci 13(2):67–74
105. Gerrano AS, Labuschagne MT, Van Biljon A, Shargie NG (2016) Quantification of mineral composition and total protein content in sorghum [*Sorghum bicolor* (L.) Moench] genotypes. Cereal Res Commun 44(2):272–285
106. Hassan AB, Osman GAM, Rushdi MAH, Eltayeb MM, Diab EE (2009) Effect of gamma irradiation on the nutritional quality of maize cultivars (*Zea mays*) and sorghum (*Sorghum bicolor*) grains. Pakistan J Nutr 8:167–171
107. Towo E, Matuscheck E, Svanberg U (2006) Fermentation and enzyme treatment of tannin sorghum gruels: effects on phenolic compounds, phytate and in vitro accessible iron. Food Chem 94:369–376
108. Wu G, Johnson SK, Bornman JF, Bennett SJ, Singh V, Simic A, Fang Z (2016) Effects of genotype and growth temperature on the contents of tannin, phytate and in vitro iron availability of sorghum grains. PLoS One 11 (2):e0148712. https://doi.org/10.1371/journal.pone.0148712
109. Wu G, Ashton J, Simic A, Fang Z, Johnson SK (2018) Mineral availability is modified by tannin and phytate content in sorghum flaked breakfast cereals. Food Res Int 103:509–514
110. Idris WH, Rahaman SMA, ElMaki HB, Babiker EE, El Tinay AH (2007) Effect of malt pretreatment on HCl extractability of calcium, phosphorus and iron of sorghum (*Sorghum bicolor*) cultivars. Int J Food Sci Technol 42:194–199
111. Badigannavar A, Girish G, Ramachandran V, Ganapathi TR (2016) Genotypic variation for seed protein and mineral content among post-rainy season-grown sorghum genotypes. Crop Journal 4:61–67
112. Bandara YMAY, Tesso TT, Bean SR, Dowell FE, Little CR (2017) Impacts of fungal stalk rot pathogens on physicochemical properties of sorghum grain. Plant Dis 101:2059–2065
113. Kumari M, Platel K (2017) Effect of sulfur-containing spices on the bioaccessibility of trace minerals from selected cereals and pulses. J Sci Food Agric 97:2842–2848
114. Asikin Y, Wada K, Imai Y, Kawamoto Y, Mizu M, Mutsuura M, Takahashi M (2017) Compositions, taste characteristics, volatile profiles, and antioxidant activities of sweet sorghum (*Sorghum bicolor* L.) and sugarcane (*Saccharum officinarum* L.) syrups. J Food Meas Charact. https://doi.org/10.1007/s11694-017-9703-2
115. Yang L, Yan Q, Cao Y, Zhang H (2012) Determination of mineral elements of some coarse grains by microwave digestion with inductively coupled plasma atomic emission spectrometry. E-Journal of Chem 9(1):93–98
116. Proietti I, Mantovani A, Mouquet-Rivier C, Guyot JP (2013) Modulation of chelating factors, trace minerals and their estimated bioavailability in Italian and African sorghum (*Sorghum bicolor* (L.) Moench) porridges. Int J Food Sci Technol 48:1526–1532
117. Ndimba R, Cloete K, Mehlo L, Kossmann J, Mtshali C, Pineda-Vargas C (2017) Using ICP and micro-PIXE to investigate possible

differences in the mineral composition of genetically modified versus wild-type sorghum grain. Nucl Instr Meth B 404:121–124
118. Oboh G, Amusan TV (2009) Nutritive value and antioxidant properties of cereal gruels produced from fermented maize and sorghum. Food Biotechnol 23:17–31
119. Hemalatha S, Platel K, Srinivasan K (2007) Zinc and iron contents and their bioaccessibility in cereals and pulses consumed in India. Food Chem 102:1328–1336
120. Llopart EE, Drago SR (2016) Physicochemical properties of sorghum and technological aptitude for popping. Nutritional changes after popping. LWT–Food Sci Technol 71:316–322
121. Okoth JK, Ochola SA, Gikonyo NK, Makokha A (2017) Development of a nutrient-dense complementary food using amaranth-sorghum grains. Food Sci Nutr 5(1):86–93
122. Ragaee S, Abdel-Aal EM, Noaman M (2006) Antioxidant activity and nutrient composition of selected cereals for food use. Food Chem 98:32–38
123. Hager AS, Wolter A, Jacob F, Zannini E, Arendt EK (2012) Nutritional properties and ultra-structure of commercial gluten free flours from different botanical sources compared to wheat flours. J Cereal Sci 56:239–247
124. Kruger J, Taylor JRN, Du X, De Moura FF, Lönnerdal B, Oelofse A (2013) Effect of phytate reduction of sorghum, through genetic modification, on iron and zinc availability as assessed by an in vitro dialysability bioaccessibility assay, Caco-2 cell uptake assay, and suckling rat pup absorption model. Food Chem 141:1019–1025
125. Paiva CL, Queiroz VAV, Simeone MLF, Schaffert RE, De Oliveira AC, Da Silva CS (2017) Mineral content of sorghum genotypes and the influence of water stress. Food Chem 214:400–405
126. Heaney RP (2001) Factors influencing the measurement of bioavailability, taking calcium as a model. J Nutr 131:1344–1348
127. Carbonell-Capella JM, Buniowska M, Barba FJ, Esteve MJ, Frigola A (2014) Analytical methods for determining bioavailability and bioaccessibility of bioactive compounds from fruits and vegetables: a review. Compr Rev Food Sci Food Saf 13:155–171
128. Chauhan BM, Mahjan L (1988) Effect of natural fermentation on the extractability of minerals from pearl millet flour. J Food Sci 53:1576–1577
129. Luten J, Crews H, Flynn A, Dael PV, Kastenmayer P, Hurrel R, Deelstra H, Shen LH, Fairweathertait S, Hickson K, Farre R, Schlemmer U, Frohlich W (1996) Interlaboratory trial on the determination of the in vitro iron dialysability from food. J Sci Food Agric 72:415–424
130. Drago SR, Binaghi MJ, Valencia ME (2005) Effect of gastric digestion pH on iron, zinc and calcium availability from preterm and term starting infant formulas. J Food Sci 70:107–112

Chapter 8

Tannin Analysis in Sorghum Grains

Linda Dykes

Abstract

Sorghum (*Sorghum bicolor* (L.) Moench) contains a wide variety of phenolic compounds and their levels and types depend on the genotype. Sorghums with a pigmented testa contain condensed tannins, which are concentrated in the pigmented testa. In this chapter, three methods of tannin analysis are described: (1) Chlorox bleach test; (2) modified vanillin-HCl assay; and (3) normal-phase high-performance liquid chromatography (NP-HPLC) with fluorescence detection. The Chlorox bleach test is a rapid and inexpensive method that identifies non-tannin and tannin sorghums. The modified vanillin-HCl assay is a colorimetric assay that provides relative tannin values among sorghum samples. The normal-phase HPLC with fluorescence detection separates and quantifies condensed tannins according to their degree of polymerization and confirms the presence or absence of tannins.

Key words Sorghum, Condensed tannins, Chlorox bleach test, Vanillin-HCl assay, Normal-phase HPLC, Fluorescence detection

1 Introduction

Sorghum (*Sorghum bicolor* (L.) Moench) contains a wide variety of phenolic compounds and their levels and types depend on the genotype [1–7]. Sorghums with a pigmented testa contain condensed tannins, which are the main phenolic compounds in those genotypes and are concentrated in the pigmented testa (Fig. 1) [1, 2]. Sorghum varieties are divided into three groups based on their genetics and chemical analyses [9]. Type I sorghums do not have a pigmented testa and contain low levels of phenolic compounds and no tannins. Type II and III sorghums both have a pigmented testa and contain tannins but the tannins in Type II sorghums are extracted with acidified methanol (1% HCl) while those in Type III sorghums are extracted with either methanol or acidified methanol [8, 10]. Typically, Type II sorghums contain lower tannin levels than Type III sorghums (Table 1).

Several methods to determine the presence of tannins in sorghum exist. The Chlorox bleach test (also known simply as the bleach procedure or test) is a rapid and inexpensive qualitative test

Zuo-Yu Zhao and Jeff Dahlberg (eds.), *Sorghum: Methods and Protocols*, Methods in Molecular Biology, vol. 1931, https://doi.org/10.1007/978-1-4939-9039-9_8, © Springer Science+Business Media, LLC, part of Springer Nature 2019

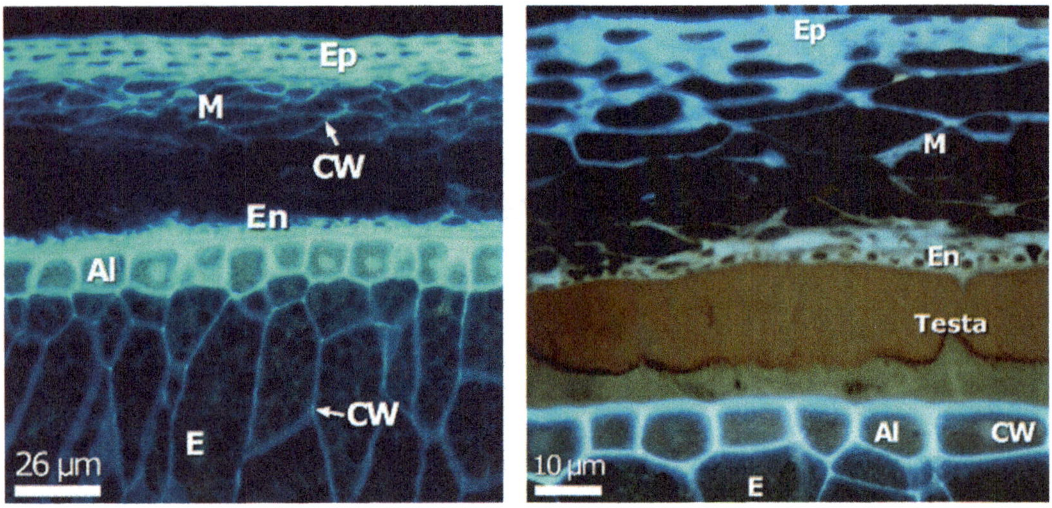

Fig. 1 Fluorescence photomicrograph of cross-sections of a non-tannin (left) and a tannin sorghum kernel (right). *Al* aleurone; *CW* cell wall; *E* endosperm; *En* endocarp; *Ep* epicarp; *M* mesocarp (Reprinted from Dykes and Rooney [8] with permission from Elsevier)

Table 1
Tannin concentrations (mg catechin equivalents/g) in Type I, II, and III sorghums determined by the vanillin-HCl assay

	Tannins	References
Type I:		
02CA4796	0.1[a]	[3]
BTx3197	0.5[a]	[11]
BTx378	1.4[a]	[11]
Tx2911	0.3[a]	[3]
Tx3362	0.7[a]	[6]
Type II:		
Hegari	6.6	[11]
SC109-14E	6.4	[11]
TAM-256	11.4	[11]
SC719-11E	10.9	[3]
Type III:		
Dobbs	20.1	[11]
(Hegari x Dobbs) OP5	18.3	[11]
SC103	28.2	[2]
Sumac	50.1	[2]
Black PI Tall	48.7	[6]

[a]Value is below the detection limit

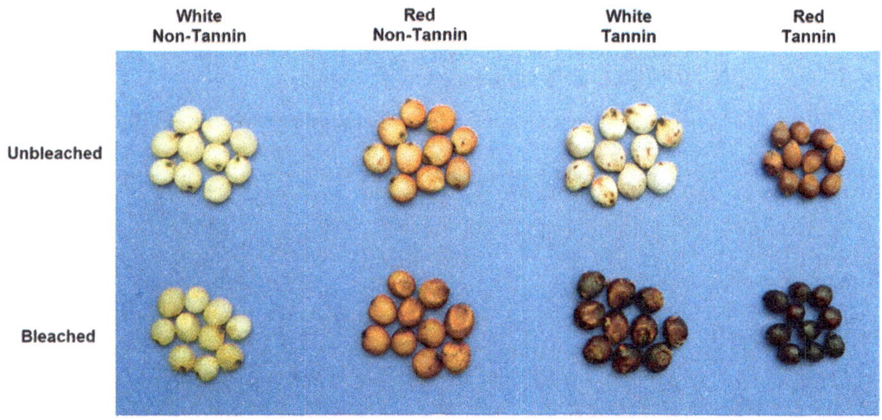

Fig. 2 Unbleached and bleached kernels of white non-tannin, red non-tannin, white tannin, and red tannin sorghums

that identifies non-tannin and tannin sorghums [12–15]. Sorghum kernels are immersed in bleach with alkali to dissolve the pericarp to reveal the presence or absence of a pigmented testa. For tannin sorghums, the kernels turn black, while non-tannin sorghums do not turn black (Fig. 2). The main disadvantage of this test is it does not clearly differentiate between Type II and III sorghums, nor does it determine their tannin levels.

Colorimetric methods, such as the modified vanillin-HCl assay [10], are rapid and inexpensive quantitative tests compared to other methods requiring expensive equipment. The vanillin-HCl assay involves the condensation of vanillin with monomeric flavanols and their oligomers to form a red adduct that absorbs at 500 nm [8, 10]. Despite being widely used for condensed tannin determination, the vanillin-HCl assay has disadvantages. The assay does not measure tannin levels accurately due to lack of appropriate standards [8, 16]. The preparation of a pure tannin standard from sorghum, quebracho, or other tannin-containing material is difficult and time-consuming [8, 17, 18]. The standard mostly used for this assay is catechin but its use results in over-estimation of tannins [10]. Also, Type I sorghums contain non-tannin phenolics, such as leucoanthocyanidins or flavan-3,4-diols, that react with vanillin, giving low "tannin" values [3, 19] (Table 1). Despite its disadvantages, the vanillin-HCl assay is a useful method to obtain relative tannin values among sorghum samples.

HPCL with UV-vis, photodiode array, or fluorescence detection separates, identifies, and quantifies sorghum phenolics, including condensed tannins [1, 4–8, 16, 20, 21]. Reversed-phase HPLC separates condensed tannins up to tetramers (DP4) and their separation is not according to their degree of polymerization [8, 16]. NP-HPLC with fluorescence detection separates and quantifies condensed tannins according to their degree of polymerization

Table 2
Procyanidin composition (mg/g) of tannin sorghum grains determined by NP-HPLC with fluorescence detection

Component	Hi Tannin	Sumac
Monomers	0.01	0.18
DP2[a]	0.09	0.40
DP3	0.12	0.51
DP4	0.21	0.69
DP5	0.26	0.74
DP6	0.49	1.10
DP7	0.38	0.79
DP8	0.38	0.74
DP9	0.63	1.17
DP10	0.31	0.55
Polymers	17.67	15.09
Total	20.50	21.97
% Oligomers[b]	14.03	31.31

Source: Awika et al. [1]
[a]DP: degree of polymerization
[b]Oligomers: DP $<$ 10

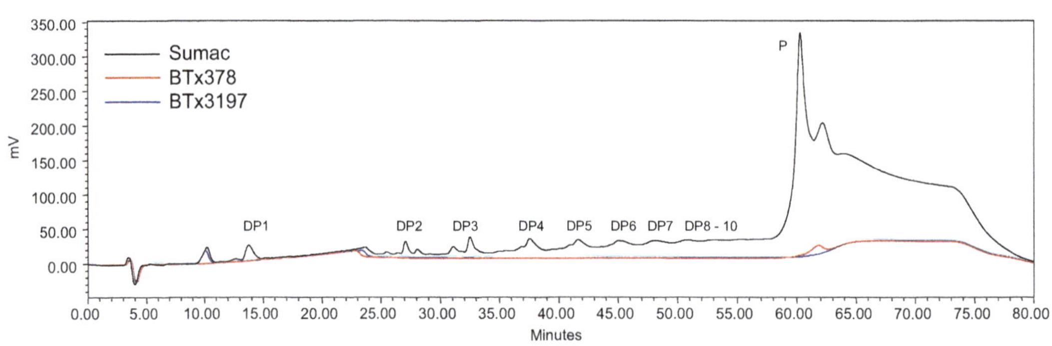

Fig. 3 NP-HPLC profiles of proanthocyanidins in tannin sorghum (Sumac) grains compared to non-tannin sorghum (BTx378 and BTx3197) grains. *DP* degree of polymerization, *P* polymers, with DP $>$ 10 (Reprinted from Dykes and Rooney [8] with permission from Elsevier)

(Table 2) [1, 8, 20, 21]. This method successfully separates condensed tannins up to decamers (DP10) and tannins with higher levels of polymerization are shown as one single peak (Fig. 3). This method also confirms the absence of tannins in Type I or non-tannin sorghums. One disadvantage of this method is the lack of commercial standards for oligomeric and polymeric tannins. These standards must be prepared from tannin-containing materials. Oligomeric and polymeric tannins can be purified from cocoa beans and blueberries, respectively [20, 21].

In this chapter, three methods for the identification and/or quantification of tannins are described: (1) Chlorox bleach test [12, 14]; (2) modified vanillin-HCl assay [10]; and (3) NP-HPLC with fluorescence detection [1, 21].

2 Materials

For all analyses, sorghum samples should consist of sound and whole sorghum kernels. Kernels should be free of glumes and all blemished, damaged, or broken kernels must be removed. For the Chlorox bleach test, whole sorghum kernels are used. For the modified vanillin-HCl assay and NP-HPLC analysis, sorghum kernels are first ground using either a coffee grinder or a grinding mill. Samples must be ground until the particles can pass through a #30 sieve (595 μm). Prepare all solutions using analytical grade reagents and HPLC-grade or ultra-pure water (18 MΩ-cm at 25 °C) unless otherwise noted.

2.1 Chlorox Bleach Test

1. Bleaching reagent: dissolve 5 g of NaOH in 100 mL of 5.25% NaOCl (or commercial/household chlorine bleach) (*see* **Note 1**).
2. 100 mL glass beakers.
3. Aluminum foil.
4. Tea strainer.
5. Paper towels.
6. Non-tannin and tannin sorghum kernels to be used as standards (*see* **Note 2**).

2.2 Modified Vanillin-HCl Assay

1. Analytical balance (4 decimal places).
2. 1% HCl in methanol: using a 1000 mL volumetric flask, add 10 mL of HCl to 500 mL of methanol. Make up to 1000 mL with methanol and mix well.
3. 4% HCl in methanol: using a 100 mL volumetric flask, add 4 mL of HCl to 50 mL of methanol. Make up to 100 mL with methanol and mix well.
4. 8% HCl in methanol: using a 100 mL volumetric flask, add 8 mL of HCl to 50 mL of methanol. Make up to 100 mL with methanol and mix well.
5. Vanillin reagent: Weigh 1.00 g of vanillin and transfer to a 100 mL volumetric flask containing 50 mL of methanol. Make up to 100 mL with methanol, mix well, and transfer to

a 250 mL glass bottle or Erlenmeyer flask. Add 100 mL of the 8% HCl in methanol solution to the 250 mL Erlenmeyer flask and mix well. The vanillin reagent must be prepared daily.

6. Catechin standard (1000 ppm): using a 100 mL volumetric flask, add 100.0 mg of catechin to 50 mL of methanol. Make up to 100 mL with methanol and mix well (*see* **Note 3**).
7. Centrifuge tubes (15- or 50 mL) with caps.
8. Glass test tubes ($\geq$10 mL).
9. Manual or automatic pipettes: 1 mL, 5 mL, and 10 mL.
10. Vortex mixer.
11. Water bath set at 30 °C.
12. Timer.
13. Centrifuge set at 4000 × *g*.
14. Polystyrene disposable cuvettes.
15. UV/Vis spectrophotometer set at 500 nm.

2.3 NP-HPLC with Fluorescence Detection

1. Analytical balance (4 decimal places).
2. Methanol.
3. Dichloromethane.
4. 50% acetic acid in water: to 500 mL of water, add 500 mL of acetic acid. Mix well.
5. Acetone/water/acetic acid (70:29.5:0.5, v/v): to 700 mL of acetone, add 295 mL of water and 5 mL of acetic acid. Mix well.
6. 30% methanol in water: to 700 mL of water, add 300 mL of methanol. Mix well.
7. 70% acetone in water: to 700 mL of acetone, add 300 mL of water. Mix well.
8. Centrifuge tubes (15- or 50 mL) with caps.
9. Centrifuge tube shaker.
10. Centrifuge set at 4000 × *g*.
11. Rotary evaporator or concentrator.
12. Solid-phase extraction tubes with frits (12–15 mL).
13. Sephadex LH-20, equilibrated in water: equilibrate the Sephadex LH-20 in water according to the manufacturer's directions. Pour the Sephadex LH-20 slurry into the solid-phase extraction tubes. The final level of the slurry should be around 5–8 mm from the top of the tube (*see* **Note 4**).
14. Collection tubes (50- to 100 mL) for the collection of fractions during the solid-phase extraction procedures.
15. Syringes (3 mL) .

16. Syringe filters: nylon, 0.2 or 0.45 μm, 13 mm diameter.
17. HPLC vials.
18. HPLC column: Silica column, 250 × 46 mm, 5 μm (e.g., Luna Silica, Phenomenex, Torrance, CA).
19. HPLC system consisting of quaternary pump, column heater, auto-sampler, and fluorescence detector.
20. Standards: Catechin; (−)-epicatechin; procyanidin B-1; procyanidin B-2; oligomeric tannins (DP3-DP10) purified from cocoa beans as described by Adamson et al. [20]; polymeric tannins (>DP10) purified from blueberries as described by Gu et al. [21].

3 Methods

3.1 Chlorox Bleach Test

This test should be done in duplicates.

1. Count 100 whole sorghum kernels and transfer them to a 100 mL beaker.
2. Add enough bleaching reagent to fully immerse sorghum kernels (~15 mL) and cover the top of the beaker with aluminum foil (*see* **Note 5**).
3. Swirl the contents in the beaker to thoroughly wet the kernels and let it sit for 20 min at room temperature. Swirl the contents in the beaker every 5 min.
4. Drain the sorghum and bleaching reagent mixture into a tea strainer and thoroughly rinse the sorghum kernels using tap water (*see* **Note 6**).
5. Place the kernels on a paper towel and gently pat the kernels to remove excess water.
6. Let the sorghum kernels dry at room temperature.
7. Evaluate the sorghum kernels. Kernels that are completely black have tannins and kernels that are white or lightly colored have no tannins (Fig. 2). Count the number of kernels that are completely black (*see* **Note 7**).
8. Express results as percentage of tannin sorghums. Duplicates should not differ by more than 5.0% [12, 14].

3.2 Modified Vanillin-HCl Assay

3.2.1 Standard Curve Preparation

1. Before sample analysis, run a standard curve using the catechin solution (1000 ppm). To prepare the standard curve, add 0.0, 0.2, 0.4, 0.6, 0.8, and 1.0 mL of the catechin solution into test tubes and then dilute to 1.0 mL with methanol.
2. Place all tubes in the water bath (30 °C). Set the timer to 20 min but do not start the timer yet.

3. Add 5 mL of vanillin reagent to the first tube and start the timer immediately.
4. Add 5 mL of vanillin reagent to each remaining tube at 15 s intervals (see **Note 8**).
5. After the 20 min incubation period is complete, measure the absorbance of each tube at 500 nm (*see* **Note 9**). Zero the spectrophotometer with the methanol blank.
6. Using Microsoft Excel (or other equivalent software), determine the slope of the line using catechin concentrations (0, 0.2, 0.4, 0.6, 0.8, and 1.0 mg/mL) as the x-axis and the absorbance values as the *y*-axis. Determine the r^2 of the curve by running a linear regression. The r^2 value should be 0.995 or higher. If it is lower, repeat **steps 1–6** until a higher r^2 value is obtained.

3.2.2 Sample Analysis

1. Weigh three replicates of 0.3 g each of freshly ground sorghum sample and transfer them into centrifuge tubes. For samples that are known to have a high tannin concentration, reduce the weight to 0.15 g.
2. Add 8 mL of 1% HCl in methanol to each tube. Mix the contents on a vortex mixer for 10 s and place each tube in the water bath for 20 min. After the first half of the incubation period (10 min), vortex each tube for 10 s and place it back into the water bath for the remaining incubation period (*see* **Note 10**).
3. Remove each tube exactly after 20 min of incubation. Mix each tube on the vortex mixer for 10 s as it is recovered from the water bath.
4. Centrifuge the samples at 4000 × *g* for 10 min.
5. Decant the supernatant carefully to avoid ground sample to be transferred into the supernatant.
6. Take two 1 mL aliquots from the supernatant, place each aliquot into a separate test tube. One tube is labeled as the "sample" tube and the other tube is labeled as the "blank" tube. Place all tubes into the water bath (30 °C). Set timer to 20 min but do not start the timer yet.
7. While in the water bath for 20 min, add 5 mL of the vanillin reagent to the "sample" tube in each pair. Start the timer once the vanillin reagent is added to the first "sample" tube. Continue to add the vanillin reagent to each "sample" tube at 15-s intervals (*see* **Note 8**).
8. Add 5 mL of 4% HCl in methanol to the "blank" tube in each pair at 15 s intervals (*see* **Note 8**).
9. Measure the absorbance of each "sample" and "blank" tube exactly after 20 min at 500 nm (*see* **Note 9**). Zero the spectrophotometer with a methanol blank.

10. For the final determination of tannin concentration, subtract the value of the "blank" from that of the "sample." Calculate tannin concentration using the following equation:

$$\text{Tannin concentration (mg catechin equiv./g)} = \frac{V * A/m}{W}$$

where:

V = Volume of extract in mL (in this case 8 mL).
A = Absorbance at 500 nm (absorbance of "sample"—absorbance of "blank").
m = slope of the standard curve.
W = Weight of sample (in this case 0.15 or 0.30 g).
Non-tannin sorghums give values of less than 2.0 mg catechin equiv./g (Table 1). When in doubt, perform the Chlorox bleach test to confirm the presence or absence of a pigmented testa.

3.3 NP-HPLC with Fluorescence Detection

1. Weigh three replicates of 1.0 g each of freshly ground sorghum sample and place into centrifuge tubes. For samples that are known to have a very high tannin concentration, reduce the weight to 0.5 g.
2. Add 10 mL of acetone/water/acetic acid (70:29.5:0.5, v/v) and place the tubes on a shaker for 2 h at high speed.
3. Centrifuge the mixture at 4000 × *g* for 10 min.
4. Concentrate 7.5 mL of the extract until all the acetone is removed using a rotary evaporator or concentrator.
5. Bring up the volume of the concentrated extract to 6 mL with water and load it to a Sephadex LH-20 column that is already equilibrated in water.
6. After the sample is loaded, wash the column with 40 mL of 30% methanol (v/v) to remove the sugars and other phenols.
7. Add 70 mL of 70% acetone (v/v) and collect the eluent, which contains the condensed tannins.
8. Concentrate the eluent containing the condensed tannins to dryness and dissolve the residue with 70% acetone (v/v) to a final volume of 5 mL. Filter the extract using a nylon filter (0.45 μm, 13 mm) prior to HPLC analysis.
9. Analyze the extracts on the HPLC system using catechin, (−)-epicatechin, procyanidin B-1 and B-2, oligomeric tannins (DP3-DP10), and polymeric tannins (>DP10) as external standards. The silica HPLC column should first be equilibrated at 37 °C with 86% dichloromethane (Solvent A), 14.0% methanol (Solvent B), and 4% acetic acid in water (Solvent C). The fluorescence detector should be set at the following wavelengths: excitation—276 nm; emission—316 nm. Once the

column is equilibrated, inject 20 μL of extract and separate the condensed tannins using the following gradient at a flow rate of 1 mL/min: 0–20 min, 14.0–23.6% B; 20–50 min, 23.6–35.0% B; 50–55 min, 35.0–86.0% B; 55–65 min, 86.0% B; 65–70 min, 86.0–14.0% B; 70–80 min, 14% B. Solvent C (4%) is kept constant throughout the gradient (*see* **Note 11**).

10. Identify the tannin components (monomers, DP2, DP3, ..., DP10, polymeric tannins) based on the retention time of each standard and quantify each component by comparing peak areas with that of the calibration curve of each standard. For oligomeric tannins, the areas of the components of the same degree of polymerization (DP) are integrated and quantified individually. The area beyond DP10 is integrated and quantified as polymeric tannins. Express the results of each component as μg/g (*see* **Note 12**).

4 Notes

1. If using commercial/laundry bleach, make sure it is fresh. The use of "old" bleach may give erroneous results.
2. One should have, at a minimum, the following sorghum standards: (1) white pericarp, non-tannin; (2) red pericarp, non-tannin; (3) white pericarp, tannin; and (4) red pericarp, tannin.
3. The catechin standard solution can be stored in the freezer (−20 °C) when not being used.
4. Do not let the Sephadex LH-20 slurry dry during the packing procedure. Make sure the water level is above (>3 mm) the slurry during this process.
5. Adding too much bleaching reagent may cause over-bleaching the sorghum kernels and thus give false-negative results [14].
6. Make sure the water flow is not too strong. A strong water flow may cause removal of the pigmented testa.
7. The bleached tannin sorghum kernels are mostly black with the exception of the germ where it is dark brown. Some non-tannin sorghums may show small to large black spots after bleaching, which may be due to kernel damage from diseases, weathering, or insect bites [13, 15].
8. The interval time to add the reagents can be longer (>15 s). However, since the reaction is time-dependent, the addition of the reagents must be exactly at the same time interval. This applies for both standards and samples. If the reagents were added at 15 s intervals for the generation of the standard curve, the reagents must also be added at 15 s intervals for sample

analysis. Failure to do so will cause erroneous absorbance readings.

9. Do not remove all test tubes from the water bath at once. The remaining test tubes should stay in the water bath until they are ready to be read on the spectrophotometer after 20 min from the time the reagent had been added. Since the reaction is time-dependent, do not go back and re-read the absorbance of the "sample" or "blank" as it will give significantly different absorbance readings.
10. Do not remove all test tubes from the water bath at once. The remaining test tubes should stay in the water bath during this process.
11. For HPLC systems equipped with only a binary pump, the mobile phase and gradient can be modified as follows: dichloromethane, methanol, water, and acetic acid (82:14:2:2 v/v) (Solvent A); methanol, water, and acetic acid (96:2:2 v/v) (Solvent B) [20]. After the column is equilibrated with 0.0% B, separate the condensed tannins using the following gradient: 0–20 min, 0.0–11.7% B; 20–50 min, 11.7–25.6% B; 50–55 min, 25.6–87.8% B; 55–65 min, 87.8% B; 65–70 min, 87.8–0.0% B; 70–80 min, 0% B. All other conditions (i.e., column temperature, flow rate, injection volume) remain the same.
12. According to Gu et al. [21], valley-to-valley integration was found to severely underestimate condensed tannin levels and it was recommended to integrate using the flat baseline method: draw a flat baseline from the beginning to the end of the run. Draw perpendicular lines from the lowest point of the valley between peaks to the flat baseline.

Acknowledgment

Special thanks to Dr. Tom Herald, USDA-ARS, Center for Grain and Animal Health Research, Manhattan, KS, for providing the sorghum samples for the Chlorox bleach test procedure.

Disclaimer: Mention of trade names or commercial products in this publication is solely for the purpose of providing specific information and does not imply recommendation or endorsement by the U.S. Department of Agriculture. USDA is an equal opportunity provider and employer.

References

1. Awika JM, Dykes L, Gu L, Rooney LW, Prior RL (2003) Processing of sorghum (*Sorghum bicolor*) and sorghum products alters procyanidin oligomer and polymer distribution and content. J Agric Food Chem 51:5516–5521
2. Awika JM, McDonough CM, Rooney LW (2005) Decorticating sorghum to concentrate healthy phytochemicals. J Agric Food Chem 53:6230–6234
3. Dykes L, Rooney LW, Waniska RD, Rooney WL (2005) Phenolic compounds and antioxidant activity of sorghum grains of varying genotypes. J Agric Food Chem 53:6813–6818
4. Dykes L, Seitz L, Rooney WL, Rooney LW (2009) Flavonoid composition of red sorghum genotypes. Food Chem 116:313–317
5. Dykes L, Peterson GC, Rooney WL, Rooney LW (2011) Flavonoid composition of lemon-yellow sorghum genotypes. Food Chem 128:173–179
6. Dykes L, Rooney WL, Rooney LW (2013) Evaluation of phenolics and antioxidant activity of black sorghum hybrids. J Cereal Sci 58:278–283
7. Hahn DH, Faubion JM, Rooney LW (1983) Sorghum phenolic acids, their high performance liquid chromatography separation and their relation to fungal resistance. Cereal Chem 60:255–259
8. Dykes L, Rooney LW (2006) Sorghum and millet phenols and antioxidants. J Cereal Sci 44:236–251
9. Rooney LW, Miller FR (1982) Variation in the structure and kernel characteristics of sorghum. In: Mertin JV (ed) Proceedings of the International Symposium on Sorghum Grain Quality. International Crops Research Institute for the Semi-Arid Tropics (ICRISAT), 28–31 Oct 1982, Patancheru, AP
10. Price ML, Van Scoyoc S, Butler LG (1978) A critical evaluation of the vanillin reaction as an assay for tannin in sorghum grain. J Agric Food Chem 26:1214–1218
11. Boren B, Waniska RD (1992) Sorghum seed color as an indicator of tannin content. J Appl Poult Res 1:117–121
12. International Association for Cereal Science and Technology (2017) Standard Method 177: Detection of tannin sorghum grain by the bleach test. Approved 2012. International Association for Cereal Science and Technology, Vienna
13. Grain Inspection, Packers and Stockyards Administration (2013) Sorghum. In: Grain inspection handbook II. https://www.gipsa.usda.gov/fgis/gradingprocedures.aspx. Accessed Oct 2017
14. Taylor JRN, Taylor J (2008) Five simple methods for the determination of sorghum grain end-use quality (with adaptations for those without laboratory facilities). INTSORMIL Scientific Publications 17. USAID, INTSORMIL, University of Pretoria, Pretoria, South Africa
15. Waniska RD, Hugo LF, Rooney LW (1992) Practical methods to determine the presence of tannins in sorghum. J Appl Poult Res 1:122–128
16. Schoffield P, Mbugua DM, Pell AN (2001) Analysis of condensed tannins: a review. Anim Feed Sci Technol 91:21–40
17. Butler LG (1989) Sorghum polyphenols. In: Cheeke PR (ed) Toxicants of plant origin, Phenolics, vol IV. CRC Press, Inc, Boca Raton, FL
18. Hagerman AE, Butler LG (1980) Condensed tannin purification and characterization of tannin-associated proteins. J Agric Food Chem 28:947–952
19. Hahn DH, Rooney LW (1986) Effect of genotype on tannins and phenols of sorghum. Cereal Chem 63:4–8
20. Adamson GE, Lazarus SA, Mitchell AE, Prior RL, Cao G, Jacobs PH, Kremers BG, Hammerstone JF, Rucker RB, Ritter KA, Schmitz HH (1999) HPLC method for the quantification of proycanidins in cocoa and chocolate samples and correlation to total antioxidant capacity. J Agric Food Chem 47:4184–4188
21. Gu L, Kelm M, Hammerstonne JF, Beecher G, Cunningham D, Vannozzi S, Prior RL (2002) Fractionation of polymeric procyanidins from lowbush blueberry and quantification of procyanidins in selected foods with an optimized normal-phase HPLC-MS fluorescent detection method. J Agric Food Chem 50:4852–4860

Chapter 9

Sorghum Phytochemicals and Their Potential Impact on Human Health

Linda Dykes

Abstract

Sorghum contains a wide array of phytochemicals and their levels are affected by the genotype. Phytochemicals identified in sorghum include phenolic acids, flavonoids, condensed tannins, polycosanols, phytosterols, stilbenes, and phenolamides. Most of these phytochemicals are concentrated in the bran fraction and have been shown to have several potential health benefits, which include antidiabetic, cholesterol-lowering, anti-inflammatory, and anticancer properties. This chapter gives an overview of sorghum genetics relevant to phytochemicals, phytochemicals identified in sorghum grain, and their potential health benefits.

Key words Sorghum, Phytochemicals, Phenolics, Phenolic acids, Flavonoids, 3-Deoxyanthocyanins, Condensed tannins, Health benefits

1 Introduction

Sorghum (*Sorghum bicolor* (L.) Moench) is the fifth major cereal in the world after wheat, maize, rice, and barley [1]. It originated in Africa and it is mainly grown in semiarid to arid regions [2]. Sorghum is primarily used as food in Africa and Asia and as feed in Western countries. However, due to its gluten-free characteristics, sorghum is increasingly used as food in gluten-free markets in the United States. Sorghum is processed into several food products and beverage such as bread, tortillas, porridges, couscous, snack foods (e.g., popped, puffed), and beer [3]. All sorghums contain a wide variety of phytochemicals and their compositions are affected by the genotype [4]. These phytochemicals are concentrated in the bran fraction and they are reported to have several potential health benefits, which include antidiabetic, cholesterol-lowering, anti-inflammatory, and anticancer properties [4–7]. This chapter gives an overview of sorghum genetics relevant to phytochemicals, phytochemicals identified in sorghum grain, and their potential health benefits.

Zuo-Yu Zhao and Jeff Dahlberg (eds.), *Sorghum: Methods and Protocols*, Methods in Molecular Biology, vol. 1931, https://doi.org/10.1007/978-1-4939-9039-9_9,

2 Overview of Sorghum Genetics Relevant to Phenolics and Tannins

Sorghum pericarp color is controlled by the *R* and *Y* genes (Fig. 1) [4, 9]. A pericarp is white when *Y* is homozygous recessive (*rryy* or *R_yy*), whereas a yellow (or lemon-yellow) pericarp has recessive *R* and dominant *Y* genes (*rrY_*). When both *R* and *Y* genes are dominant, the pericarp is red. The "red" color varies and includes orange, light red, bright red, light brown, dark brown, and black. Black sorghum is a special red sorghum with both dominant *R* and *Y* genes that turns black when grown in the presence of sunlight (Fig. 2) [10]. The intensifier gene *I* affects the intensity of the pericarp color and is clearly visible in red sorghums [4, 9]. Red sorghums with the intensifier gene appear bright red compared to those without the gene [9].

Secondary plant color is controlled by the *P* and *Q* genes, which is also associated with glume color [9]. Plants with *P_Q_* and *P_qq* genes produce purple- and red-pigmented plants, respectively, while recessive *ppqq* genes produce tan-pigmented plants. Secondary plant color can affect the appearance of the grain. Sorghums with the *P_Q_* and *P_qq* genes have purple/red stains that are leached from the purple/red glumes onto the pericarp when grown under humid conditions or when rainfall occurs [9].

The presence or absence of a pigmented testa (Fig. 3) is controlled by the $B_1\_B_2\_$ genes and sorghums with a pigmented testa must have both dominant genes ($B_1\_B_2\_$) [9]. Sorghums with the $B_1\_B_2\_$ genes have condensed tannins, which contribute to their astringency and some resistance to molds and deterioration of the grain [11]. These genotypes are also referred to as "bird proof" or "bird resistant" since they provide some resistance to bird predation in the field [4]. When given a choice, birds prefer non-tannin sorghums but they do consume tannin sorghums when no other food is available [4]. In sorghum nurseries with non-tannin white, non-tannin red, and tannin sorghums (both Types II and III), birds were found to eat sorghum in the following order: (1) non-tannin

Fig. 1 Genetics of sorghum pericarp color (*R* and *Y* genes): (**A**) White (*rryy* or *R_yy*); (**B**) lemon-yellow (*rrY_*); (**C**) red (*R_Y_*); and (**D**) black (*R_Y_*). Images adapted from Dykes [8]

Fig. 2 Tx430 Black sorghum grains from panicles that were (**A**) covered and (**B**) uncovered during their development in College Station, TX, USA in 2004. Reproduced from Dykes et al. [10] with permission from Elsevier

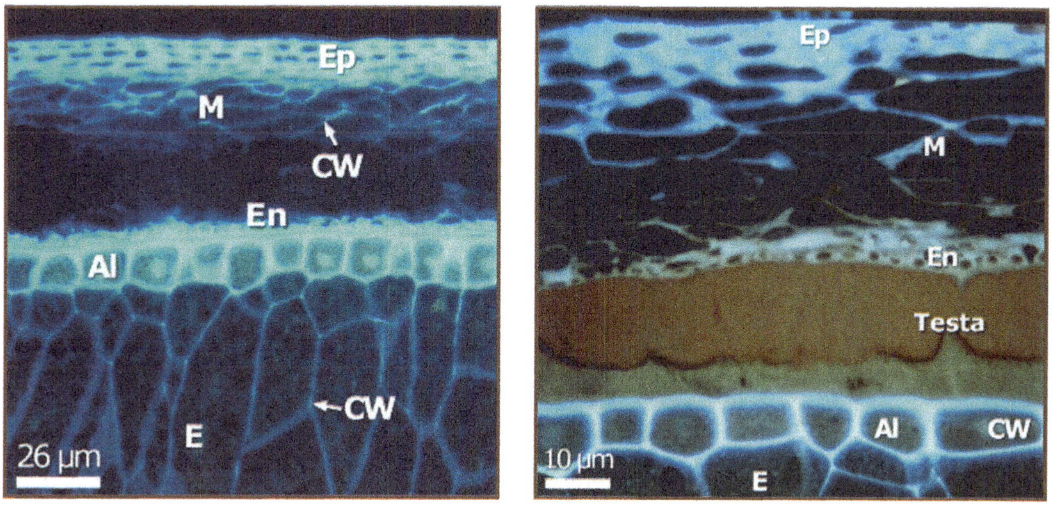

Fig. 3 Fluorescence photomicrograph of cross-sections of a non-tannin (left) and a tannin sorghum kernel (right). *Al* aleurone, *CW* cell wall, *E* endosperm, *En* endocarp, *Ep* epicarp, *M* mesocarp (Reprinted from Dykes and Rooney [4] with permission from Elsevier)

white sorghum; (2) non-tannin red sorghum; (3) Type II sorghum; and (4) Type III sorghum [4, 12]. Pericarp color and its intensity are not a reliable indicator of tannins in sorghums [13], and it is erroneously believed that all sorghums with a red or brown pericarp have tannins [12]. Sorghums with a white, lemon-yellow, red, brown, or black pericarp may or may not have tannins depending upon the presence of a pigmented testa.

The spreader gene *S_* controls the presence of brown pigments, possibly tannins, in the epicarp and endocarp when a pigmented testa is present [4]. Tannin sorghums with a white pericarp and the spreader gene *S_* appear tan-colored or light brown while those without the spreader gene (*ss*) appear white. Tannin sorghums without the spreader gene ($B_1\_B_2\_ss$) are known as Type II sorghums while those with the spreader gene ($B_1\_B_2\_S\_$) are known as Type III sorghums [9]. The extractability of condensed tannins in Type II and III sorghums differs. The condensed tannins in Type II sorghums ($B_1\_B_2\_ss$) are extracted with acidified methanol (1% HCl methanol) while those in Type III sorghums ($B_1\_B_2\_S\_$) are extracted with either methanol or acidified methanol [4]. The difference in extractability is due to the fact that tannins in Type II and III sorghums are stored differently in the testa layer. In Type II sorghums, tannins are stored in vesicles within the testa layer, whereas the tannins in type III sorghums are stored along the cell walls of the testa and some are present in the pericarp [14]. This may explain why acid (e.g., HCl) is required to disrupt the structure of the vesicles to release the tannins in Type II sorghums [14]. In general, condensed tannin concentrations are highest in sorghums containing dominant $B_1\_B_2\_SS$ genes and these sorghums have high bird and mold resistance [4, 13].

Pericarp thickness is controlled by the *Z* gene [4, 9]. A pericarp is thick when the genes are homozygous recessive (*zz*) and thin when they are dominant (*ZZ*). Sorghums with a thick pericarp have significantly far more starch granules in the mesocarp than those with a thin pericarp [15]. Pericarp thickness may affect the appearance of the grain. In general, sorghums with a thin pericarp have a "pearly" appearance while those with a thick pericarp have a "chalky" appearance [16]. Some white sorghums (*rryy* or *R_yy*) with a thin pericarp and yellow endosperm appear yellow in color and could erroneously be classified as lemon-yellow sorghums. Thus, care must be taken when evaluating these genotypes. Pericarp thickness may also have an effect on phenolic concentrations. It has been reported that sorghum genotypes with a thick pericarp have higher phenolic concentrations than those with a thin pericarp [16].

3 Phytochemical Compounds in Sorghum Grain

3.1 Phenolic Acids

All sorghum genotypes contain phenolic acids, which are located in the pericarp, testa, aleurone layer, and endosperm [17]. Phenolic acids consist of two classes: hydroxybenzoic and hydroxycinnamic acids. Hydroxybenzoic acids are directly derived from benzoic acid and they include gallic, protocatechuic, *p*-hydroxybenzoic, vanillic, and syringic, acids. The hydroxycinnamic acids have a C6-C3 structure and they include caffeic, *p*-coumaric, ferulic, and sinapic acids.

Sorghum phenolic acids are found in free and bound forms [18]. Free phenolic acids are located in the pericarp, testa, and aleurone layer and they are extracted in organic (e.g., methanol) or aqueous organic solvents (e.g., 70 or 80% methanol)

[18–20]. Bound phenolic acids are esterified to the cell walls and are extracted either in alkaline solution (e.g., 2 *M* NaOH) or in boiling 2 *M* HCl [18–20]. Phenolic acids in sorghum are present mostly in bound form with ferulic acid being dominant (24–76%) [18, 19]. Gallic, sinapic, and diferulic acids have been detected only in bound forms [18, 19]. Protocatechuic, *p*-hydoxybenzoic, vanillic, caffeic, *p*-coumaric, and cinnamic acids have been reported in both free and bound forms [18]. Grain hardness affects phenolic acid concentrations where hard sorghum varieties have higher phenolic acid concentrations (310–383 μg/g) than soft varieties (107–163 μg/g) [19].

Several glycerol esters of phenolic acids have also been identified in sorghum. They include 2-*O*-caffeoyl-glycerol, 1-*O*-caffeoyl-glycerol, 2-*O*-coumaroyl-glycerol, 1,3-*O*-dicaffeoyl-glycerol, 1,3-coumaroyl-caffeoyl-glycerol, 1,3-coumaroyl-feruloyl-glycerol, 1,3-*O*-dicoumaroyl-glycerol, and 1,3-*O*-diferuloyl-glycerol [20–22]. These compounds have been reported in sorghums with white and red/brown pericarps [20–22].

3.2 Flavonoids

Many sorghum flavonoids have been isolated and identified over the years (Table 1), which include 3-deoxyanthocyanins, flavan-4-ols, flavanones, flavones, and flavonols (Fig. 4). Anthocyanins are pigments that contribute to the reds, blues, and purples, in fruits, vegetables, and flowers. Sorghum anthocyanins are unique since, unlike the anthocyanidins (e.g., cyanidin, delphinidin, malvidin, pelargonidin, petunidin, peonidin), they lack the hydroxyl group at the 3-position of the C-ring and thus are called 3-deoxyanthocyanins. Sorghum is the only dietary source of those compounds [27]. The two major 3-deoxyanthocyanidins are the yellow apigeninidin and the orange luteolinidin. The lack of the hydroxyl group at the 3-position of the C-ring is reported to increase pigment stability at higher pH and temperature [29, 35–37], which make these pigments good candidates for natural food colorants. The 3-deoxyanthocyanins are also phytoalexins since they are produced as a response to mold invasion or other stresses in sorghum [2, 28]. Apigeninidin, luteolinidin, 5-methoxyluteolinidin, and 7-methoxyapigeninidin are the four major 3-deoxyanthocyanidins identified and quantified in sorghum [10, 23–27]. Other sorghum 3-deoxyanthocyanins identified include apigeninidin-5-*O*-glucoside, luteolinidin-5-*O*-glucoside, 5-methoxyluteolinidin-7-*O*-glucoside, and 7-methoxyapigeninidin-5-*O*-glucoside (Table 1) [27].

Sorghum 3-deoxyanthocyanins are located in the bran fraction and their concentrations are 3–4 times higher than in the grain [30]. Sorghums with a black pericarp have the highest concentrations of 3-deoxyanthocyanins (329–1054 μg/g) and their concentrations are increased in darker grains [24]. However, it cannot be concluded that high concentrations of 3-deoxyanthocyanins lead to

Table 1
Flavonoids identified in sorghum grain

Compound	References
3-Deoxyanthocyanins:	
7-Methoxyapigeninidin	[10, 22–27]
7-Methoxyapigeninidin-5-*O*-glucoside	[27]
5-Methoxyluteolinidin	[10, 23–28]
5-Methoxyluteolinidin-7-*O*-glucoside	[27]
Apigeninidin	[10, 22–30]
Apigeninidin-5-*O*-glucoside	[27, 31]
Luteolinidin	[10, 22–30]
Luteolinidin-5-*O*-glucoside	[27]
Dihydroflavonols:	
Dihydrokaempferol	[21]
Taxifolin	[20, 21]
Taxifolin-7-*O*-glucoside	[32]
Flavan-3-ols:	
7-Methoxycatechin	[21]
Catechin	[20, 21]
Catechin hexoside	[21]
Flavan-4-ols:	
Apiforol	[33]
Luteoforol	[34]
Flavanones:	
Eriodictyol	[10, 20, 21, 23–26]
Eriodictyol-5-*O*-glucoside	[32]
Eriodictyol-7-*O*-glucoside	[20]
Homoeriodictyol	[21]
Naringenin	[10, 20, 21, 23–26, 32]
Naringenin-7-*O*-glucoside	[20, 22]
Flavones:	
7-Methoxyapigenin	[22]
7-Methoxyluteolin	[22]
Apigenin	[10, 20–26]
Apigenin-6-*C*-glucoside	[21]
Apigenin-4′-*O*-glucoside	[21]
Apigenin-7-*O*-glucoside	[21]
Chrysoeriol hexoside	[21]
Luteolin	[10, 20–26]
Tricin	[21]
Tricin hexoside	[21]
Flavonols:	
Isorhamnetin	[21]
Kaempferol-3-rutinoside-7-glucuronide	[31]
Quercetin	[21]
Quercetin hexoside	[21]

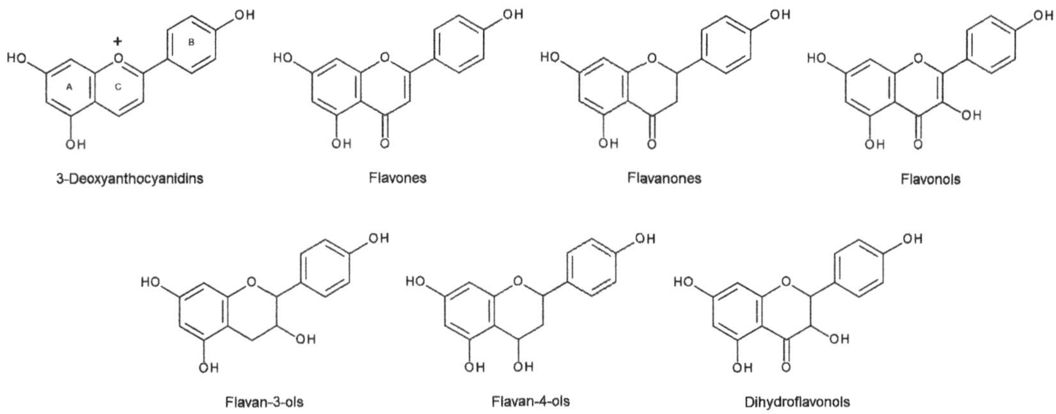

Fig. 4 Chemical structure of flavonoids in sorghum grain

a darker or black pericarp since it is unknown as to what compounds contribute to its black color [24]. Secondary plant color also affects 3-deoxyanthocyanin concentration, as well as their composition. Sorghums with purple or red secondary plant color are reported to have higher concentrations of 3-deoxyanthocyanins (8–680 μg/g) than those with tan secondary plant color (nd-1.5 μg/g) [10, 23]. Apigeninidin and 7-methoxyapigeninidin are reported to be the main 3-deoxyanthocyanidins in sorghums with red secondary plant color which, combined, account for 66–89% of the total [10, 23]. On the other hand, luteolinidin and 5-methoxyluteolinidin are reported to be the main 3-deoxyanthocyanidins in sorghums with purple secondary plant color which, combined, account for 57–81% of the total [10, 23].

Flavan-4-ols (e.g., luteoforol, apiforol) are a group of compounds that are produced from flavanones (e.g., eriodictyol, naringenin) and may be precursors of 3-deoxyanthocyanins [38]. Sorghums with a black pericarp have the highest flavan-4-ol concentrations, followed by those with a red pericarp and purple/red secondary plant color [16, 39]. However, not all sorghums with a red pericarp have detectable flavan-4-ols [39, 40]. Some genotypes with a white pericarp are reported to have flavan-4-ols as well [16, 39–43]. Pericarp thickness may affect flavan-4-ol concentrations. One study reported that sorghums with purple/red secondary plant color and a thick pericarp had higher concentrations of flavan-4-ols (4.3–9.3 abs/mL/g) than those with purple/red secondary plant color and a thin pericarp (3–4 abs/mL/g) and those with tan secondary plant color (2–3 abs/mL/g) [16]. Flavan-4-ols may play a role in mold resistance as several studies report a correlation between flavan-4-ols concentration and mold resistance in sorghums [40–44]; however, these findings have been ineffective in selecting sorghums with mold resistance [4].

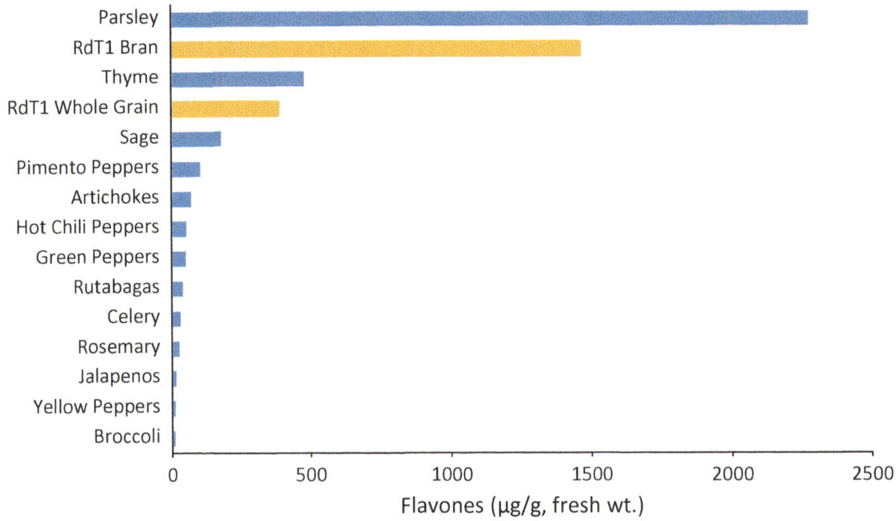

Fig. 5 Comparison of flavone concentrations in red sorghum with tan secondary plant color (RdT1) with those from common food sources. Reproduced from Dykes and Rooney [7] with permission from Food Science Publisher

Flavones (Fig. 4) are a group of flavonoids that are generally associated with herbs and vegetables. Luteolin and apigenin are two main flavones reported in sorghum [10, 20–24]. Other flavones identified include 7-methoxyapigenin, 7-methoxyluteolin, apigenin-6-*C*-glucoside, apigenin-4′-*O*-glucoside, apigenin-7-*O*-glucoside, chrysoeriol hexoside, and tricin [21, 22] (Table 1). Flavones are located in the pericarp and their concentrations increase almost four-folds in the bran fraction [10]. Pericarp and secondary plant colors affect flavone concentrations. These compounds are predominant in red or lemon-yellow sorghums with tan secondary plant color (60–394 µg/g) [10, 23, 26], which make these genotypes a good source of flavones (Fig. 5). Sorghums with a white pericarp and tan secondary plant color also contain flavones but their concentrations (1–65 µg/g) are comparable to those with purple or red secondary plant color (nd-67 µg/g) [10, 23, 26].

Flavanones (Fig. 4) are another group of flavonoids and they are mostly associated with citrus fruits. Eriodictyol and naringenin are the two major flavanones quantified in sorghum [10, 20, 23, 24, 26]. Other flavanones reported include eriodictyol-5-*O*-glucoside, eriodictyol-7-*O*-glucoside, homoerioedictyol, and naringenin-7-*O*-glucoside (Table 1) [20–22]. Flavanones are also located in the pericarp and their concentrations are affected by pericarp color [24]. Sorghums with a lemon-yellow pericarp (*rrY_*) have higher flavanone concentrations (308–1823 µg/g) than other genotypes (nd-382 µg/g) [23–26]. Decorticating lemon-yellow sorghums to produce bran increases flavanone concentrations four-folds [23]. The high flavanone concentrations

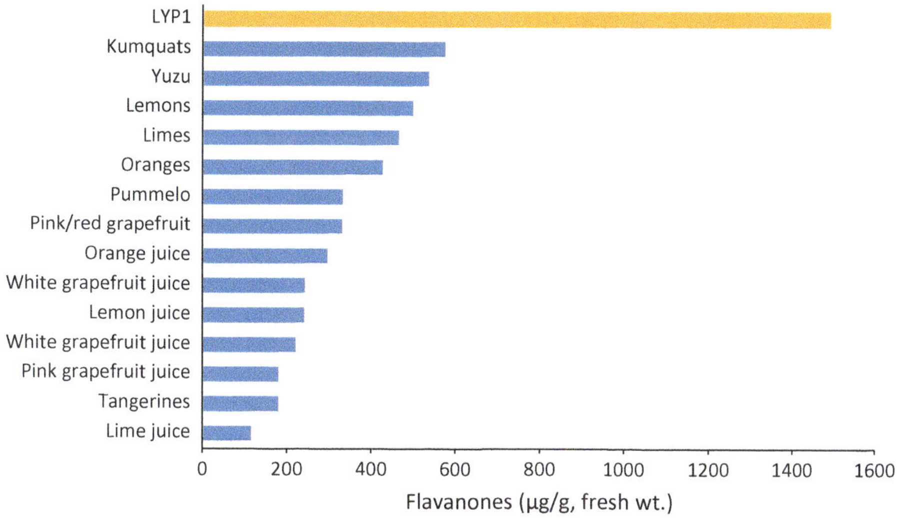

Fig. 6 Comparison of flavanone concentrations in lemon-yellow sorghum grain (LYP1) with those from citrus fruits. Reproduced from Dykes and Rooney [7] with permission from Food Science Publisher

make these genotypes a good source of these compounds since their concentrations are higher than those found in citrus fruits on an as-is-basis (Fig. 6).

Other flavonoids identified in sorghum include the flavonols quercetin [21] and kaempferol-3-rutinoside-7-glucuronide [31]. The dihydroflavonols taxifolin and taxifolin-7-*O*-glucoside have also been isolated [20, 21, 32].

3.3 Condensed Tannins

Sorghums with the $B_1\_B_2\_$ gene contain tannins, which are the major phenolic compounds in those genotypes [17]. These compounds contribute to some resistance to molds and deterioration of the grain [11]. In general, type II and III sorghums have tannin concentrations of 0.6–1.6 and 1.1–3.7 mg/100 mg catechin equivalents, respectively [13]. The tannins in these sorghums are of the condensed types and they are also known as proanthocyanidins, or procyanidins. Tannic acid, a hydrolyzable tannin, has never been found in sorghum even though some studies report tannin concentrations as tannic acid equivalents/g [4, 12].

Proanthocyanidins or condensed tannins are mainly polymers of flavan-3-ols units and are grouped into two types based on their interflavan linkages: A-type and B-type. A-type proanthocyanidins (Fig. 7) consist of flavan-3-ol units linked by C4 → C8 interflavan bonds and by an additional ether bond between C2 → C7, which have been identified mostly in cranberries [45, 46]. B-type proanthocyanidins (Fig. 7) consist of polymerized flavan-3-ol and/or flavan-3,4-diol units linked by C4 → C8 interflavan bonds. The proanthocyanidins in tannin sorghums are mostly of the B-type with (−)-epicatechin as extension units and catechin as

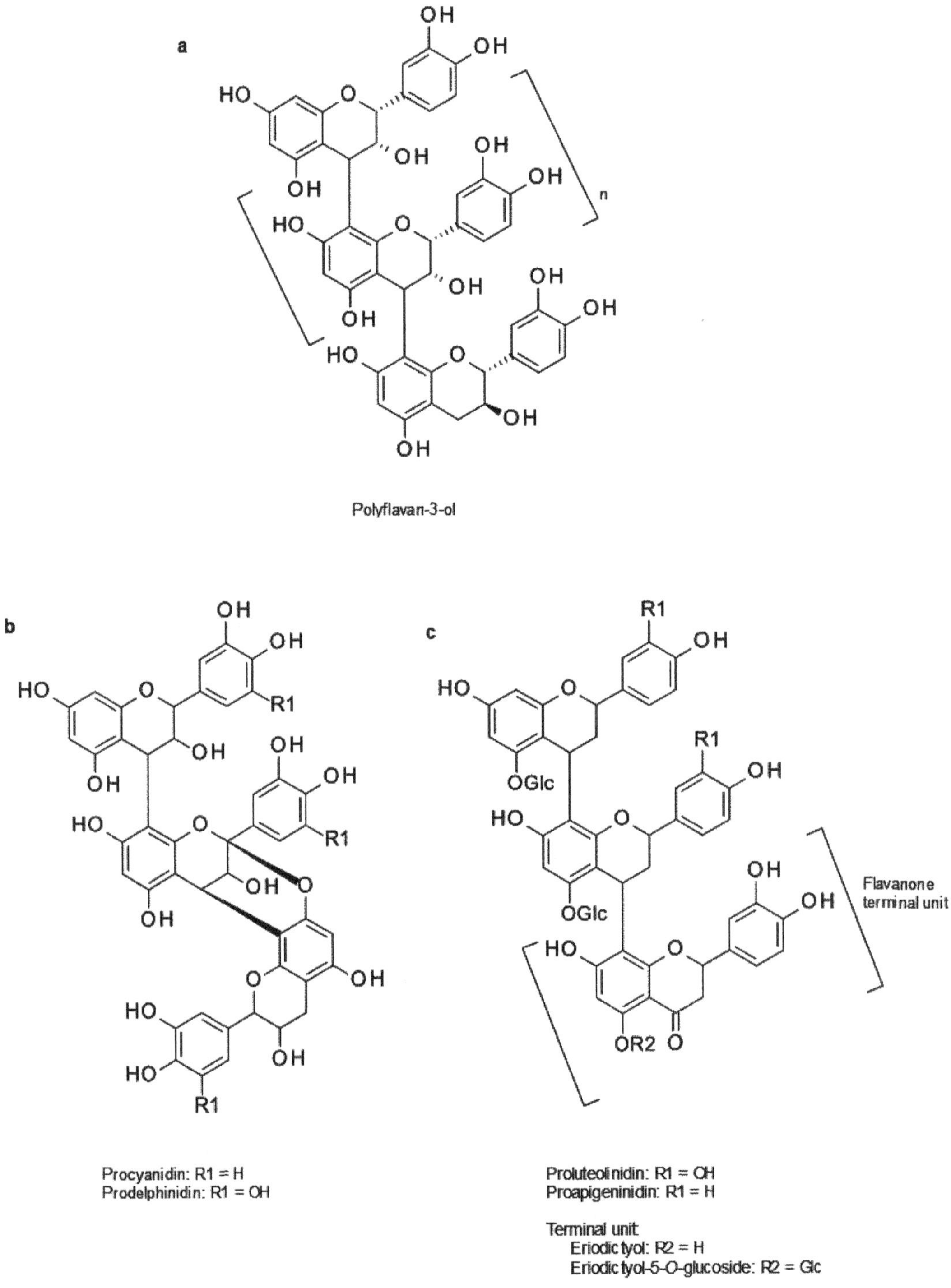

Fig. 7 Structures of proanthocyanidins reported in sorghum. (**a**) Polyflavan-3-ol with B-type interflavan linkages; (**b**) Heteropolyflavan-3-ols with A- and B-type interflavan linkages; (**c**) Glucosylated heteropolyflavans with a flavanone as the terminal unit. Reproduced from Dykes and Rooney [4] with permission from Elsevier

terminal units [46, 47]. Diversity of sorghum proanthocyanidins has been reported [4, 48]. Heteropolyflavan-3-ols with both A- and B-type interflavan linkages with catechin/epicatechin or gallocatechin/epigallocatechin as extension and terminal units (Fig. 7) have been identified [48, 49]. Glucosylated heteropolyflavans with either proluteolinidin or proapigeninidin flavan as extension units and the flavonones eriodictyol or eriodictyol-5-*O*-*β*-glucoside as terminal units (Fig. 7) have also been identified [32, 48].

3.4 Other Phytochemicals Identified in Sorghum Grain

Polycosanols and phytosterols are compounds associated with the lipid fraction of sorghum grain [6]. Reported polycosanol concentrations in polished and unpolished sorghums are 10 and 75 mg/100 mg, respectively, which are significantly higher than those reported in brown maize (0.2 mg/100 mg), rice (2 mg/100 mg), and unpolished purple rice (4 mg/100 mg) [50]. Campesterol, stigmasterol, and β-sitosterol have been identified in sorghum with β-sitosterol being dominant [51, 52]. The stilbenes *trans*-resveratrol and *trans*-piceid are also reported in red sorghum with *trans*-piceid being dominant ($\leq$1 mg/kg) [53]. In addition, the phenolamides N^1, N^4-dicaffeoyl spermidine, N^1, N^8-dicaffeoyl spermidine, and N^1, N^8-caffeoyl-feruloyl spermidine have been identified in white and red/brown sorghums [21].

4 Potential Contribution of Sorghum and Their Phytochemicals to Health

4.1 Sorghum Phenolics and Their Antioxidant Activity Potential

Sorghum has consistently been shown to have antioxidant activity potential in vitro, which is contributed by phenolic compounds [16, 39, 54]. Sorghums with a pigmented testa for the presence of condensed tannins have the highest antioxidant activity potential among all genotypes [16, 39, 54]. They approach or exceed the antioxidant levels of fruits and vegetables (Fig. 8). The elevated antioxidant activity potential of tannin sorghums is not surprising since condensed tannins have been shown to be more potent antioxidants than monomeric phenolic compounds [55]. Decortication of sorghum to produce bran increases sorghum phenolics and antioxidant activity potential 3–5 times over the original grain (Fig. 8) [54, 56]. The antioxidant activity potential of sorghum extracts is strongly correlated to total phenols [16, 39, 54] and condensed tannins [24, 39]. A strong correlation between antioxidant activity and flavan-4-ol levels ($r = 0.88$) among non-tannin sorghums with a red pericarp has been reported as well [16]. However, this correlation has not been found in tannin sorghums with a black pericarp [24].

4.2 Antidiabetic Properties

Antidiabetic properties of sorghum have been reported in several studies. For example, tannin and black sorghum bran are reported to inhibit protein glycation and thus have potential use for diabetic

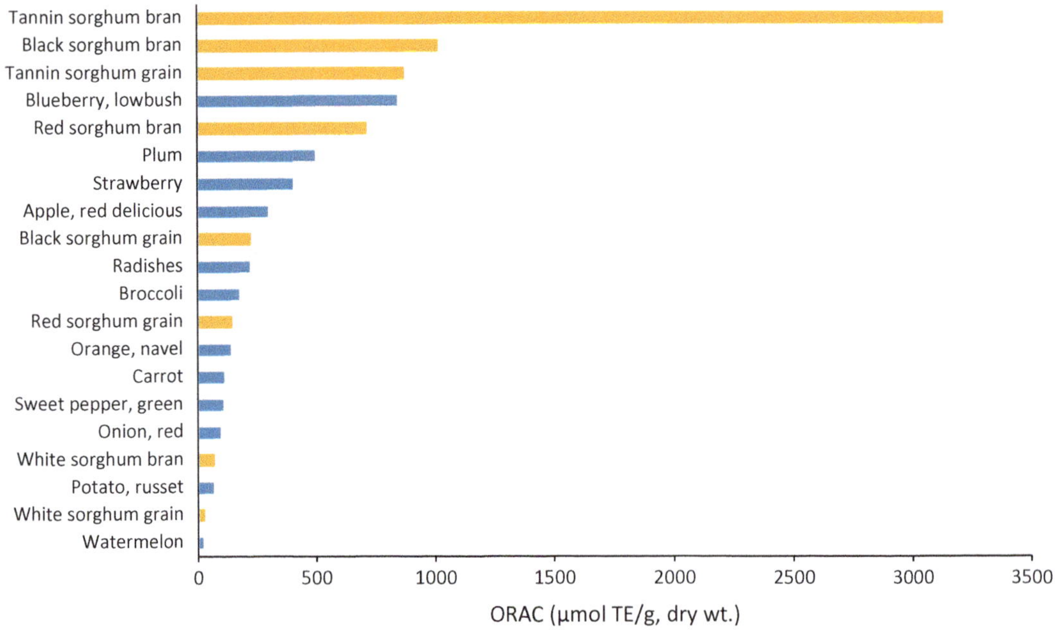

Fig. 8 Antioxidant activity (ORAC) of sorghum grain and bran compared to common fruits and vegetables. Data obtained from Dykes and Rooney [4]

or other metabolic syndrome patients [57]. In another study, rats that were fed a high-fat-fructose diet with tannin whole sorghum flour had decreased fasting glucose, improved glucose tolerance, insulin resistance, and reduced insulin secretion [58]. The antidiabetic properties of sorghum extracts were also investigated. One in vitro study demonstrated that sorghum extracts had elevated inhibitory activities against α-glucosidase and inhibitory degradation of starch by pancreatic and salivary α-amylase [59]. Park et al. [60] reported that mice fed a high-fat diet containing 1% sorghum extract had reduced serum glucose and insulin concentrations [60]. It was suggested the hypoglycemic effect of the sorghum extract was due to increased PPAR-γ expression in adipose tissue, which then increased adiponectin expression and decreased TNF-α expression [60]. In another study, diabetic rats that were orally administered sorghum extracts had lower serum glucose concentration, which was reported to be a result of the inhibition of hepatic gluconeogenesis via suppression of PEPCK and p38 expressions while increasing p-AMPK expression [61].

4.3 Cholesterol-Lowering Properties

Heart disease is the leading cause of death in the United States, killing about 610,000 people annually [62]. One of the key risk factors for heart disease is high blood cholesterol, and people with high blood cholesterol double their risks of heart disease [62]. Thus, lowering blood cholesterol is one of the main strategies to decrease the risks of heart of disease. Several reports on the cholesterol-lowering properties of sorghum exist. In one study, guinea pigs fed low tannin sorghum had significantly lower cholesterol levels than those fed whole wheat, rolled oats, or pearl millet [63]. Sorghum grain lipid extracts were found to decrease plasma non-HDL cholesterol, liver cholesteryl esters, and cholesterol absorption in vivo, and it was suggested the cholesterol reduction was contributed by phytosterols [64]. In another report, mice and diabetic rats fed a high-fat diet containing 0.5–1% sorghum extracts had lower levels of serum total cholesterol, LDL-cholesterol, and triglycerides compared to those fed a high-fat diet [60]. Kim et al. [65] demonstrated the oral administration of sorghum extracts in mice suppressed HMGCR, SREBP2, and FAS expressions and caused an increase in p-AMK expression, which reduced hepatic cholesterol biosynthesis. They also demonstrated that sorghum extracts increased CYP7A1 expression, which increased hepatic cholesterol metabolism via bile acid biosynthesis [65]. Phytochemical components that contributed to cholesterol reduction were not identified in those three reports.

4.4 Anti-inflammatory Properties

Chronic inflammatory conditions, such as inflammatory joint disease and skin aging, are increased by hyaluronidase activity [66]. One study reported that tannin sorghum reduced hyaluronidase activity in vitro [66]. In a separate study using mice, a black sorghum bran extract was found to inhibit the secretion of both TNF-α and IL-1β, two pro-inflammatory cytokines implicated in the pathogenesis of numerous inflammatory conditions [67]. In addition, the study also demonstrated that both black and tannin sorghum bran extracts also significantly reduced edema in inflamed ears [67]. Moraes et al. [68] reported that rats fed a hyperlipidic diet containing non-tannin red sorghum flour exhibited lower concentrations of TBARS in the liver and TNF-α in the epidymal adipose tissues, which suggested that sorghum can reduce oxidative stress and inflammation associated with obesity. Sorghum brans rich in condensed tannins and 3-deoxyanthocyanins may also be beneficial in the intervention of inflammatory bowel disease such as ulcerative colitis. In an in vivo study, sorghum brans rich in condensed tannins and 3-deoxyanthocyanins were found to have the potential to protect against alterations observed during dextran sodium sulfate (DSS)-colitis to include microbial diversity and richness, as well as dysbiosis [69]. A recent study reported these same sorghum brans could also have the potential to suppress symptoms reported in DDS-colitis such as weight loss and bloody diarrhea [70].

4.5 Anticancer Properties

Anticancer properties of sorghum and their phenolic components have been examined. Two studies reported that populations consuming sorghum and millet had lower incidences of esophageal cancer compared to those consuming wheat or maize

[71, 72]. Turner et al. [73] demonstrated that black and tannin sorghum brans could have the potential to reduce colon carcinogenesis. In that study, rats fed diets containing black or tannin sorghum bran had fewer aberrant crypts than those that were fed diets containing cellulose or white sorghum bran [73]. It was suggested the reduction of aberrant crypts could be due to the elevated antioxidant activity potential of black and tannin sorghum brans, which resulted in the reduction of oxidative stress involved in colon cancer development [73].

Anticancer properties of sorghum phenolic extracts were investigated but most of the studies were performed in vitro. In one report, a sorghum extract rich in phenolics was demonstrated to inhibit the proliferation and colonization of ovarian cancer cells and to chemosensitize those cells to the chemotherapy drug paclitaxel [74]. In a recent study, Ryu et al. [75] reported a sorghum extract containing phenolic acids and flavonoids decreased the proliferation of PC3M prostate cancer cells, increased apoptosis rate, induced cell cycle arrest, and reduced cell migration and expression of MMP2 and MMP9. In the same study, the phenolic extract was also found to reduce growth and metastasis of prostate cancer in vivo without major side effects such as weight loss [75]. In another report, extracts rich in condensed tannins and 3-deoxyanthocyanins were shown to reduce the proliferation of both colon HT-29 and esophageal (OE33) cancer cell lines [76]. Gomez-Cordoves et al. [77] demonstrated that sorghum condensed tannins increased melanogenic activity without increasing total melanin and reduced the formation of human melanoma colony cells. In a recent study, sorghum bran extracts rich in condensed tannin were also found to inhibit proliferation and migration of HepG2 liver cancer cell lines through the activation of the AMPKα pathway and inhibition of MAPK pathway [78]. Sorghum 3-deoxyanthocyanins were found to have anticancer properties in leukemia (HL-60), hepatoma (HepG2 and Hepa1c1c7), and colon cancer (HT-29) cell lines [79, 80]. These compounds were also found to inhibit the proliferation of MCF-7 breast cancer cell lines [81], which were reported to be mediated by the stimulation of the P^{53} gene and the downregulation of the Bcl-2 gene [82]. Woo et al. [83] found the 3-deoxyanthocyaninidin apigeninidin induced apoptosis in leukemia HL-60 cancer cells by the activation of the pro-apoptotic Bcl-2 proteins Bak and Bax, mitochondrial membrane potential loss, release of mitochondrial cytochrome *c* and AIF into the cytoplasm, and activation of enzymes caspase-9 and caspase-3.

4.6 Health Potential of Sorghum Flavones and Flavanones

Since some sorghum genotypes are rich sources of flavones and flavanones, they may have potential health applications that should be investigated [7]. For instance, flavones were reported to have anticancer, anti-inflammatory, anti-allergic, estrogenic, and analgesic properties [22, 84–89]. These compounds were also demonstrated as effective vascular relaxation agents, and they could be used for the treatment of corneal neovascularization [90, 91].

Lemon-yellow sorghums are a rich source of flavanones and potential health properties of these genotypes should also be explored. Several potential health benefits of flavanones were reported and reviewed in the literature [7, 23]. Potential health benefits for eriodictyol include anti-inflammatory properties [92], prevention of pulmonary valve insufficiency [93], protection of skin cells from UV-induced damage [94], and reduction in the development of age-related macular degeneration [95]. Potential health benefits for naringenin include cholesterol-lowering [96, 97] and anti-ulcer properties [98]. Naringenin was also reported as an expectorant [99] and as a chemopreventive agent against neurodegenerative disease [100].

5 Conclusion

Sorghum contains a wide variety of phytochemicals and their levels and compositions are affected by the genotype. Studies on the potential health benefits of sorghum rich in phytochemicals have significantly increased in the past 15 years showing promising results. However, these studies have been performed either in vitro or in animal models. Human trials investigating the health-promoting properties of these sorghum genotypes are much needed.

Acknowledgment

Disclaimer: Mention of trade names or commercial products in this publication is solely for the purpose of providing specific information and does not imply recommendation or endorsement by the U.S. Department of Agriculture. USDA is an equal opportunity provider and employer.

References

1. FAO (2017) FAOSTAT, Crops production database, Yearly production. http://faostat.fao.org. Accessed Nov 2017
2. Waniska RD, Rooney LW (2000) Structure and chemistry of the sorghum caryopsis. In: Smith CW, Frederiksen RA (eds) Sorghum: origin, history, technology, and production. John Wiley and Sons, Inc, New York
3. Rooney LW, Serna-Saldivar SO (2000) Sorghum. In: Kulp K, Ponte JG Jr (eds) Handbook of cereal science and technology, 2nd edn. Marcel Dekker, Inc., New York
4. Dykes L, Rooney LW (2006) Sorghum and millet phenols and antioxidants. J Cereal Sci 44:236–251
5. Althwab S, Carr TP, Weller CL, Dweikat IM, Schlegel V (2015) Advances in grain sorghum

and its co-products as a human health promoting dietary system. Food Res Int 77:349–359
6. Cardoso LDM, Pinheiro SS, Martino HSD, Pinheiro-Sant'Ana HM (2017) Sorghum (*Sorghum bicolor* L.): Nutrients, bioactive compounds, and potential impact on human health. Crit Rev Food Sci Nutr 57:372–390
7. Dykes L, Rooney LW (2010) Special sorghums for health foods. In: Martirrosyan DM, Abate N (eds) Functional foods for chronic diseases, vol 5. Food Science Publisher, Richardson, TX
8. Dykes L (2008) Flavonoid composition and antioxidant activity of pigmented sorghums of varying genotypes. Dissertation, Texas A&M University
9. Rooney LW, Miller FR (1982) Variation in the structure and kernel characteristics of sorghum. In: Mertin JV (ed) Proceedings of the international symposium on sorghum grain quality. International Crops Research Institute for the Semi-Arid Tropics (ICRISAT), October 28–31, 1982, Patancheru, AP
10. Dykes L, Seitz L, Rooney WL, Rooney LW (2009) Flavonoid composition of red sorghum genotypes. Food Chem 116:313–317
11. Waniska RD, Poe JH, Bandyopadhyay R (1989) Effects of growth conditions on grain molding and phenols in sorghum caryopsis. J Cereal Sci 10:217–225
12. Rooney LW (2005) Ten myths about tannins in sorghums. Int Sorgh Mill Newsl 46:3–5
13. Boren B, Waniska RD (1992) Sorghum seed color as an indicator of tannin content. J Appl Poult Res 1:117–121
14. Earp CF, McDonough CM, Awika J, Rooney LW (2004) Testa development in the caryopsis of *Sorghum bicolor* (L.) Moench. J Cereal Sci 39:303–311
15. Earp CF, McDonough CM, Rooney LW (2004) Microscopy of pericarp development in the caryopsis of *Sorghum bicolor* (L.) Moench. J Cereal Sci 39:21–27
16. Dykes L, Rooney LW, Waniska RD, Rooney WL (2005) Phenolic compounds and antioxidant activity of sorghum grains of varying genotypes. J Agric Food Chem 53:6813–6818
17. Hahn DH, Rooney LW, Earp CF (1984) Tannins and phenols of sorghum. Cereal Foods World 29:776–779
18. Hahn DH, Faubion JM, Rooney LW (1983) Sorghum phenolic acids, their high performance liquid chromatography separation and their relation to fungal resistance. Cereal Chem 60:255–259
19. Chiremba C, Taylor JRN, Rooney LW, Beta T (2012) Phenolic acid content of sorghum and maize cultivars in varying hardness. Food Chem 134:81–88
20. Svensson L, Sekwati-Monang B, Lutz DL, Schieber A, Gänzle M (2010) Phenolic acids and flavonoids in nonfermented and fermented red sorghum (*Sorghum bicolor* (L.) Moench). J Agric Food Chem 58:9214–9220
21. Kang J, Price WE, Ashton J, Tapsell LC, Johnson S (2016) Identification and characterization of phenolic compounds in hydromethanolic extracts of sorghum wholegrains by LC-ESI-MS. Food Chem 211:215–226
22. Yang L, Allred KF, Geera B, Allred CD, Awika JM (2012) Sorghum phenolics demonstrate estrogenic action and induce apoptosis in nonmalignant colonocytes. Nutr Cancer 64:419–427
23. Dykes L, Peterson GC, Rooney WL, Rooney LW (2011) Flavonoid composition of lemon-yellow sorghum genotypes. Food Chem 128:173–179
24. Dykes L, Rooney WL, Rooney LW (2013) Evaluation of phenolics and antioxidant activity of black sorghum hybrids. J Cereal Sci 58:278–283
25. Taleon V, Dykes L, Rooney WL, Rooney LW (2012) Effect of genotype and environment on flavonoid concentration and profile of black sorghum grains. J Cereal Sci 56:470–475
26. Taleon V, Dykes L, Rooney WL, Rooney LW (2014) Environmental effect on flavonoid concentrations and profiles of red and lemon-yellow sorghum grains. J Food Compost Anal 34:178–185
27. Wu X, Prior RL (2005) Identification and characterization of anthocyanins by high-performance liquid chromatography-electrospray ionization-tandem mass spectrometry in common foods in the United States: vegetables, nuts, and grains. J Agric Food Chem 53:3101–3113
28. Lo S-CC, De Verdier K, Nicholson RL (1999) Accumulation of 3-deoxyanthocyanidin phytoalexins and resistance to *Colletotrichum sublineolum* in sorghum. Physiol Mol Plant Pathol 55:263–273
29. Awika JM, Rooney LW, Waniska RD (2004) Properties of 3-deoxyanthocyanins from sorghum. J Agric Food Chem 52:4388–4394
30. Awika JM, Rooney LW, Waniska RD (2004) Anthocyanins from black sorghum and their antioxidant properties. Food Chem 90:293–301

31. Nip WK, Burns EE (1969) Pigment characterization in grain sorghum. I. Red varieties. Cereal Chem 46:490–495
32. Gujer R, Magnolato D, Self R (1986) Glucosylated flavonoids and other phenolic compounds from sorghum. Phytochemistry 25:1431–1436
33. Watterson JJ, Butler LG (1983) Occurrence of an unusual leucoanthocyanidin and absence of proanthocyanidins in sorghum leaves. J Agric Food Chem 31:41–45
34. Bate-Smith EC (1969) Luteoforol (3′,4,4′,5,7-pentahydroxyflavan) in *Sorghum Vulgare* L. Phytochemistry 8:1803–1810
35. Mazza G, Brouillard R (1987) Color stability and structural transformations of cyanidin 3,5-diglucoside and four 3-deoxyanthocyanins in aqueous solutions. J Agric Food Chem 35:422–426
36. Sweeny JG, Iacobucci GA (1983) Effect of substitution on the stability of 3-deoxyanthocyanidins in aqueous solutions. J Agric Food Chem 31:531–533
37. Yang L, Dykes L, Awika JM (2014) Thermal stability of 3-deoxyanthocyanidin pigments. Food Chem 160:246–254
38. Wharton PS, Nicholson RL (2000) Temporal synthesis and radiolabelling of the sorghum 3-deoxyanthocyanidin phytoalexins and the anthocyanin, cyanidin 3-dimalonyl glucoside. New Phytol 145:457–469
39. Dicko MH, Gruppen H, Traore AS, Van Berkel WJH, Voragen AGJ (2005) Evaluation of the effect of germination on phenolic compounds and antioxidant activities in sorghum varieties. J Agric Food Chem 53:2581–2588
40. Audilakshmi S, Stenhouse JW, Reddy TP, Prasad MVR (1999) Grain mould resistance and associated characters of sorghum genotypes. Euphytica 107:91–103
41. Jambunathan R, Kherdekar MS, Bandyopadhyay R (1990) Flavan-4-ols concentration in mold-susceptible and mold-resistant sorghum at different stages of grain development. J Agric Food Chem 38:545–548
42. Jambunathan R, Kherdekar MS, Vaidya MS (1991) Ergosterol concentration in mold-susceptible and mold-resistant sorghum at different stages of grain development and its relationship to flavan-4-ols. J Agric Food Chem 39:1866–1870
43. Menkir A, Ejeta G, Butler L, Melakeberhan A (1996) Physical and chemical kernel properties associated with resistance to grain mold in sorghum. Cereal Chem 73:613–617
44. Melake-Berhan A, Butler LG, Ejeta G, Menkir A (1996) Grain mold resistance and polyphenol accumulation in sorghum. J Agric Food Chem 44:2428–2434
45. Foo LY, Lu Y, Howell AB, Vorsa N (2000) A-type proanthocyanidin trimers from cranberry that inhibit adherence of uropathogenic P-fimbriated *Escherichia coli*. J Nat Prod 63:1225–1228
46. Gu L, Kelm MA, Hammerstone JF, Beecher G, Holden J, Haytowitz D, Prior RL (2003) Screening of foods containing proanthocyanidins and their structural characterization using LC-MS/MS and thiolytic degradation. J Agric Food Chem 51:7513–7521
47. Gu L, Kelm M, Hammerstone JF, Beecher G, Cunningham D, Vannozzi S, Prior RL (2002) Fractionation of polymeric procyanidins from lowbush blueberry and quantification of procyanidins in selected foods with an optimized normal-phase HPLC-MS fluorescent detection method. J Agric Food Chem 50:4852–4860
48. Krueger CG, Vestling MM, Reed JD (2003) Matrix-assisted laser desorption/ionization time-of-flight mass spectrometry of heteropolyflavan-3-ols and glucosylated heteropolyflavans in sorghum (*Sorghum bicolor* (L.) Moench). J Agric Food Chem 51:538–543
49. Brandon MJ, Foo LY, Porter L, Meredith P (1982) Proanthocyanidins of barley and sorghum; composition as a function of maturity of barley ears. Phytochemistry 21:2953–2957
50. Hwang KT, Kim JE, Weller CL (2005) Policosanol contents and compositions in wax-like materials extracted from selected cereals of Korean origin. Cereal Chem 82:242–245
51. Leguizamón C, Weller CL, Schlegel VL, Carr TP (2009) Plant sterol and policosanol characterization of hexane extracts from grain sorghum, corn and their DDGS. J Am Oil Chem Soc 86:707–716
52. Wang L, Weller CL, Schlegel VL, Carr TP, Cuppett SL (2007) Comparison of supercritical CO_2 and hexane extraction of lipids from sorghum distillers grains. Eur J Lipid Sci Technol 109:567–574
53. Bröhan M, Jerkovic V, Collin S (2011) Potentiality of red sorghum for producing stilbenoid-enriched beers with high antioxidant activity. J Agric Food Chem 59:4088–4094
54. Awika JM, Rooney LW, Wu X, Prior RL, Cisneros-Zevallos L (2003) Screening methods to measure antioxidant activity of sorghum (*Sorghum bicolor*) and sorghum products. J Agric Food Chem 51:6657–6662

55. Hagerman AE, Riedl KM, Jones GA, Sovik KN, Ritchard NT, Hartzfeld PW, Riechel TL (1998) High molecular weight plant polyphenolics (tannins) as biological antioxidants. J Agric Food Chem 46:1887–1892
56. Awika JM, McDonough CM, Rooney LW (2005) Decorticating sorghum to concentrate healthy phytochemicals. J Agric Food Chem 53:6230–6234
57. Farrar JL, Hartle DK, Hargrove JL, Greenspan P (2008) A novel nutraceutical property of select sorghum (*Sorghum bicolor*) brans: inhibition of protein glycation. Phytother Res 22:1052–1056
58. Moraes EA, Marineli RDS, Lenquiste SA, Queiroz VAV, Camargo RL, Borck PC, Carneiro EM, Júnior MRM (2017) Whole sorghum flour improves glucose tolerance, insulin resistance and preserved pancreatic islets function in obesity diet-induced rats. J Funct Foods. https://doi.org/10.1016/j.jff.2017.03.047
59. Kim J-S, Hyun TK, Kim M-J (2011) The inhibitory effects of ethanol extracts from sorghum, foxtail millet and proso millet on α-glucosidase and α-amylase activities. Food Chem 124:1647–1651
60. Park JH, Lee SH, Chung I-M, Park Y (2012) Sorghum extract exerts an anti-diabetic effect by improving insulin sensitivity via PPAR-γ in mice fed a high-fat diet. Nutr Res Pract 6:322–327
61. Kim J, Park Y (2012) Anti-diabetic effect of sorghum extract on hepatic gluconeogenesis of streptozotocin-induced diabetic rats. Nutr Metab 9:106. https://doi.org/10.1186/1743-7075-9-106
62. Center for Disease Control and Prevention (2017) Heart disease and stroke fact sheets. https://www.cdc.gov/dhdsp/data_statistics/fact_sheets. Accessed Dec 2017
63. Klopfenstein CF, Varriano-Marston E, Hoseney RC (1981) Cholesterol-lowering effect of sorghum diet in guinea pigs. Nutr Rep Int 24:621–627
64. Carr TP, Weller CL, Schlegel VL, Cuppett SL, Guderian DM Jr, Johnson KR (2005) Grain sorghum lipid extract reduces cholesterol absorption and plasma non-HDL cholesterol concentration in hamsters. J Nutr 135:2236–2240
65. Kim E, Kim S, Park Y (2015) Sorghum extract exerts cholesterol-lowering effects through the regulation of hepatic cholesterol metabolism in hypercholesterolemic mice. Int J Food Sci Nutr 66:308–313
66. Bralley E, Greenspan P, Hargrove JL, Hartle DK (2008) Inhibition of hyaluronidase activity by select sorghum brans. J Med Food 11:307–312
67. Burdette A, Garner PL, Mayer EP, Hargrove JL, Hartle DK, Greenspan P (2010) Anti-inflammatory activity of select sorghum (*Sorghum bicolor*) brans. J Med Food 13:879–887
68. Moraes EA, Natal DIG, Queiroz VAV, Schaffert RE, Cecon PR, de Paula SO, Benjamin LA, Ribeiro SMR, Martino HSD (2012) Sorghum genotype may reduce low-grade inflammatory response and oxidative stress and maintains jejunum morphology of rats fed a hyperlipidic diet. Food Res Int 49:553–559
69. Ritchie L, Sturino JM, Carroll RJ, Rooney LW, Azcarate-Peril MA, Turner ND (2015) Polyphenol-rich sorghum brans alter colon microbiota and impact species diversity and species richness after multiple bouts of dextran sodium sulfate-induced colitis. FEMS Microbiol Ecol. https://doi.org/10.1093/femsec/fiv008
70. Ritchie LE, Taddeo SS, Weeks BR, Carroll RJ, Dykes L, Rooney LW, Turner ND (2017) Impact of novel sorghum bran diets on DSS-induced colitis. Nutrients. https://doi.org/10.3390/nu9040330
71. Van Rensburg SJ (1981) Epidemiological and dietary evidence for a specific nutritional predisposition to esophageal cancer. J Natl Cancer Inst 67:243–251
72. Chen F, Cole P, Mi Z, Xing L-Y (1993) Corn and wheat-flour consumption and mortality from esophageal cancer in Shanxi, China. Int J Cancer 53:902–906
73. Turner ND, Diaz A, Taddeo SS, Vanamala J, McDonough CM, Dykes L, Murphy ME, Carroll RJ, Rooney LW (2006) Bran from black or brown sorghum suppresses colon carcinogenesis. FASEB J 20:A599
74. Dia VP, Pangloli P, Jones L, McClure A, Patel A (2016) Phytochemical concentrations and biological activities of *Sorghum bicolor* alcoholic extracts. Food Funct 7:3410–3420
75. Ryu J-M, Jang GY, Woo KS, Kim TM, Jeong HS, Kim DJ (2017) Effects of sorghum ethyl-acetate extract on PC3M prostate cancer cell tumorigenicity. J Funct Foods 37:449–459
76. Awika JM, Yang L, Browning JD, Faraj A (2009) Comparative antioxidant, antiproliferative and phase II enzyme inducing

potential of sorghum (*Sorghum bicolor*) varieties. Lebensm Wiss Technol 42:1041–1046

77. Gomez-Cordoves C, Bartoleme B, Vieira W, Virador VM (2001) Effects of wine phenolics and sorghum tannins on tyrosinase activity and growth of melanoma cells. J Agric Food Chem 49:1620–1624
78. Zhu Y, Shi Z, Yao Y, Hao Y, Ren G (2017) Antioxidant and anti-cancer activities of proanthocyanidins-rich extracts from three varieties of sorghum (*Sorghum bicolor*) bran. Food Agric Immunol 28:1530–1543
79. Shih C-H, Siu S-O, Ng R, Wong E, Chiu LCM, Chu IK, Lo C (2007) Quantitative analysis of anticancer 3-deoxyanthocyanidins in infected sorghum seedlings. J Agric Food Chem 55:254–259
80. Yang L, Browning JD, Awika JM (2009) Sorghum 3-deoxyanthocyanins possess strong phase II enzyme inducer activity and cancer cell growth inhibition properties. J Agric Food Chem 57:1797–1804
81. Devi PS, Kumar MS, Das SM (2011) Evaluation of antiproliferative activity of red sorghum bran anthocyanin on a human breast cancer cell line (MCF-7). Int J Breast Cancer. https://doi.org/10.4061/2011/891481
82. Suganyadevi P, Saravanakumar KM, Mohandas S (2013) The antiproliferative activity of 3-deoxyanthocyanins extracted from red sorghum (*Sorghum bicolor*) bran through P53-dependent and Bcl-2 gene expression in breast cancer cell line. Life Sci 92:379–382
83. Woo HJ, Oh IT, Lee JY, Jun DY, Seu MC, Woo KS, Nam MH, Kim YH (2012) Apigeninidin induces apoptosis through activation of Bak and Bax and subsequent mediation of mitochondrial damage in human promyelocytic leukemia HL-60 cells. Process Biochem 47:1861–1871
84. Block LC, Santo ARS, De Souza MM, Scheidt C, Yunes RA, Santos MA, Monache FD, Filho VC (1998) Chemical and pharmacological examination of antinociceptive constituents of *Wedelia paludosa*. J Ethnopharmacol 61:85–89
85. Hirano T, Higa S, Arimitsu J, Naka T, Shima Y, Ohshima S, Fujimoto M, Yamadori T, Kawase I, Tanaka T (2004) Flavonoids such as luteolin, fisetin, and apigenin are inhibitors of interleukin-4 and interleukin-13 production by activated human basophils. Int Arch Allergy Immunol 134:135–140
86. Horinaka M, Yoshida T, Shiraishi T, Nakata S, Wakada M, Nakanishi R, Nishino H, Matsui H, Sakai T (2005) Luteolin induces apoptosis via death receptor 5 upregulation in human malignant tumor cells. Oncogene 24:7180–7189
87. Matsui J, Kiyokawa N, Takenouchi H, Taguchi T, Suzuki K, Shiozawa Y, Saito M, Tang W-R, Katagiri YU, Okita H, Fujimoto J (2005) Dietary bioflavonoids induce apoptosis in human leukemia cells. Leuk Res 29:573–581
88. Cherng J-M, Shieh D-E, Chiang W, Chang M-Y, Chiang L-C (2007) Chemopreventive effects of minor dietary constituents in common foods on human cancer cells. Biosci Biotechnol Biochem 71:1500–1504
89. Ziyan L, Yongmei Z, Nan Z, Ning T, Baolin L (2007) Evaluation of the anti-inflammatory activity of luteolin in experimental animal models. Planta Med 73:221–226
90. Joussen AM, Rohrschneider K, Reichling J, Kirchhof B, Kruse FE (2000) Treatment of corneal neovascularization with dietary isoflavonoids and flavonoids. Exp Eye Res 71:483–487
91. Xu YC, Leung SWS, Yeung DKY, Hu LH, Chen GH, Che CM, Man RYK (2007) Structure-activity relationships of flavonoids for vascular relaxation in porcine coronary artery. Phytochemistry 68:1179–1188
92. Zhang XF, Hung TM, Phuong PT, Ngoc TM, Min B-S, Song K-S, Seong YH, Bae K (2006) Anti-inflammatory activity of flavonoids from *Populus davidiana*. Arch Pharm Res 29:1102–1108
93. Aviado DM, Bacalzo LV Jr, Belej MA (1974) Prevention of acute pulmonary insufficiency by eriodictyol. J Pharmacol Exp Ther 189:157–166
94. Lee E-R, Kim J-H, Kang Y-J, Cho S-G (2007) The anti-apoptotic and anti-oxidant effect of eriodictyol on UV-induced apoptosis in keratinocytes. Biol Pharm Bull 30:32–37
95. Hanneken A, Lin F-F, Johnson J, Maher P (2006) Flavonoids protect human retinal pigment epithelial cells from oxidative-stress-induced death. Invest Ophthalmol Vis Sci 47:3164–3177
96. Borradaile NM, Carroll KK, Kurowska EM (1999) Regulation of HepG2 cell apolipoprotein B metabolism by the citrus flavanones hesperitin and naringenin. Lipids 34:591–598

97. Wilcox LJ, Borradaile NM, de Dreu LE, Huff MW (2001) Secretion of hepatocyte apoB is inhibited by the flavonoids, naringenin and hesperitin, via reduced activity and expression of ACAT2 and MTP. J Lipid Res 42:725–734
98. Martin MJ, Motilva V, Alarcón de la Lastra C (1993) Quercetin and naringenin: Effects on ulcer formation and gastric secretion in rats. Phytother Res 7:150–153
99. Lin B-Q, Li P-B, Wang Y-G, Peng W, Wu Z, Su W-W, Ji H (2008) The expectorant activity of naringenin. Pulm Pharmacol Ther 21:259–263
100. Heo HJ, Kim D-O, Shin SC, Kim MJ, Kim BG, Shin D-H (2004) Effect of antioxidant flavanone, naringenin, from *Citrus junos* on neuroprotection. J Agric Food Chem 52:1520–1525

Chapter 10

Identification and Quantification of Carotenoids and Tocochromanols in Sorghum Grain by High-Performance Liquid Chromatography

Darwin Ortiz and Mario G. Ferruzzi

Abstract

Carotenoids and tocochromanols are lipid-soluble secondary plant metabolites that are essential for the normal functioning of plants and, in some cases function as a source of vitamin A and E in humans. Enhancement of the provitamin A carotenoid and tocochromanols levels in sorghum and other cereal grains through traditional breeding and transgenic methods has increased in recent years with interest in biofortification of grains to combat micronutrient deficiencies in developing countries. With this increase in research, reliable methodology for the extraction, identification, and quantification of individual carotenoids and tocochromanols species from sorghum and other cereal grains is essential. Here, we describe a basic method for extraction of carotenoid and tocochromanols adapted to sorghum grain and chromatographic condition for separation, identification, and quantification of individual carotenoid using High-Performance Liquid Chromatography (HPLC) system.

Key words Carotenoids, Vitamin A, Vitamin E, Lutein, Zeaxanthin, β-carotene, Tocopherols, Tocotrienols

1 Introduction

Carotenoids are natural pigments, responsible for the yellow, orange, and red coloration of many commonly consumed fruits, vegetables, and grains. The rich color of these molecules is a result of their tetraterpene structure with an extensive system of conjugated double bonds [1]. Carotenoids are generally classified into two major groups: carotenes, which are strictly hydrocarbons, that are either cyclized such as β-carotene, or linear, such as lycopene; xanthophylls, or oxycarotenoids, which contain oxygen that include lutein, zeaxanthin, and b-cryptoxanthin [2]. Dietary carotenoids are those carotenoids in the food chain, and their number has been estimated to be approximately 40–50; however, only 24 of them have been reported in human plasma and tissues [3, 4]. Consumption of these phytochemicals has been associated with specific

Zuo-Yu Zhao and Jeff Dahlberg (eds.), *Sorghum: Methods and Protocols*, Methods in Molecular Biology, vol. 1931, https://doi.org/10.1007/978-1-4939-9039-9_10,

health benefits such as reduced risk of cardiovascular diseases, certain type of cancer, Alzheimer's disease, and eye-related maculopathy [5–8]. Lutein, zeaxanthin, and the lutein metabolite meso-zeaxanthin accumulate in the macula of the human eye and are vital for ocular health [9]. Many health benefits are attributed to carotenoids and their role in modulating inflammatory response [10], and oxidative stress [11]; ability to regulate gene transcription, gap junction communication, phase II enzyme-inducing activity, and ability to enhance immune function [12]. While promising, provitamin A activity remains the best-known role of carotenoids in human health [13]. Nutritionally, α-, β-, γ-carotene as well as α- and β-cryptoxanthin are considered provitamin A carotenoids (pVACs). These carotenoids have at least one unsubstituted β-ionone ring, which when cleaved in the intestine or liver leads to the production of retinol (vitamin A) [14]. With dietary retinol forms being derived primarily from animal or supplemental sources, pVACs remain the main plant-derived dietary form and therefore critical to delivery of vitamin A to many at-risk populations globally. As such, efforts to develop and implement new sorghum genotypes with enhanced pVAC levels are under way to increase the availability of provitamin A and other beneficial carotenoids in the human diet.

Similar to carotenoids, tocochromanols are a group of amphipathic, lipid-soluble organic molecules composed of a polar moiety derived from tyrosine and a hydrophobic polyprenyl side chain originating from the isoprenoid pathway [15]. Tocochromanols are grouped into two classes, tocopherols characterized by a phytyl-derived side chain whereas tocotrienols are those with a geranylgeranylderived side chain. The four different forms of tocopherols and tocotrienols (α-, β-, γ and δ-) differ by the degree of methylation of the polar moiety [15]. As such, four tocopherols and four tocotrienols constitute natural vitamin E forms found in food. However, the humans preferentially use α-tocopherol as vitamin E and only this form can reverse deficiency symptoms and, by extension, meet human vitamin E requirements [16]. α-tocopherol's major vitamin functions are to maintain the integrity of long-chain polyunsaturated fatty acids in the membrane cells and thus maintain their bioactivity acting as a potent peroxyl radical scavenger [17]. Vitamin E is an essential nutrient in the human diet that can only be produced by photosynthetic organisms. Diets that are low in α-tocopherol may be at risk for increased oxidative damage because they have insufficient protection from oxidative stress [16]. While α-tocopherol is the most relevant vitamin E compound, it is not the main form consumed since the composition of most major crops is dominated by γ-tocopherol [18]. Therefore, several strategies have been undertaken to improve vitamin E content in crops. In addition, enhancing the levels of vitamin E might enhance the accumulation and stability of provitamin A carotenoids in biofortified sorghum [19]. Hence, a method for simultaneous analysis of vitamin E and carotenoids isomers is essential in plant breeding programs; therefore, here it is described an optimized method for the identification and quantification of carotenoids and tocochromanols in sorghum grain by HPLC.

2 Materials

Prepare all solutions using ultrapure water (prepared by purifying deionized water, to attain a sensitivity of 18 MΩ-cm at 25 °C). All sample preparations and extractions must be performed under low light conditions or yellow/red light to minimize potential for photo-isomerization reaction.

1. Extraction solvents: Acetone and Methyl tert-butyl ether. It is recommended their use in American Chemical Society (ACS)-grade and fresh or the addition of butylated hydroxytoluene (BHT) (0.1%) as a stabilizer agent.
2. Mobile phase solvents: Methanol and Ethyl Acetate. It is recommended their use in HPLC-grade.
3. Ammonium acetate: 1.0 M Ammonium acetate must be prepared using double distilled water and adjusted to pH 4.6 with glacial acetic acid.
4. Saturated saline solution. Solubility of NaCl in water is 36 g per 100 mL at 20 °C. Prepare the saturate saline solution by slowly adding sodium chloride (ACS-grade) to a double distilled water and stir the mixture thoroughly.

3 Methods

3.1 Sample Preparation

1. Collection of sorghum grain from field plots should be conducted based on appropriate sampling and statistical plan. Sufficient samples should be collected to ensure reflection of natural variation and replication of analysis.
2. In the laboratory, take collected grain head and dehull them. Pile the grains (approximately 500 g–1 kg) evenly on a clean surface, flatten the pile and spread into a circle.
3. Make a cross, dividing the circle into four roughly equal parts. Discard two diametrically opposite quarters and remix the remaining two quarters.
4. Repeat the quartering procedure until the amount is reduced to approximately 125–250 g.
5. Grind the grains to <0.5 mm and keep the sorghum meal in a tightly closed container stored frozen at −20 °C.

3.2 Extraction

1. Keep ground sorghum meal on ice while weighing and covered (*see* **Note 1**).
2. Weigh ~600 mg of ground sorghum flour in a 15 mL Falcon tube, polypropylene. Keep weighed samples on regular ice in covered styrofoam cooler. Keep covered to reduce light.
3. Spike samples at this point with 80 μL of internal standards β-apo-8-carotenal and echinenone (resolubilized in 1 mL Petroleum ether). Run 2 spikes and one control samples for each extraction set. Also, spikes should be in a range of 1X–3X in concentration to the highest carotenoid present. Inject 10 μL.
4. Add 1.0 mL of saturated saline solution. Vortex to mix. Place on ice in covered cooler for 10 min.
5. Add 5 mL of chilled acetone (with 0.1% butylated hydroxytoluene). Vortex to mix. Agitate for 5 min on rotary mixer. Wrap sample rack in foil to reduce light. Incubate for 5 min on ice in covered ice cooler (*see* **Note 2**).
6. Centrifuge samples at 3000 RPM (1734 × *g*) for 3 min cooled to 4 °C.
7. Remove samples from centrifuge and transfer acetone into a second 15 mL test tube.
8. Evaporate acetone fraction under nitrogen gas.
9. Add a 5 mL aliquot of chilled acetone to grain sample. Vortex to mix pellet. Agitate for 5 min on rotary mixer (Vortex-Genie 2, 3000 RPM). Incubate for 5 min on ice. Centrifuge at 3000 RPM (1734 × *g*) for 3 min at 4 °C. Add this second acetone fraction to the initial acetone fraction under nitrogen gas.
10. Add 2 mL of methyl tert-butyl ether (with 0.1% butylated hydroxytoluene) to ground sample. Vortex to mix pellet. Agitate for 5 min on rotary mixer. Incubate for 5 min on ice in a covered cooler. Vortex for 5 min on a rotary mixer (Vortex-Genie 2, 3000 RPM).
11. Centrifuge samples at 3000 RPM (1734 × *g*) for 3 min at 4 °C.
12. Mix the methyl tert-butyl ether (lipophilic) fraction with the acetone (aqueous) fraction that was previously evaporated under nitrogen gas (*see* **Note 3**).
13. Add a 1 mL of methyl tert-butyl ether and gently add 1 mL saturated saline solution. Gently mix the mixture and wait until a very well-defined phase separation occurs in the mixture.
14. Collect the upper phase with a Pasteur glass pipette avoiding the lower phase and transfer it to a new 15 mL test tube. Repeat this **step 3** more times or until no visible color appears

in the upper phase of the mixture. Place the new 15 mL test tube with all the lipophilic fractions under nitrogen gas to completely evaporate the extracts.

15. Redissolve the dried extract in 2 mL of ethyl acetate:methanol (1:1) for immediate analysis. Vortex briefly to mix sample and syringe filter through 0.45 mm filters prior to injection.

3.3 Chromatographic Separation

Carotenoids and tocochromanols analysis is recommended to be performed in a HPLC couple with a multichannel UV-Visible detector or Photodiode Array Detector (PDA or DAD detector). A PDA is preferred as it allows for the acquisition of the UV-Visible spectra in-line, assisting in the identification of main carotenoids and tocopherols and providing a means of verifying peak purity (absence of interfering compounds).

Column type:	YMC C30 3 μm 2.0 mm × 150 mm column, with a YMC carotenoid guard column (2.0 × 23 mm)
Column temperature:	40 °C
Mobile phase:	Binary mobile phase system. Phase A: Methanol:1 M ammonium acetate (98:2 v/v). Phase B: Ethyl acetate
Flow:	0.37 mL/min
Detection:	UV-Visible or PDA, 295 nm, 348 nm, and 450 nm.
Injection:	10 μL
Internal standard:	*trans*-β-apo-8-carotenal (CAS Number: 1107-26-2), echinenone (CAS Number: 80348-65-8), and DL α-tocopherol acetate (CAS Number: 7695-91-2)
Elution gradient:	0 min 80:15 v/v, phase A:phase B; 6 min 20:80 v/v, phase A:phase B; 8 min 0:100 v/v, phase A:phase B; 12 min 0:100 v/v, phase A:phase B, 14 min 80:15 v/v, phase A:phase B
Post-injection time:	3 min to equilibrate between runs

3.4 Identification

While more advanced mass spectroscopy methods have been recently applied to the characterization of carotenoids and tocochromanols in foods [20, 21], tentative identification is most commonly carried out by co-chromatography with authentic standards and comparison of ultraviolet and visible absorption spectrum. The commercial availability of many all-trans carotenoid standards has made this possible. Maximum absorption peak (λmax) and the shape (spectral fine structure) of the spectrum can be used as a tool for carotenoids identification (Fig. 1) (*see* **Note 4**). Characteristics λmax values of major carotenoid species in sorghum are presented in Table 1. A typical carotenoids separation from transgenic sorghum meal is shown in Fig. 2.

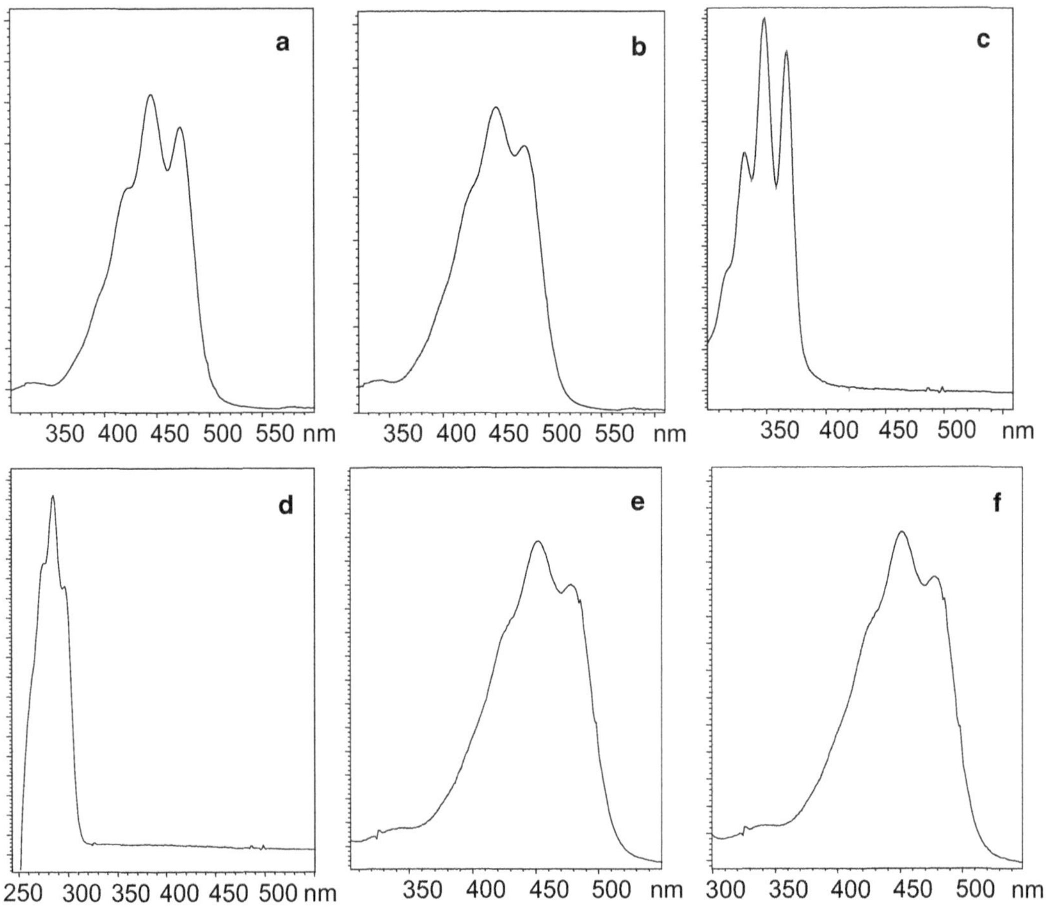

Fig. 1 UV-Visible absorption spectra collected in-line during LC analysis of carotenoid species in sorghum meal. Peak identification: lutein (**a**), zeaxanthin (**b**), phytofluene (**c**), phytoene (**d**), all-trans-β-carotene (**e**), β-cryptoxanthin (**f**)

3.5 Quantification

Carotenoids and tocochromanols in solution obey the Beer-Lamber law; therefore, their absorbance is directly proportional to the concentration. Thus, carotenoids and tocochromanols are typically quantified spectrometrically for generation of stock solutions suitable for calibration curves and compatible with typically UV-Visible detection. In this regard, HPLC quantification is most commonly carried out by generation of an external calibration curve with the respective standard. Concentrations of standard solutions are determined by visible absorption spectrometry. Table 2 shows the absorption coefficients of the major carotenoids and tocochromanols present in sorghum meal. Following preparation of standards, a calibration curve is constructed by injecting a range of standard concentrations (see below) on the HPLC from which the peak areas are plotted against the concentration. Using linear regression, a calibration curve can be established from which carotenoid concentrations can be calculated from peak areas obtained from sample analysis.

Table 1
Ultraviolet and visible absorption data for major carotenoid species sorghum meal

Carotenoid	Solvent	λmax, nm[a]			% III/II[b]
α-Carotene	Acetone	424	448	476	55
	Chloroform	433	457	484	
	Ethanol	423	444	473	
	Hexane, petroleum ether	422	445	473	55
β-Carotene	Acetone	(429)	452	478	15
	Chloroform	(435)	461	485	
	Ethanol	(425)	450	478	25
	Hexane, petroleum ether	(425)	450	477	25
α-Cryptoxanthin	Chloroform	435	459	487	
	Ethanol	423	446	473	60
	Hexane	421	445	475	60
β-Cryptoxanthin	Chloroform	(435)	459	485	
	Ethanol	(428)	450	478	27
	Petroleum ether	(425)	449	476	25
Lutein	Chloroform	435	458	485	
	Ethanol	422	445	474	60
	Petroleum ether	421	445	474	60
Phytoene	Hexane, petroleum ether	(276)	286	(297)	
Phytofluene	Hexane, petroleum ether	331	348	367	
Zeaxanthin	Acetone	(430)	452	479	
	Chloroform	(433)	462	493	
	Ethanol	(428)	450	478	26
	Petroleum ether	(424)	449	476	25

Adapted from: Rodriguez-Amaya et al. [22]
[a]Parenthesis indicate a shoulder
[b]Ratio of the height of the longest-wavelength absorption peak, designated III, and that of the middle absorption peak, designated II, taking the minimum between the two peaks as baseline, multiplied by 100

3.5.1 External Calibration Procedure

1. Turn on UV/vis spectrophotometer lamps.
2. Dissolve a small quantity (~0.5–1 mg) of authentic standard into a culture tube.
3. Add 10–15 mL of the appropriate solvent according (Table 2) to and sonicate for 5 min.
4. Filter with 0.45 μm syringe filter and label tube A.

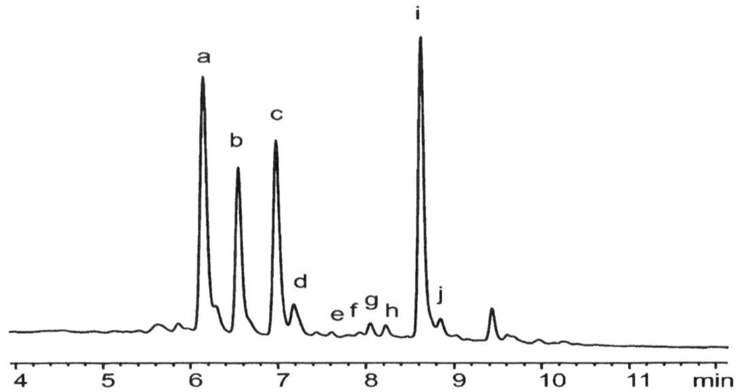

Fig. 2 Chromatogram of major carotenoid species in transgenic sorghum meal (Hemi203-A) analyzed by HPLC with diode-array detection at 450 nm. Peak identification: lutein (**a**), zeaxanthin (**b**), β-apo-8′-carotenal internal standard (**c**), α-cryptoxanthin (**d**), β-cryptoxanthin (**e**), 15-cis-β-carotene (**f**), 13-cis-β-carotene (**g**), α-carotene (**h**), all-trans-β-carotene (**i**), 9-cis-β-carotene (**j**). Reprinted with permission from Lipkie et al. [23] (Copyright 2017, American Chemical Society)

Table 2
UV absorptivities of carotenoids and tocochromanols

Carotenoid	Wavelength (nm)	$A^{1\%}_{1\ cm}$	Solvent
β-carotene	450	2592	Petroleum Ether
Lycopene	470	3450	Petroleum Ether
α-carotene	445	2710	Hexanes
β-cryptoxanthin	450	2460	Hexanes
α-cryptoxanthin	445	2636	Hexanes
Lutein	445	2550	Ethanol
Zeaxanthin	449	2348	Petroleum Ether
Phytoene	286	1250	Petroleum Ether
Phytofluene	348	1350	Petroleum Ether
Tocochromanols	Wavelength (nm)	$A^{1\%}_{1cm}$	Solvent
α-Tocopherol	292	75.8	Ethanol
β-Tocopherol	296	89.4	Ethanol
γ-Tocopherol	298	91.4	Ethanol
δ-Tocopherol	298	87.3	Ethanol
α-Tocotrienol	290	91	Ethanol
β-Tocotrienol	295	87.5	Ethanol
γ-Tocotrienol	298	103	Ethanol
δ-Tocotrienol	298	83	Ethanol

Source: Britton G [24] and Podda et al. [25]

Table 3
Schematic dilution procedure for curve calibration preparation

Sample	Dilution	Sample added	Solvent
A	1		
B	1/2	5 mL A	5 mL
C	1/10	2 mL B	8 mL
D	1/50	2 mL C	8 mL
E	1/100	1 mL C volumetrically	In 10 mL volumetric flasks
F	1/500	1 mL D volumetrically	In 10 mL volumetric flasks
G	1/1000	1 mL E volumetrically	In 10 mL volumetric flasks

5. Read absorbance, adjust to ~1 with appropriate solvent.
6. Dilute following the scheme (Table 3) for A–D. Vortex and cap.
7. Read absorbance of three different 1 mL aliquots each with new cuvette.
8. Prepare E–G (Table 3) volumetrically using volumetric flask and volumetric pipets.
9. If solvent is hexane transfer 2 mL volumetrically of each sample into a new culture tube.
   (a) Dry with nitrogen.
   (b) Resolubilized volumetrically 1 mL EtOAc and 1 mL MeOH.
10. If methanol, transfer directly to HPC vial.
11. Inject 10 μL.

4 Notes

1. Extraction should be carried out under a yellow light to avoid photoisomerization reactions. Solvents used during the extraction procedure must be chilled and stabilized with butylated hydroxytoluene (BHT) (0.1%).
2. Since water was added at the beginning of the extraction process, and water is soluble in the acetone fraction, therefore the resulting fraction will be a mixture of acetone and water.
3. Recommendation. Add the methyl tert-butyl ether fraction to the acetone fraction placed previously under nitrogen after you notice that carotenoids begin to float on the surface of the solution or foaming formation on the surface of the solution. This phenomenon is an indication that most of the acetone in

the mixture has been evaporated and the composition of the remaining solution is mainly water.

4. Identification of carotenoids based solely on the retention times/co-chromatographic with standards or the absorption spectra may lead to erroneous conclusions [22]. However, carotenoids with well-known structure can be conclusively identified by the combined and judicious use of chromatographic behavior, UV-visible absorption spectra, and for the xanthophyll's, specific group chemical reactions [26].

References

1. Boon CS, McClements DJ, Weiss J, Decker EA (2010) Factors influencing the chemical stability of carotenoids in foods. Crit Rev Food Sci Nutr 50:515–532. https://doi.org/10.1080/10408390802565888
2. IUPAC (1974) Nomenclature of carotenoids. Pure Appl Chem 41:405–431
3. Khachik F, Beecher GR, Goli MB, Lusby WR, Daitch CE (1992) Separation and quantification of carotenoids in human plasma. Methods Enzymol 213:205–219
4. Ross AC (2012) In: Ross AC, Caballero B, Cousins RJ, Tucker KL, Ziegler TR (eds) "Vitamin A," in modern nutrition in health and disease. Wolters Kluwer Health/Lippincott Williams & Wilkins, Philadelphia, pp 260–277
5. Kesse-Guyot E, Andreeva VA, Ducros V, Jeandel C, Julia C, Hercberg S, Galian P (2014) Carotenoid-rich dietary patterns during midlife and subsequent cognitive function. BJN 111:915–923. https://doi.org/10.1017/S0007114513003188
6. Liu S, Lee IM, Ajani U, Cole SR, Buring JE, Manson JE (2001) Intake of vegetables rich in carotenoids and risk of coronary heart disease in men: the Physicians' health study. Int J Epidemiol 30:130–135. https://doi.org/10.1093/ije/30.1.130
7. Takata Y, Xiang YB, Yang G, Li H, Gao J, Cai H, Gao YT, Zheng W, Shu XO (2013) Intakes of fruits, vegetables, and related vitamins and lung cancer risk: results from the Shanghai Men's health study (2002-2009). Nutr Cancer 65:51–61. https://doi.org/10.1080/01635581.2013.741757
8. Vaicaitiene R, Luksiene DK, Paunksnis A, Cerniauskiene LR, Domarkiene S, Cimbalas A (2003) Age-related maculopathy and consumption of fresh vegetables and fruits in urban elderly. Medicina (Kaunas) 39:1231–1236
9. Bernstein PS, Li B, Vachali PP, Gorusupudi A, Shyam R, Henriksen BS, Nolan JM (2016) Lutein, zeaxanthin, and meso-zeaxanthin: the basic and clinical science underlying carotenoid-based nutritional interventions against ocular disease. Prog Retin Eye Res 50:34–66. https://doi.org/10.1016/j.preteyeres.2015.10.003
10. Rubin LP, Ross AC, Stephensen CB, Bohn T, Tanumihardjo SA (2017) Metabolic effects of inflammation on vitamin A and carotenoids in humans and animal models. Adv Nutr 8:197–212. https://doi.org/10.3945/an.116.014167
11. Fiedor J, Burda K (2014) Potential role of carotenoids as antioxidants in human health and disease. Nutrients 6:466–488. https://doi.org/10.3390/nu6020466
12. Cooperstone JL, Schwartz SJ (2016) Recent insights into health benefits of carotenoids. In: Carle R, Schweiggert R (eds) Handbook on natural pigments in food and beverages. Elsevier Ltd, Amsterdam, pp 473–497. https://doi.org/10.1016/B978-0-08-100371-8.00020-8
13. Tanumihardjo SA, Nestel P (2005) Beta-carotene-rich orange-fleshed sweet potato improves the vitamin A status of primary school children assessed with the modified-relative-dose-response test. Am J Clin Nutr 81:1080–1087
14. Harrison EH (2012) Mechanisms involved in the intestinal absorption of dietary vitamin A and provitamin A carotenoids. Biochim Biophys Acta 1821:70–77. https://doi.org/10.1016/j.bbalip.2011.06.002
15. Mène-Saffrané L, DellaPenna D (2010) Biosynthesis, regulation and functions of tocochromanols in plants. Plant Physiol Biochem 48(5):301–309. https://doi.org/10.1016/j.plaphy.2009.11.004
16. Traber MG (2013) Chapter 4. Vitamin E. In: Zempleni J, Suttie JW, Gregory JF III, Stover

PJ (eds) Handbook of vitamins, 5th edn. CRC Press, Boca Raton, FL. https://doi.org/10.1201/b15413

17. Traber MG, Atkinson J (2007) Vitamin E, antioxidant and nothing more. Free Radic Biol Med 43(1):4–15. https://doi.org/10.1016/j.freeradbiomed.2007.03.024
18. Mène-Saffrané L, Pellaud S (2017) Current strategies for vitamin E biofortification of crops. Curr Opin Biotechnol 44:189–197
19. Che P, Zhao ZY, Glassman K, Dolde D, Hu TX, Jones TJ, Gruis DF, Obukosia S, Wambugu F, Albertsen MC (2016) Elevated vitamin E content improves all-trans β-carotene accumulation and stability in biofortified sorghum. PNAS 113(39):11040–11045
20. Van-Breemen RB, Dong L, Pajkovic ND (2012) Atmospheric pressure chemical ionization tandem mass spectrometry of carotenoids. Int J Mass Spectrom 312:163–172. https://doi.org/10.1016/j.ijms.2011.07.030
21. Albahrani AA, Rotarou V, Roche PJ, Greaves RF (2016) A simultaneous quantitative method for vitamins A, D and E in human serum using liquid chromatography-tandem mass spectrometry. J Steroid Biochem Mol Biol 159:41–53. https://doi.org/10.1016/j.jsbmb.2016.02.019
22. Rodriguez-Amaya DB (2001) A guide to carotenoid analysis in foods. ILSI human nutrition institute. One Thomas circle. NW, Washington DC, pp 20005–25802
23. Lipkie TE, de Moura FF, Zhao ZY, Albertsen MC, Che P, Glassman K, Ferruzzi MG (2013) Bioaccessibility of carotenoids from transgenic provitamin A biofortified sorghum. J Agric Food Chem 61(24):5764–5771. https://doi.org/10.1021/jf305361s
24. Britton G (1995) UV/visible spectroscopy. In: Britton G, Liaaen-Jensen S, Pfander H (eds) Carotenoids: spec- troscopy, vol 1B. Birkhäuser Verlag, Basel, pp 13–63
25. Podda M, Weber C, Traber MG, Packer L (1996) Simultaneous determination of tissue tocopherols, tocotrienols, ubiquinols, and ubiquinones. J Lipid Res 37:893–901
26. Azevedo-Meleiro CH, Rodriguez-Amaya DB (2004) Confirmation of the identity of the carotenoids of tropical fruits by HPLC-DAD and HPLC-MS. J Food Compos Anal 17:385–396

Part III

Biotechnologies

Chapter 11

Gene Editing in Sorghum Through *Agrobacterium*

Jeffry D. Sander

Abstract

The application of CRISPR/Cas to introduce targeted genomic edits is powering research and discovery across the genetic frontier. Applying CRISPR/Cas in sorghum can facilitate the study of gene function and unlock our understanding of this robust crop that serves as a staple for some of the most food insecure regions on the planet. When paired with recent advances in sorghum tissue culture and *Agrobacteria* technology, CRISPR/Cas can be used to introduce desirable changes and natural genetic variations directly into agriculturally relevant sorghum lines facilitating product development. This chapter describes CRISPR/Cas gene editing and provides high-level strategies and expectations for applying this technology using *Agrobacterium* in sorghum.

Key words Cas9, Sorghum, CRISPR, CRISPR-Cas, CRISPR/Cas, Gene editing, *Agrobacterium*

1 Introduction

The development of programmable CRISPR/Cas (CRISPR stands for clustered regularly interspersed short palindromic repeats and Cas stands for CRISPR associated)-enabled gene editing is breaking ground across fields in the life sciences. Its capacity to precisely introduce targeted changes into the genome of virtually any organism has transcended academic discovery and moved into the therapeutic, commercial, and agricultural arenas [1, 2]. CRISPR/Cas is particularly suited to agriculture where traditional breeding has a long history of capturing variation identical to outcomes that can be achieved with gene editing.

CRISPR/Cas gene editing systems are derived from a family of bacterial immune systems that evolved to protect bacteria from invading viruses. These CRISPR/Cas systems employ nuclease enzymes that complex with an RNA to recognize and cleave specific DNA sequences present within the genomes of invading viruses.

The most widely used CRISPR/Cas system for gene editing comprises a complex of the Cas9 nuclease from *Streptococcus pyogenes* (spyCas9) and a single synthetic guide RNA (gRNA)

Zuo-Yu Zhao and Jeff Dahlberg (eds.), *Sorghum: Methods and Protocols*, Methods in Molecular Biology, vol. 1931,
https://doi.org/10.1007/978-1-4939-9039-9_11,

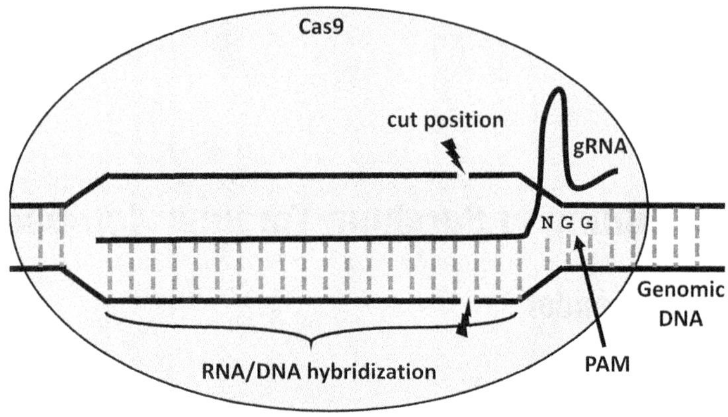

Fig. 1 Schematic of a spyCas9 ribo-protein cleaving double-stranded DNA

comprising approximately 100 nucleotides (Fig. 1). The spyCas9 protein recognizes the DNA sequence 5′-NGG-3′ referred to as the Protospacer Adjacent Motif (PAM). The 5′ end of the complexed gRNA referred to as the spacer hybridizes with ~20 nucleotides of genomic DNA immediately 5′ of the PAM sequence via standard Watson-Crick base pairing. Upon target recognition, the Cas9 enzyme undergoes an allosteric shift and cleaves both the stands of the genomic DNA three base pairs upstream of the NGG PAM sequence. Altering the 5′ sequence of the gRNA enables this system to be reprogrammed to cut virtually genomic DNA sequence addressed by the PAM sequence "NGG."

Double-stranded breaks in genomic DNA such as those created by Cas9 are natural and common place occurrences in the cells of all living organisms. These naturally occurring breaks originate from many sources at random locations throughout the genome and are subsequently repaired by the cell. The dominant repair processes in sorghum and most other eukaryotes are imperfect and can result in small insertions and deletions (indels) at the sites of these breaks. Simultaneous double-stranded breaks in contiguous sequence can result in the loss or inversion of the intervening sequence. A common outcome of this variation is to inactivate genes. This is highly valuable for studying gene function and can have direct agricultural applications. Analogous to these naturally occurring breaks, Cas9 can cleave the genome to create targeted breaks in desirable locations thus enabling efficient creation of frameshift and dropout edits through imperfect repair in these targets—a great leap forward compared to traditional screening methods. To date gene editing has been demonstrated for numerous applications including improved disease resistance [3, 4], increased shelf life, altered starch content [5, 6], improved health profile of oil in seeds [7], reduced acrylamide in processed chips, [5], and reduction of allergenic proteins including gluten [8].

Double-stranded breaks can also be repaired through less active pathways that utilize a DNA template harboring sequence identical to that flanking the break to precisely copy across the broken region. These template-based edits can be used to replicate natural variation such as single nucleotide polymorphisms (SNPs) by introducing them within a DNA template provided concurrently with DNA break generation. Alternatively, this process can be used to improve the delivery of transgenic traits by targeting them to genomic locations amenable to stacking via introgression. Because these template repairs compete with the more active template-independent pathways, it may require screening hundreds or even thousands of regenerated plants to identify one perfectly repaired using the template. Template-based repair has not yet been demonstrated in sorghum; however, it has been demonstrated in many other diverse organisms across the plant and animal kingdoms including maize, a close relative of sorghum [9].

Early examples of gene editing in plants [10] demonstrated potential for gene editing by expressing the Cas9 protein and associated gRNA in non-regenerable cell cultures, or model organisms selected for their ability to be readily transformed. Recent developments in monocot transformation have enabled efficient regeneration in product lines of agriculturally relevant monocots including maize and sorghum [11–15]. This in combination with genome editing provides a powerful new tool to conduct plant research and facilitate the development of product germplasm.

To conduct CRISPR/Cas-mediated gene editing, the following are required: (1) annotated genomic sequence of the target gene and full DNA sequence of the sorghum genome—ideally from the sorghum inbred being edited, (2) bioinformatics software tools, (3) capability of vector construction and sorghum transformation, (4) molecular biology tools to sample leaf tissue, purify DNA, amplify DNA fragments and performing molecular analyses, (5) controlled environments for growing and propagating sorghum.

2 Materials

2.1 DNA Sequence Information

1. Annotated DNA sequence of the target region from a target sorghum inbred, such as TX430.
2. High-quality sequence of the full sorghum genome. It is recommended that sequence is for the exact sorghum line and specific seed source being used in the gene editing experiment.

2.2 Bioinformatics Tools

1. Bioinformatics software capable of comprehensively screening the sorghum genome for DNA sequence motifs sharing high degrees of similarity with the Cas9 target [16] (*see* **Note 1**).
2. Bioinformatics software such as BWA [17] that aligns amplicon reads from CRISPR/Cas treated plants with the wild type sequence to characterize edited alleles.

2.3 Sorghum Transformation Protocol and Facilities

1. T-DNA vectors expressing spyCas9 and gRNA(s).
2. Sorghum transformation protocols and associated tissue culture facilities.
3. Greenhouse and associated materials for growing and propagating sorghum.

2.4 Molecular Biology Tools and Facilities

1. PCR assays designed to the various components of the T-DNA and accompanying vectors.
2. PCR for amplification of genomic edits.
3. DNA sequencing capability.
4. Capillary electrophoresis equipment or comparable equipment for running agarose gels.
5. Southern-by-sequencing [18] (or comparable method) and oligo capture library designed to capture sequence fragments from T-DNA expressing vectors.

3 Methods

The general methods of gene editing mediated by CRISPR/Cas9 in sorghum include sequence analysis, vector construction, *Agrobacterium*-mediated sorghum transformation, molecular analysis to identify edits, backcrossing, and molecular analysis to segregate away T-DNA components from the intended edit.

Che et al. [14] described the application of ternary vector in sorghum transformation resulting in broad sorghum genotypes capable for transformation. A step-by-step protocol of *Agrobacterium*-mediated sorghum transformation is available [15] and Anand et al. describe the application of ternary vector in sorghum in this book. This chapter focuses on the methods specific to CRISPR/Cas9-mediated gene editing.

3.1 Design Experiment

Select an experimental approach that best achieves the intended outcome.

1. Frameshift edit: A small insertion or deletion created by cleaving a single genomic locus (Fig. 2). This approach is commonly used to turn off a gene by creating a frame shift in protein coding sequence (*see* **Note 2**).

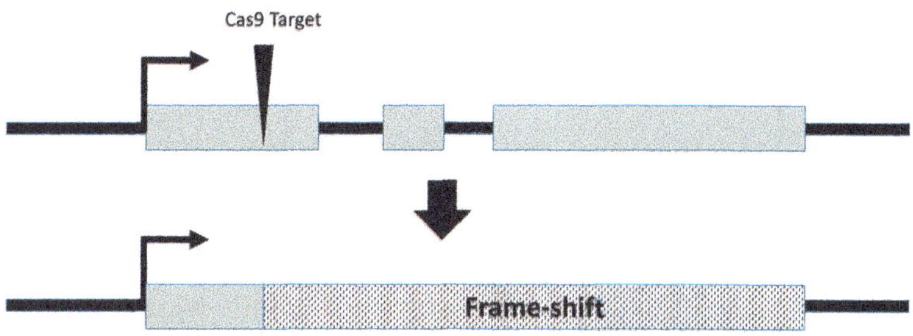

Fig. 2 Schematic of a frameshift edit disrupting the coding sequence of a gene. The start of transcription is denoted by a right facing arrow. The cut site is represented by a black triangle. Exons are shown as solid gray bars. Out of frame sequence is represented by white bar with black dashes

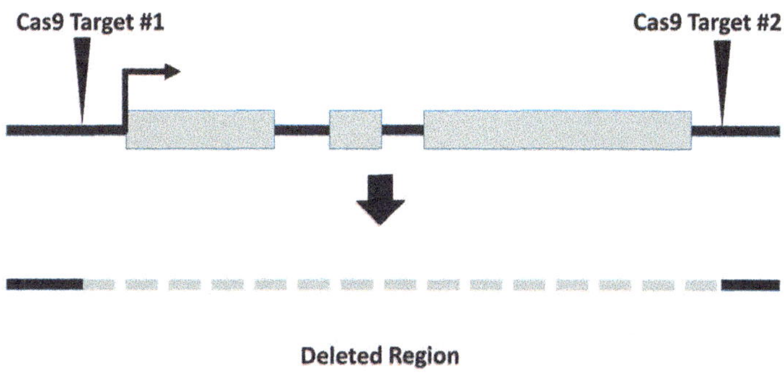

Fig. 3 Schematic of a dropout edit creating an absence variant. The start of transcription is denoted by a right facing arrow. Cut sites are represented by black triangles. Exons are shown as solid gray bars. Deleted sequence is represented by a dashed gray line

2. Dropout edit: A dropout of a region created by simultaneous cleaving and loss of the intervening sequence (Fig. 3). This approach is commonly used to turn off a gene by removing the entire open reading frame (*see* **Note 3**).
3. Template edit: Template-based edits are used to make precise base pair changes or insert sequence into a genomic DNA. Template-based editing is achieved by including a donor DNA with the desired sequence changes flanked by sequence sharing identity with the genomic sequence flanking a nuclease generated break (Fig. 4) (*see* **Note 4**). It is expected that this can be achieved in sorghum. However, template-based edits have not yet been demonstrated with *Agrobacterium*-mediated gene editing in sorghum and are not covered in this chapter.

3.2 Screen Sorghum Genome for Potential CRISPR/Cas9 Targets

1. Screen the genomic DNA sequence of the experimental sorghum line with bioinformatics software to identify all canonical spyCas9 targets in the region of interest following the motif 5′-N20NGG-3′ (Fig. 5). These targets can be located on either

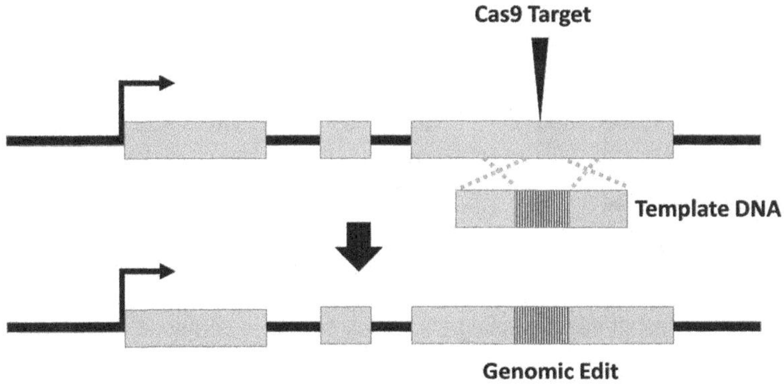

Fig. 4 Schematic of a template-based edit. The start of transcription is denoted by a right facing arrow. The cut site is represented by a black triangle. Exons are shown as solid gray bars. Sequence variation in the template is represented as a striped bar

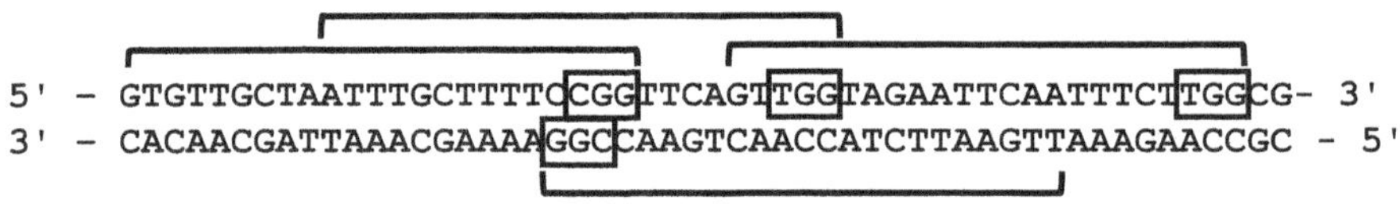

Fig. 5 Sites that can be targeted by spyCas9 on genomic double-stranded DNA sequence. PAM sequences are noted by boxes. Full target sequences are marked by horizontal brackets

strand of the double-stranded genomic DNA (*see* **Notes 1** and **5**).

2. Identify the 20 nucleotides upstream of the 5′ of the PAM sequence at the target site in sorghum genome. These 20 nucleotides along with the PAM comprise the full CRISPR/Cas9 target sequence.

3.3 Characterize the Distinctness of Each Potential Target Sequence

1. Characterize the distinctness of each potential target sequence by performing a comprehensive informatics search of the sorghum genome for sites that share a high degree of identity with the target site and have a canonical 5′-NGG-3′ PAM sequence or 5′-NAG-3′ PAM that is recognized by spyCas9 with reduced efficiency (*see* **Note 1**). Perform the off-target analyses using the precise sorghum inbred being edited and from the exact seed source when possible.
2. Pick targets that achieve the desired outcome while maintaining the greatest orthogonality to background genomic sequences. Avoid using targets that have similar genomic sequences including sites with two or fewer total mismatches or up to four mismatches clustered primarily in the 5′ region of the spacer region of the gRNA. These regions have increased likelihood for the potential to be co-edited alongside the

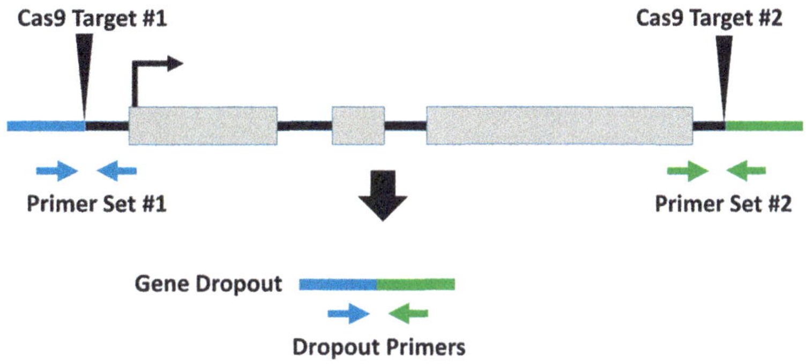

Fig. 6 Depiction of PCR assays developed to characterize dropout mutations. Blue inward facing arrows represent primers that amplify Cas9 Target #1. Green inward facing arrows represent primers that amplify Cas9 Target #2. The outer blue and green arrows combine to form a third PCR assay that amplifies the novel junction represented by adjacent blue and green bars

intended targets [19–22]. The best sites will also have a minimal number of targets with 3 or 4 total mismatches. Sometimes it may not be possible to avoid picking targets with other highly similar genomic sequences. For example, it can be difficult to find distinct targets within the coding sequence of closely related gene families. In those instances, develop PCR assays for the potential off-target regions as described in Subheading 3.4 to characterize and if necessary segregate away unwanted edits as described in Subheading 3.7.

3.4 Develop PCR Assays That Amplify Genomic Targets

1. Design PCR primers to amplify the genomic sequence around the gRNA target site. The designs for dropout editing comprise two target sites that use the same PCR conditions to amplify both the target sites (*see* **Note 6**). The PCR assays designed in this step will be used later to identify and characterize edited alleles in modified plants. In the case of a dropout design, the distal primers from the individual assays will be used to identify dropout edits (Fig. 6).
2. Validate the integrity of the PCR product using the following three steps: (1) conduct DNA amplification by these PCR primers with wild type (WT) sorghum genomic DNA, (2) run the DNA fragment on agarose gel, and (3) sequence the PCR amplified fragment. If a single DNA fragment appeared on the gel with the expected size and contains the expected sequence, then successful PCR design has been achieved.

3.5 Construct Expression Vectors

1. Design gRNAs to the selected target sites in sorghum genome. The DNA sequence encoding the gRNA is driven by a maize U6 promoter [9]. A guanine in the final base of the U6

G<u>NNNNNNNNNNNNNNNNNNN</u>GTTTTAGAGCTAGAAATAGCAAGTTAAAATA
AGGCTAGTCCGTTATCAACTTGAAAAAGTGGCACCGAGTCGGTGCT**TTTTTT**

Fig. 7 DNA sequence encoding gRNA and terminating poly-T sequence (bold). Programmable DNA hybridization region is underlined. The spacer sequence is represented by N's. The boxed G is both the final nucleotide of the U6 promoter and the first nucleotide encoding the DNA hybridization region of the gRNA [23]

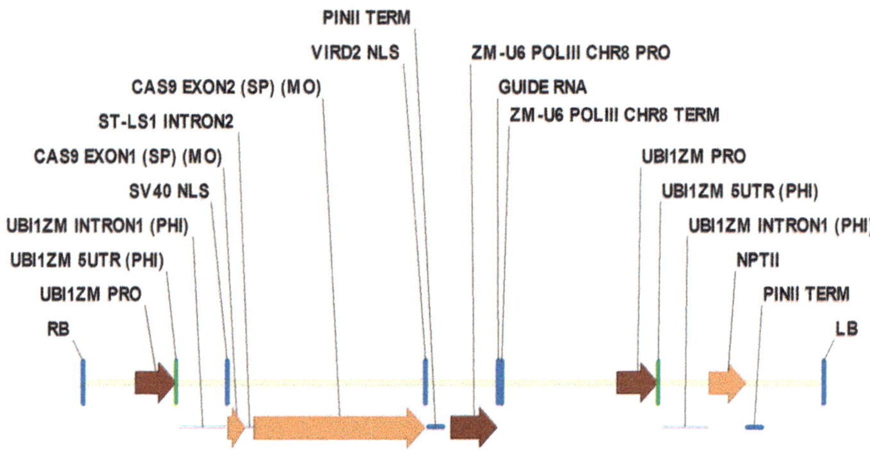

Fig. 8 Schematic of T-DNA expressing Cas9 and gRNA used by Che et al. [14]

promoter is the first nucleotide in the spacer sequence of the gRNA (Fig. 7). The spacer sequence is the region of the gRNA hybridizing with the target site in sorghum genome. If the 5′ nucleotide in this 20-nucleotide spacer sequence is not a G, change it to a G. This will maximize expression of the gRNA from the U6 promoter with minimal or no impact to activity at the target site (*see* **Notes 7** and **8**).

2. Clone the DNA sequence expressing the spyCas9, gRNA, and a selection marker gene for plant transformation such as NPTII into the T-DNA region (Fig. 8) in a ternary vector system ([14] and Chapter 13).
3. For dropout edits, clone two gRNA expression cassettes into the T-DNA (Fig. 9).
4. Transform the vector and accompanying helper vectors [14] into an *Agrobacterium* strain, such as LBA4404 (*see* **Note 9**).

3.6 Transform Sorghum and Regenerate Plants

Conduct sorghum transformation following the well-established step-by-step protocol [15] with the *Agrobacterium* vector described above for gene editing and generate T0 plants.

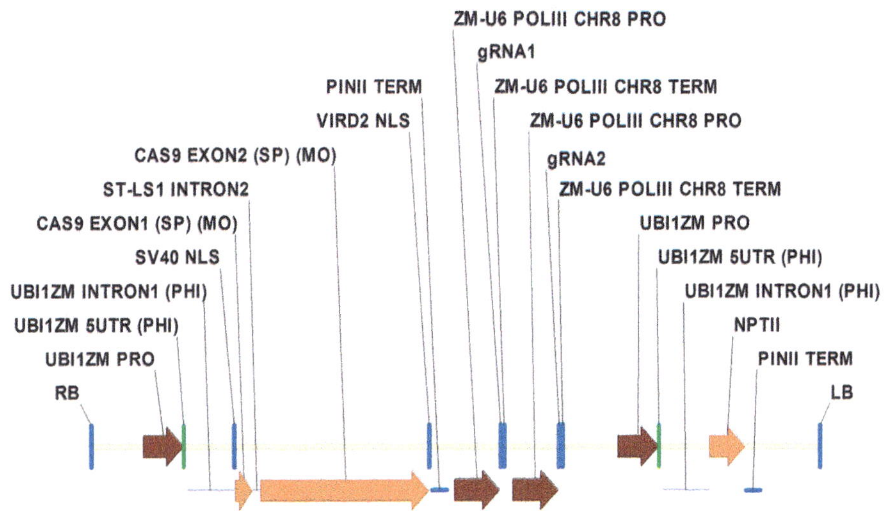

Fig. 9 Schematic of T-DNA expressing Cas9 and two gRNAs used to create a dropout edit

Example: Homozygous edit

Reference	-	GGCGCCCACCGTGATGGCTCGAACCAAGCACCAGGCCGTGAGGAAGC	Reads
Sequence1	-	GGCGCCCACCGTGATGGCTCGAACC------------------AAGC	x54364
Sequence2	-	GGCGCCCACCGTGATGGCTCGAACCAAGCACCAGGC-GTGAGGAAGC	x46911

Example: Hemizygous edit

Reference	-	GGCGCCCACCGTGATGGCTCGAACCAAGCACCAGGCCGTGAGGAAGC	Reads
Sequence1	-	GGCGCCCACCGTGATGGCTCGAACCAAGCACCA-----TGAGGAAGC	x50944
Sequence2	-	GGCGCCCACCGTGATGGCTCGAACCAAGCACCAGGCCGTGAGGAAGC	x47618

Fig. 10 Example Illumina sequence data amplified from DNA harvested from frameshift edits for homozygous and hemizygous edits. Deleted bases are represented by dashes

3.7 Analyze Sequence and Segregate Edits from T-DNA

1. Isolate genomic DNA from individual T0 plants and WT sorghum plants for molecular analysis (*see* **Note 10**).
2. PCR amplify the target region(s) with sorghum genomic DNA from these T0 plants and WT plants using the PCR primers designed in Subheading 3.4.
3. Identify T0 plants with edited sequences through the following steps: (1) sequence amplicons using a single molecule sequencing technology such as Illumina, (2) align amplicon sequences to the WT sorghum reference sequence using software such as BWA (Fig. 10), (3) select the T0 plants with desired edits for propagation (*see* **Note 11**).
4. For dropout edits, use spanning primers from individual targets (Fig. 6). Run the amplicons from the spanning primers on an agarose gel and identify bands consistent with the expected size

of a successful dropout before proceeding with sequencing (*see* **Notes 12** and **13**).

5. Self-pollinate these edited T0 plants to generate a population of T1 seeds that are segregating the edits from the accompany T-DNA.
6. Plant 16–32 T1 seeds from each segregating seed population derived from an individual T0 plant. These will be segregating for both the Cas9 machinery and the selective marker gene away from the edits.
7. Isolate genomic DNA from individual T1 plants at approximately 2–3 leaf stage for analysis.
8. Use PCR assays designed against the T-DNA region and accompanying vectors [11] to identify the T1 plants of absence of the entire T-DNA including the Cas9 and gRNA components as well as the selection marker.
9. Using the amplicon and sequencing strategy from **step 3** in this section to identify the plants having the desired edits in T1 generation.
10. Maintain and propagate those T1 plants that are negative for T-DNA integration and have only the desired edits.
11. Perform Southern-by-sequencing [18] (or alternative high-fidelity screen) to verify no DNA from any of the exogenous vector sequences used in the process remains integrated in the selected plants.
12. Pollinate and harvest the T2 seeds only from these T1 plants that have the intended edit and are verified to have no insertion of T-DNA or vector backbone.

4 Notes

1. BLAST searches can provide a crude picture of off-targets including if the sequence is highly repetitive. However, this approach will miss many closely related sequences and is insufficient for selecting targets. Many tools exist that scan genomes for related sequences [16]; however, as of the writing of this chapter, most do not support sorghum. CRISPRdirect [24] is a publicly available resource that does support sorghum and can facilitate this process identifying potential targets and how many times the exact site occurs as well as the 12mer and 8mer [24]. However, this approach does not take into context the sequence of the entire site and therefore should be used with care. For example, this strategy will miss highly similar sites that vary by a single nucleotide in the 8–12 bases near the PAM. It will also incorrectly count sites that match exactly in

the sequence near the PAM but are mismatched at many or all positions in the 8–12 bases distal to the PAM and therefore are not closely related.

2. Most spyCas9 gRNAs (>90%) designed lacking a poly-T pol III terminating motif in the target sequence will be active when expressed from the promoter in T-DNA. In line with this, it is recommended to avoid targets with 4 or more adjacent T's. Efficient *Agrobacterium* delivery systems typically result with high mutagenesis rates with most (50–90%) of T0s being edited with indels. Frameshift edits generated using spyCas9 commonly comprise a single bp insertion or deletion. Larger deletions are possible; however, they are typically limited to less than a few dozen bp and occur in fewer than 30% of alleles. Insertions can also comprise other DNA present in the cell at the time of repair including T-DNA or the sorghum genomic DNA. Homozygous edits at both alleles are common. These edits can be different at each allele and the frequency can vary by target sites from more than 70% of plants having homozygous edits to less than 30%. The precise makeup of edits varies across target sites and the role of the gene in cellular processes. For example, homozygous mutations in essential genes will be uncommon and will result in more in-frame insertions and deletions that conserve gene function. Chimeras can occur and its frequency varies by target sites. One must wait until the T1 generation to be fully confident that the desired edit is stably inherited.
3. For a successful dropout, both gRNA complexes must cleave their intended target sites. When using a robust expression and transformation system, greater than 90% of individual gRNAs actively modify their targets. Precise deletions can occur between the individual break points. It is common for additional sequence flanking the cut sites to be removed during this process as well. Dropouts occur at a lower and more highly variable rate than frameshifts, typically less than 20% of plants have a dropout. The spyCas9 system has been used to efficiently delete regions on the order of 10,000 bp in sorghum. Much larger dropouts have been demonstrated in other crops. However, for most dropouts these much larger deletions are generally expected to occur less frequently than smaller deletions. When dropouts are not achieved, the individual gRNA targets will typically create indels at their individual cut sites.
4. The CRISPR/Cas-induced cut should be as close to the location of the desired edit as possible and preferably fewer than 20 nucleotides away from the break. The length of each flanking arm identical to the target site should be more than 400 bp. Although template-based editing has not yet been reported in sorghum using *Agrobacterium*, it has been demonstrated using

bombardment in maize with rates typically lower than 1 plant per 100 plants screened. Success rates for template-based editing are variable and site-dependent.

5. Additional CRISPR/Cas systems have been discovered [25, 26] and engineered [27, 28]. These alternative CRISPR/Cas systems can target different sequence motifs, require different cognate gRNAs and may have different activities and behaviors as compared with spyCas9.
6. The insertion and deletion edits will start from the cut site within the target site and extend out in one or both directions. Therefore, one should avoid designing the PCR priming sites too close to the edit because this can result in an inability to amplify and characterize edits that extend into either of the primer regions. Amplicons should also be sufficiently small to accommodate the selected sequencing technology.
7. If the 5′ is not a guanine, the off-target screen should ideally be assessed against a gRNA with a guanine at this position.
8. The NGG PAM region is recognized by the spyCas9 protein and should not be included in the gRNA sequence.
9. Vector integrity should be confirmed by sequencing DNA purified from the transformed *Agrobacterium* strain before proceeding to transformation.
10. Clean the sampling equipment between plants to minimize contamination from other plants that could be amplified and provide a false signal in downstream PCR steps.
11. A high percentage of reads for more than two sequences or an imbalance in observed WT and edited alleles may represent that the regenerated plant is a chimera. Large insertions will typically not be amplified because they will be outcompeted in the PCR process by the smaller allele lacking the insert. Plants must be assessed in the T1 generation to confirm stable inheritance.
12. Full gene dropouts create novel junctions often separated by more than several thousand base pairs that often cannot be reliably amplified. The use of flanking primers verified to amplify the individual target sites under identical conditions improves success as compared to unverified primers pairs (Fig. 6).
13. Bands with smaller than expected sizes that were not present in unedited samples may represent dropouts with addition exonuclease activity and can be moved forward.

Acknowledgments

The author would like to acknowledge the Crop Genome Engineering, Vector Construction, Molecular Characterization, and Controlled Environment groups from Corteva Agriscience™, Agriculture Division of DowDuPont™ for support and Clara Alarcon and Todd J. Jones for comments on the manuscript. This work was funded by a subaward from the CSIRO under the Capturing Heterosis for Smallholder Farmers grant from BMGF. Corteva Agriscience™, Agriculture Division of DowDuPont™ provided funding and in-kind donations.

References

1. Sander JD, Joung JK (2014) CRISPR-Cas systems for editing, regulating and targeting genomes. Nat Methods 32:347–355
2. Chilcoat D, Liu ZB, Sander JD (2017) Use of CRISPR/Cas9 for crop improvement in maize and soybean. In: Weeks D, Yang B (eds) Progress in molecular biology and translational science, vol 149. Elsevier, San Diego, pp 27–46
3. Li T, Liu B, Spalding MH, Weeks DP, Yang B (2012) High-efficiency TALEN-based gene editing produced disease-resistant rice. Nat Biotechnol 30:390–392
4. Wang Y, Cheng X, Shang Q, Zhang Y, Liu J, Gao C et al (2014) Simultaneous editing of three homeoalleles in hexaploid bread wheat confers heritable resistance to powdery mildew. Nat Biotechnol 32:947–951
5. Clasen BM, Stoddard TJ, Luo S, Demorest ZL, Li J, Cedrone F et al (2016) Improving cold storage and processing traits in potato through targeted gene knockout. Plant Biotechnol J 14 (1):169–176. https://doi.org/10.1111/pbi.12370
6. Andersson M, Turesson H, Nicola A, Fält AS, Samuelsson M, Hofvander P (2017) Efficient targeted multiallelic mutagenesis in tetraploid potato (Solanum tuberosum) by transient CRISPR-Cas9 expression in protoplasts. Plant Cell Rep 36:117–128
7. Demorest ZL, Coffman A, Baltes NJ, Stoddard TJ, Clasen BM, Luo S et al (2016) Direct stacking of sequence-specific nuclease-induced mutations to produce high oleic and low linoleic soybean oil. BMC Plant Biol 16:255
8. Sanchez-Leon S, Gil-Humanes J, Ozuna CV, Giménez MJ, Sousa C, Voytas DF et al (2018) Low-gluten, nontransgenic wheat engineered with CRISPR/Cas9. Plant Biotechnol J 16 (4):902–910
9. Svitashev S, Young JK, Schwartz C, Gao H, Falco SC, Cigan AM (2015) Targeted mutagenesis, precise gene editing, and site-specific gene insertion in maize using cas9 and guide RNA. Plant Physiol 169(2):931–945
10. Jiang W, Zhou H, Bi H, Fromm M, Yang B, Weeks DP (2013) Demonstration of CRISPR/Cas9/sgRNA-mediated targeted gene modification in Arabidopsis, tobacco, sorghum and rice. Nucleic Acids Res 41(20):e188. https://doi.org/10.1093/nar/gkt780
11. Wu E, Lenderts B, Glassman K, Berezowska-Kaniewska M, Christensen H, Asmus T et al (2014) Optimized agrobacterium-mediated sorghum transformation protocol and molecular data of transgenic sorghum plants. In Vitro Cell Dev Biol Plant 50(1):9–18
12. Cho MJ, Wu E, Kwan J, Yu M, Banh J, Linn W et al (2014) *Agrobacterium*-mediated high-frequency transformation of an elite commercial maize (Zea mays L.) inbred line. Plant Cell Rep 33:1767–1777
13. Lowe K, Wu E, Wang N, Hoerster G, Hastings C, Cho MJ et al (2016) Morphogenic regulators baby boom and wuschel improve monocot transformation. Plant Cell 28:1998–2015
14. Che P, Anand A, Wu E, Sander JD, Simon MK, Zhu W et al (2018) Developing a flexible, high-efficiency agrobacterium-mediated sorghum transformation system with broad application. Plant Biotechnol J 16(7):1388–1395. https://doi.org/10.1111/pbi.12879
15. Wu E, Zhao ZY (2017) *Agrobacterium*-mediated sorghum transformation. In: Schmidt A (ed) Plant germline development, methods and protocols, methods in molecular biology, vol 1669. Humana Press, New York, pp 355–364

16. Sangar C, Samant L, Pawar S, Dhawale P, Chowdhary AS (2016) Online and offline tools: crispr/cas off-target detection. Int J Pharm Sci Res 7(5):1889–1895
17. Li H, Durbin R (2009) Fast and accurate short read alignment with burrows-wheeler transform. Bioinformatics 25(14):1754–1760
18. Zastrow-Hayes GM, Lin H, Sigmund AL, Hoffman JL, Alarcon CM, Hayes KR et al (2015) Southern-by-sequencing: a robust screening approach for molecular characterization of genetically modified crops. Plant Genome 8(1):1–15
19. Fu Y, Foden JA, Khayber C, Maeder ML, Reyon D, Joung LK et al (2013) High-frequency off-target mutagenesis induced by CRISPR-Cas nucleases in human cells. Nat Biotechnol 31(9):822–826
20. Mali P, Aach J, Stranges PB, Esvelt KM, Moosburner M, Kosuri S et al (2013) Cas9 transcriptional activators for target specificity screening and paired nickases for cooperative genome engineering. Nat Biotechnol 31 (9):833–838
21. Pattanayak V, Lin S, Guillinger JP, Ma E, Doudna JA, Liu DR (2013) High-throughput profiling of off-target DNA cleavage reveals RNA-programmed Cas9 nuclease specificity. Nat Biotechnol 31(9):839–843
22. Hsu PD, Scott DA, Weinstein JA, Ran FA, Konermann S, Agarwala V et al (2013) DNA targeting specificity of RNA-guided Cas9 nucleases. Nat Biotechnol 31:827–832
23. Mali P, Yang L, Esvelt KM, Aach J, Guell M, DiCarlo JE et al (2013) RNA-guided human genome engineering via Cas9. Science 339:823–826
24. Naito Y, Hino K, Bono H, Ui-Tei K (2015) CRISPRdirect: software for designing CRISPR/Cas guide RNA with reduced off-target sites. Bioinformatics 31:1120–1123
25. Esvelt KM, Mali P, Braff JL, Moosburner M, Yaung SJ, Church GM (2013) Orthogonal Cas9 proteins for RNA-guided gene regulation and editing. Nat Methods 10:1116–1121
26. Zetsche B, Gootenberg JS, Abudayyeh OO, Slaymaker IM, Makarova KS, Essletzbichler P et al (2015) Cpf1 is a single RNA-guided endonuclease of a class 2 CRISPR-Cas system. Cell 163:759–771
27. Kleinstiver BP, Prew MS, Tsai SQ, Topkar VV, Nguyen NT, Zheng Z et al (2015) Engineered CRISPR-Cas9 nucleases with altered PAM specificities. Nature 523:481–485
28. Hu JH, Miller SM, Geurts MH, Tang W, Chen L, Sun N et al (2018) Evolved Cas9 variants with broad PAM compatibility and high DNA specificity. Nature 556:57–63

Chapter 12

Genome Editing by CRISPR/Cas9 in Sorghum Through Biolistic Bombardment

Guoquan Liu, Jieqing Li, and Ian D. Godwin

Abstract

The advanced biotechnology CRISPR/Cas9 as a powerful genome editing tool has displayed great potential for improving important agronomic traits such as yield and quality. It has gained momentum worldwide for gene function research of plants in recent years. As for cereals, numerous studies of CRISPR/Cas9 have been reported predominately on rice and quite a few on other cereals including maize, wheat, and barley. In contrast, there are only a couple of reports on sorghum up to date. In this chapter, the CRISPR/Cas9 system has been investigated for sorghum genome editing through biolistic bombardment. Two target genes, cinnamyl alcohol dehydrogenase (CAD) and phytoene desaturase (PDS), have been investigated by CRISPR/Cas9 though bomboarment. Successful genome editing has been achieved within the sorghum genotype Tx430. Furthermore, sequencing PCR product of transgenic plants has confirmed that the CRISPR/Cas9 successfully edited the target gene in sorghum. Both homozygosis and heterozygosis editings of CAD gene have been confirmed in T_0 primary transgenic lines through sequencing PCR products. T_1 generation of CRISPR plants has been investigated as well. The results illustrated that the edited gene has passed down to next generation. More experiments, such as optimizing promoters for guide RNA (gRNA) and Cas9 in sorghum, are under investigation. Three factors were considered crucial elements to establish an efficient CRISPR/Cas9 system for genome editing in sorghum: (1) an efficient transformation system, (2) the design of targeted gene sequence for gRNA, (3) effective expression of CRISPR components including Cas9 and gRNA.

Key words CRISPR/Cas9, Biolistic bombardment, Genome editing, Immature embryo, Tissue culture, Sorghum, Genetic transformation

1 Introduction

CRISPR, which stands for Clustered Regularly Interspaced Short Palindromic Repeats, play an important role in the prokaryotic immune system. With the help of gRNA which originates from CRISPR, nuclease Cas9, containing two domains HNH domain and RucV-like domain, recognizes and cuts exogenous DNA [1]. Two domains of Cas9 cut each side of the double-stranded DNA and generate a double-strand break (DSB). Once DSB is formed by CRISPR/Cas9, the endogenous DNA repair

Zuo-Yu Zhao and Jeff Dahlberg (eds.), *Sorghum: Methods and Protocols*, Methods in Molecular Biology, vol. 1931, https://doi.org/10.1007/978-1-4939-9039-9_12,

mechanisms are initiated [2]. There are two types of DNA repair mechanisms: nonhomologous end-joining (NHEJ) and homology-directed repair (HDR). Genome editing in plants by CRISPR/Cas9 system mostly relys on NHEJ which is error-prone for DNA repair. Hence, the outcome of successful edits is usually a knockout of the gene's expression because the repair has introduced indels resulting in a frame-shift mutation. To date, most plant genome editing is the production of null alleles. HDR, while theoretically possible, has not been very successful in plants.

The CRISPR/Cas9 holds great promise for plant research and as a precise breeding tool for rapid trait modification. Unlike traditional mutagenesis breeding via chemical or ionizing radiation means, the outcomes of successful CRISPR/Cas9 genome editing lead to a significantly more precise mutation on the target gene [3]. Chemical or radiation mutations produce large numbers of off-target effects, and the screening of large populations hampers selection of appropriate phenotypes, especially for traits that are difficult to measure [4, 5]. CRISPR/Cas9 genome editing in major cereals has been reported by a number of groups [6–9]. In contrast, the first report of genome editing in sorghum was published along with Arabidopsis, tobacco, and rice in 2013, while stably edited plants were not illustrated [10]. Recently, scientists at DuPont Pioneer have reported on stable edits in sorghum plants through an *Agrobacterium*-mediated transformation method, although experimental detail was not described [11]. Despite the endless possibility of genome editing in plants, the investigation on CRISPR/Cas9 in sorghum has been overwhelmingly lacking by comparison with CRISPR in other major cereals such as rice, maize, and wheat [8, 9, 12, 13].

The genome editing system CRISPR/Cas9 has been recognized as a revolutionary tool for plant precise breeding and improvement due to its simplicity, feasibility, and versatility [6]. Significant progress has been made for plant genome editing in the last 5 years [6, 14–17]. For example, a highly efficient ligation-independent cloning system was developed for the CRISPR/Cas9 system in plants and used for cloning multiple gRNAs. This research would enable CRISPR to target multiple genes at once. Moreover, it is reported that heat stress could significantly increase the efficiency of on-target mutagenesis in plants using CRISPR/Cas9 [14]. Recently, a DNA-free system based on CRISP/Cas9 has emerged for precise genome editing in plants [9, 16, 18, 19]. More importantly, products from the DNA-free system would be more likely acceptable to the public and be considered non-GM. In fact, the US Department of Agriculture (USDA) has made clear that CRISPR/Cas9-edited plants can be cultivated and sold free from regulation under certain circumstance [7]. This will pave the way for commercialization of plant products through accurate genome editing [16]. The genome editing has rapidly become a mainstream for plant improvement in this decade [20].

To deploy the system in plants, generally, plasmids containing Cas9 gene and gRNA expression cassettes would be designed, cloned, and then be delivered into plant cells [21]. There are two common systems to deliver DNA into plant cells, which are *Agrobacterium*-mediated and biolistic transformation systems. Two previous reports on sorghum genome editing utilized the *Agrobacterium*-mediated transformation system [10, 11]. Here, we describe the biolistic bombardment methodology that we have successfully produced stable edited plants of the sorghum inbred line, Tx430. The DNA delivery system, including tissue culture and transformation, was elaborately described in previous studies [22–24]. Using this methodology, sorghum genome editing can be achieved at feasible frequencies (25% among total transgenic lines, not published data).

We are working toward improvements in the system which does not require plant tissue culture. Ultimately, products based on CRISPR/Cas9 do not contain marker genes or other inserts from constructs for sorghum research, especially for breeding. This will rapidly produce genome edited plants in a wide range of sorghum genotypes and promote the technology as a major tool for genetic advancement in plant breeding programs. This chapter is a combination of some preliminary investigation on CRISPR/Cas9 genome editing in sorghum and a few updates on sorghum transformation through biolistic bombardment.

2 Materials

2.1 Materials for Sorghum Immature Seeds

1. Temperature-controlled physical containment glasshouse (PC2) (18–28 °C) (www.ogtr.gov.au) or field in summer season (*see* **Note 1**).
2. Pots (20 l capacity), decontaminated potting mix, slowly released fertilizers such as Osmocote, watering facility, and benches.
3. Mature sorghum seeds for planting, such as Tx430 (*see* **Note 2**).
4. Healthy immature seeds for isolating immature embryos (IEs) (*see* **Note 3**).
5. Commercial bleach, for example, White King containing 4% (m/v) sodium hypochlorite for seeds surface disinfection (*see* **Note 4**).
6. Surfactant Tween 20.
7. 70% ethanol for cleaning beaches and seeds.
8. Stock solution: 1 mol/l (M) $Ca(NO_3)_2$ for calcium deficiency at the seedling stage, especially in the winter season.

9. Sterilized water.
10. Platform shaker etc.

2.2 Materials for Sorghum Tissue Culture

1. Tissue culture room with temperature control (27 ± 1 °C).
2. Fluorescent lights with a luminescence of approximately 100 μmol s^{-1} m^{-2} (16 h/day).
3. Cabinets for store tissue culture media.
4. Marker pens.
5. Autoclave.
6. Petri dishes (90 × 25 mm, 90 × 15 mm).
7. Laminar air flow.
8. pH meter.
9. Autoclavable glassware (50, 100, 250, 500, 1000 ml).
10. Beakers in different sizes.
11. Balance ($d = 0.01$ g; $d = 0.1$ mg).
12. Forceps with fine point tips.
13. Surgical blades (size 11, 23, or 24).
14. Parafilm.
15. Filter papers (90 mm∅; 70 mm∅, Whatman®).
16. Purified water (Millipore Milli-Q® water) etc.

2.3 Stock Solutions for Media

1. 5 mM copper sulfate ($CuSO_4$) (filter sterilization).
2. 70%, 90%, and 100% ethanol.
3. 1 M potassium hydroxide (KOH).
4. 5 M KOH.
5. 1 M sodium hydroxide (NaOH).
6. 1 M hydrochloric acid (HCl).
7. Stock solutions of hormone for tissue culture [22] (Table 1).

2.4 Media for Sorghum Tissue Culture

1. MS medium: MS [24, 25] powder with Gamborg vitamins 4.44 g/l, 30 g/l sucrose, and 8 g/l agar (*see* **Note 5**).
2. Callus induction medium (CIM): MS medium plus 1 g/l L-proline, 1 g/l L-asparagine, 1 g/l potassium dihydrogen phosphate (KH_2PO_4), 1 μM $CuSO_4$, and 1 mg/l 2,4-D. The CIM can be stored in a tissue culture room in the dark for 1 month.
3. Regeneration medium: MS medium plus 1 mg/l BAP, 1 mg/l IAA, and 1 μM $CuSO_4$. This medium can be stored in a tissue culture room in the dark for 2 months.

Table 1
Plant hormone for sorghum tissue culture

Stock solution	2,4-D	BAP	IAA	IBA	NAA
Full name	2,4-Dichloro-phenoxyacetic	6-Benzy-laminopurine	Indole-acetic acid	Indole-butyric acid	α-Aphthale-neacetic acid
Molecular weight (g/Mol)	221.0	225.3	175.2	203.2	186.2
Powder storage	RT	RT	−0 °C	2–8 °C	RT
Solution storage	0–5 °C	0–5 °C	0–5 °C	0–5 °C	0–5 °C
Solvent	EtOH/1 M NaOH	1 M KOH	EtOH/1 M NaOH	EtOH/1 M NaOH	1 M NaOH
Diluent	Water	Water	Water	Water	Water
Sterilization	CA[a]	F[b]	F	F	F
Stock concentration (mg/ml)	1.0	1.0	1.0	1.0	1.0
Working concentration (mg/l)	0.01–5.0	0.1–5.0	0.01–3.0	0.1–10.0	0.1–10.0

[a]CA: Co-autoclave with other media components
[b]F: Filter sterilization (32 mm Syringe Filter with 0.2 μm Supor® Membrane)

4. Rooting medium: MS medium plus 1 mg/l NAA, 1 mg/l IAA, 1 mg/l IBA, and 1 μM $CuSO_4$. The rooting medium can be stored in a tissue culture room in the dark for 3 months.
5. Selective regeneration medium: regeneration medium plus selective reagent such as 30 mg/l geneticin (G418) (disulfate salt solution, Sigma) (*see* **Note 6**).
6. Selective rooting medium: rooting medium plus selective reagent 30 mg/l G418.
7. Osmotic medium: MS medium plus 0.2 M D-sorbitol and 0.2 M D-mannitol. The osmotic medium can be stored in a tissue culture room in the dark for 3 months.

2.5 Materials for Particle Bombardment

1. Biolistic PDS 1000/He (Bio-Rad) or particle inflow gun (PIG) system, both connecting to a vacuum pump, and a high-pressure helium tank.
2. Microcarrier holder, microcarrier, rupture disks (900, or 1100, or 1300 psi) for Bio-Rad system, and stopping screens for PDS 1000/He system.

3. Sterilized syringe filters, and sterilized baffles for PIG system.
4. Vortex shaker.
5. 1.5, and 2.0 ml microfuge tubes.
6. Benchtop centrifuges.
7. Pipettes: 2, 10, 20, 200, 1000, and 5000 μl and their respective tips.
8. Refrigerator (4 °C), and freezers (−20 °C and − 80 °C).
9. 0.6 μm gold particles (*see* **Note** 7).

2.6 Solutions for Particle Bombardment

1. Stock solution: 2.5 M calcium chloride ($CaCl_2$) (filter sterilization). It can be stored at −20 °C for up to 1 years.
2. 70%, and 100% ethanol.
3. 0.1 M spermidine. Make a fresh solution per round of bombardment (2 μl spermidine in 125 μl sterilized water).
4. 50% and 80% glycerol (autoclaved).
5. Plasmids of the target and selective marker genes at the concentration of 1 μg/μl if co-transformation is performed.

2.7 Genetic Materials for CRISPR

1. CRISPR construct including Cas9 gene and guild-DNA (gRNA) (Fig. 1). In the case of CAD CRISPR construct, the guild-DNA is the nucleotides with bold letters and the PAM is the nucleotides with bold italic underline letters

   (-TCAGTTCGA**ATGGCTGCTGAATCAGAGCA*<u>CGG</u>*-**CAACTGCAA-). Theoretically, any gene containing PAM sequence, which is 5'-NGG-3′ where "N" is any nucleobase followed by two guanines (G) nucleobases, can be used as a target gene for CRISPR/Cas9 genome editing system (Fig. 1a) [2, 3].
2. In our study of CRISPR on CAD gene, gRNA for CRISPR was driven by the effective promoter such as U3 [26]. Rice U3 promoter has been utilized and proven effective for expressing gRNA in sorghum.
3. Cas9 gene was driven by the effective maize *ubi* promoter for sorghum (Fig. 1b).
4. Selective marker gene such as the NPTII gene. Once the target gene is edited by CRISPR/Cas9, the marker gene and Cas9 gene could be selected out in further progenies. As a result, there is no vector trace in plants except that the target gene has been edited.

2.8 Bioinformatic Software/Tools

There are many CRISPR online tools to help design, evaluate, and clone guild sequences. It is recommended that at least two software should be used to design the gRNA sequence of the target gene. Two websites are highly useful and efficient for CRISPR guild

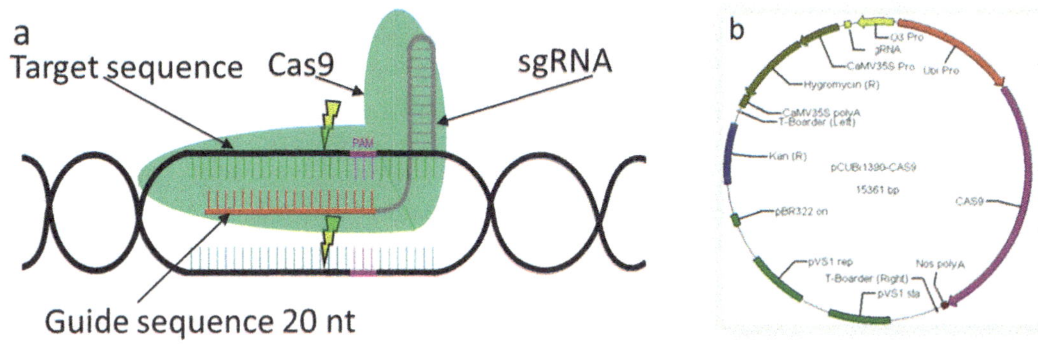

Fig. 1 Schematic diagram of CRISPR/Cas9 editing of target genes. (**a**) a sketch of CRISPR/Cas9 system; (**b**) CRISPR/Cas9 plasmid

sequence. One is CRISPR-P 2.0 (the new version CRISPR/Cas9 tool for genome editing especially for plants) [27]. Another one is CRISPOR 4.4 which will come with support for using new Cpf1 and saCas9 and xCas9 CRISPR systems [28].

3 Methods

3.1 CRISPR/Cas9 Construct Cloning

1. Design the sequence gRNA of the target gene by using the recommended online software (http://crispor.tefor.net/ or http://crispr.hzau.edu.cn/CRISPR2/).
2. Synthesize gRNA with two restriction enzymes cutting site at each end (HindIII/SamI).
3. Digest CRISPR/Cas9 construct with (HindIII/SamI) (Fig. 1b).
4. Ligate the digested vector and gRNA with T4 DNA ligase.
5. Transform ligated construct into bacteria such as DH5.
6. Extract plasmids from bacteria.
7. Sequence gRNA region to verify the correct clone.

3.2 Plant Sorghum in a Temperature-Controlled Glasshouse

1. Fill 20 l-pots with a decontaminated potting/nutrient mix (90% full).
2. Add fertilizers to the potting mix: For example, 30 g/pot of Dolomite, 10 g/pot of Super Phosphate, 10 g Osmocote®/pot.
3. Sow 4–6 sorghum seeds per pot. It is recommended that a maximum of three seedlings can be kept in each pot.
4. After germination, wait for 2 weeks before applying another round fertilizer.

5. Water plants daily or twice per day in summer season. Sorghum is famous for drought tolerance, but sufficient water helps sorghum stay in healthy and vigorous condition.
6. In the winter season, calcium deficiency may occur (*see* **Note 8**). It is necessary to apply a foliar spray of $Ca(NO_3)_2$ twice per week for the first month until the symptoms are disappearing. A stock of 1 M $Ca(NO_3)_2$ can be made and then dilute 20 times in spray bottle.
7. Apply slow release fertilizer twice per month, e.g., 5 g Osmocote®/pot.
8. Pest and disease should be monitored weekly and managed in time; otherwise, unhealthy immature seeds may result in contamination and reduce the efficiencies of both tissue culture and transformation.

3.3 Collect Sorghum Immature Seeds

1. Monitor sorghum panicle development.
2. Select healthy sorghum whole or partial panicle 12–15 days after pollination.
3. One panicle could be harvested in two or three times from top to bottom sections subsequently.
4. Collect an appropriate amount of immature seeds from panicles.

3.4 Disinfect Immature Seeds

1. Submerge immature seeds in 70% ethanol in a bottle and shake them at 200 rpm for 5 min.
2. Rinse seeds with sterilized water once in a laminar flow hood.
3. Submerge seeds in 100% commercial bleach with a few drops of Tween 20.
4. Shake it at 200 rpm for 10 min. It is not recommended that immature seeds are kept in bleach for an excessive time, which will reduce the quality of IEs for embryogenesis.
5. Rinse seeds more than five times with sterilized water until all bleach is washed off.
6. Place seeds in an aseptic petri dish in the laminar flow hood and allow drying for 20–30 min.

3.5 Isolate IEs

1. Isolate IEs with forceps and surgical blade within a laminar flow hood.
2. Place IEs with the scutellum side up on the callus induction media (maximum 25 IEs per 90 × 15 mm petri dish).
3. Seal Peti dishes with parafilm and mark them well.
4. Keep IEs in the dark to allow callus formation at 26–28 °C.

5. A fresh subculture would promote the production of embryogenic callus in 1 week after initiation.
6. After 9–11 days in the dark, IEs are ready for particle bombardment transformation.
7. Select high quality of IEs that have formed compact, globular, white embryogenic callus.
8. Place 6–8 IEs at the center of Petri dish filled with the osmotic medium for 2–3 h before bombardment.

3.6 Prepare Gold Particles

1. Weigh out 50 mg gold (0.6 μm diameter from Bio-Rad) in a 1.5 ml microfuge tube.
2. Add 1 ml 100% ethanol into the tube.
3. Vortex the tube thoroughly for 5 min.
4. Stand it for 15 min.
5. Pellet gold at 15,493 × *g* for 10 s in a benchtop centrifuge.
6. Remove ethanol, wash particles three times in 1 ml sterilized water, vortex for 1 min, allow to stand for 1 min, and then pellet gold particles in a bench top centrifuge at 825 × *g* for 10 s, and finally remove the supernatant.
7. Resuspend particles in 1 ml of autoclaved 50% glycerol. The concentration of gold particles is 50 mg/ml.
8. Allocate 50 μl aliquots into 1.5 microfuge tubes. One aliquot is generally used for 6 bombardments (aliquots can be stored at −20 °C up to 6 months).

3.7 Coat Gold Particles with DNA for 6 Bombardments

Perform the following procedure under aseptic conditions in a laminar flow hood.

1. To add $CaCl_2$ and 0.1 M spermidine together while gold particles are still suspended (Pre-drawn up solutions in two pipettors are ready for the next step).
2. Add 50 μl of 2.5 M $CaCl_2$ and 20 μl of 0.1 M spermidine into one tube at the same time.
3. Vortex the mixture for 1–2 min.
4. Precipitate for 5 Vortex tube(s) with 50 μl aliquot of gold particles thoroughly (homogenize completely, leave no visual clumps).
5. Add 10 μl of 1 μg/μl plasmid containing a selective marker, Cas9, and gRNA (for co-bombardment, 5 μg plasmid with selective gene and 5 μg plasmid Cas9, and gRNA).
6. Vortex tube(s) for 1–2 min.
7. It is important min on ice.
8. Pellet at 825 × *g* for 10 s in a bench top centrifuge.

9. Remove the supernatant.
10. Add 130 μl of 70% ethanol.
11. Vortex for 1–2 min.
12. Precipitate for 5 min on ice.
13. Pellet at 825 × *g* for 10 s in a bench top centrifuge.
14. Remove the supernatant.
15. Add 35 μl of 100% ethanol (analytical grade) and resuspend by vortex.
16. Visually confirm dispersal and immediately apply 5 μl of suspension to the center of syringe filter for each shot when using a PIG.

3.8 Proceed the PIG Delivery System

1. Set up the PIG in a laminar flow hood.
2. Spray the chamber with 70% ethanol inside and outside. Allow enough time to dry before shooting (normally 1–2 h).
3. Turn on the vacuum pump.
4. Open the knob on the helium tank to allow the gas to flow out of the cylinder.
5. Open the second knob on the helium tank and adjust helium pressure to the desired level (for example 1000 kPa for sorghum).
6. Turn the helium valve situated on top of PIG one full turn to open the valve at the optimal aperture.
7. Adjust and make sure that the timer for each shot is set to 0.05 s.
8. Mix the particle-coated DNA (which is prepared from Subheading 3.7). Add 5.0 μl of the suspension to the center of the syringe filter. Screw filter into the top of the vacuum chamber.
9. Place one petri dish with target tissues (which is prepared from Subheading 3.5) under a baffle in the chamber at the certain level and close the chamber door. The distance from the filter holder to the target cells is set at 18.5 cm for sorghum. Ensure that the vacuum release-valve to the chamber is closed.
10. Open the valve to vacuum in the chamber. Wait until the pressure gets to −90 kPa, and then close the valve.
11. Flick the firing switch to allow shot of helium to flow through the filter and shoot the particle-coated DNA into target tissues.
12. Open the release valve to release the vacuum.
13. Open the door of the chamber. Remove the baffle, take out the petri dish, and cover it with the petri dish lid.

14. Repeat **steps** from **8** to **13** for each plate which is required to be transformed with the same plasmid(s). Require separate sterilized baffles and filters if different plasmid(s) are used.
15. When all bombardment is complete, all baffles and filters should be washed, and autoclaved to be prepared for next experiment.
16. Close the bottom knob on helium tank to stop releasing helium to the tubes.
17. Flick the firing switch a number of times to release helium from tubes.
18. Close the helium valve one full turn.
19. Close the vacuum release valve to the chamber.
20. Turn off the vacuum pump.
21. Switch off the PIG system.
22. Clean the chamber and the laminar flow hood with 70% ethanol.

3.9 Recover Tissue After Bombardment

1. Keep bombarded IEs onto the osmotic medium for 3–4 h in the dark after bombardment.
2. Subculture IEs onto CIM and place the Petri dishes at 27 ± 1 °C in the dark for 3 days to recover the bombarded IEs.

3.10 Select Transgenics—Post-bombardment Selection

1. Subculture IEs from CIM onto selective regeneration medium and continue subculturing fortnightly until plantlets grow at least 3–5 cm long.
2. Subculture individual transgenic plantlet onto the selective rooting medium (2–3 plantlets from one transgenic event could share one plate) (*see* **Note 9**).
3. Keep plantlets on selective rooting medium for 3–4 weeks. In general, no further subculturing is needed.

3.11 Acclimate and Pot Out Transgenic Plantlets

1. Open lids of Petri dishes in order to hard off plantlets in a tissue culture room.
2. Add sterile water daily into Petri dishes to cover selective rooting medium.
3. Keep plantlets in the tissue culture room for 3–4 days.
4. Carefully extract plantlets from rooting medium and rinse off excess medium with water (non-sterile) from around the roots.
5. Transfer plantlets into a physical containment glasshouse with temperature control (18–28 °C).
6. Plant one transgenic event in one pot filled with potting mix.
7. Tag pots.

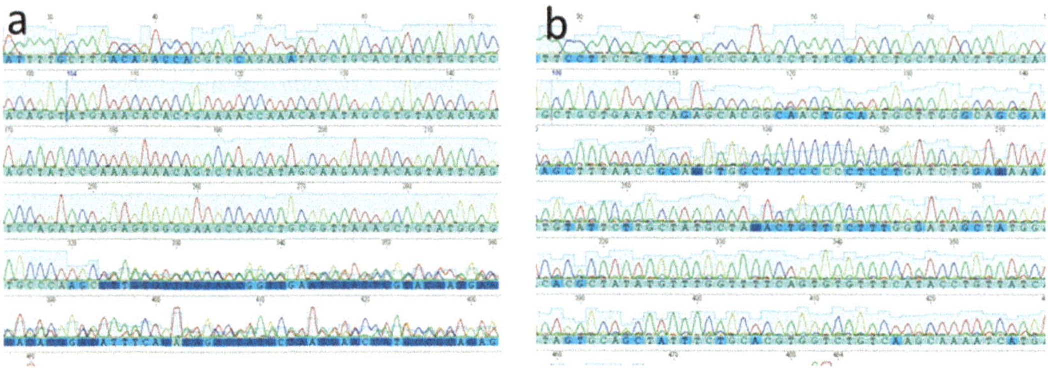

Fig. 2 Sequencing PCR products. (**a**) Homozygosis gene editing; (**b**) heterozygosis gene editing

3.12 Analyse CRISPR/Cas9 Plants

1. Design PCR primers.
2. PCR screening of target gene sequence. If the DNA fragment of insertion or deletion is long enough such as more than 20 bp, the size of CRISPR edited PCR product could be distinguished from the control by gel electrophoresis.
3. Sequence PCR product of transformed lines. Homozygosis or heterozygosis gene editing can be detected by PCR sequencing (Fig. 2). The sequencing of homozygosis editing is similar to the control and the mutation of indel happened in the target gene (Fig. 2a). The sequencing of heterozygosis editing is distinguishable from the control as it has double peaks after PAM sequence (Fig. 2b).

3.13 Eliminate Vector Trace from the Edited Individuals

1. Select the CRISPR edited lines.
2. Outcross the individual edited line with the corresponding wild type line.
3. Perform PCR screening of next generation population.
4. Select individual plants with the edited target gene but without vector trace.
5. Self-pollinate the selected plant.
6. Perform molecular analysis on next generation, such as PCR screening, Southern blotting, DNA sequencing.
7. Select homozygosis edited line without vector trace.

4 Notes

1. Seeds from glasshouse-grown plants are more reliable than those from field-grown plants because the temperature in glasshouse can be controlled, particularly in winter season while the

temperature in the field is not under control. However, there is no significant difference observed between IEs from glasshouses and field sites in terms of tissue culture in the summer season. The explants from the field are more likely vulnerable to bacteria or fungi contamination. One important environmental factor is minimum daily temperature. If the outside minimum temperature falls below 18 °C, the callus induction and callus regeneration rates will reduce dramatically for some sorghum genotypes such as SA281.

2. Tx430 is the most successful sorghum line for tissue culture and transformation in our lab. We also obtained transformants from sorghum genotypes SA281, 91419R, but the transformation efficiencies of SA281 and 91419R are much lower than that of Tx430.
3. In cereals, IEs have been demonstrated to be the most productive and reliable explant source to develop embryogenic calli. Transgenic plants have been obtained utilizing particle bombardment from cereal IEs including maize, rice, wheat, barley, sorghum, and pearl millet. In general, sorghum immature seeds are collected 12–15 days after anther emergence when the endosperm is milky and soft. Although the IE size of each variety is variable, generally, the IE size from 1.1 to 2.1 mm is more efficient to produce embryogenic callus than the others. The IE size less than 1.1 mm indicates that it is too young, whereas the IE size over 2.1 mm suggests that it is close to mature.
4. The active component of commercial bleach, sodium hypochlorite, may decrease in effectiveness after prolonged usage, leading to less success in disinfection of seed surface. It is recommended to change a fresh bottle monthly if it is not used within a month.
5. All media for sorghum tissue culture are based on MS medium with a pH adjusted to 5.7 and autoclaved at 121 °C for 15 min. 2,4-D is added to medium before autoclaving. The other hormones and $CuSO_4$ are sterilized with 0.2 μm filter and added into media post-autoclaving.
6. In general, the appropriate concentration of selection agent is variable, depending on species and genotypes. Therefore, a kill curve experiment of any specific sorghum line is vital to determine the most suitable concentration of selection agent. The ideal concentration of selection agent would be just able to prevent non-transgenic cells from developing shoots or plantlets. If the selective agent concentration is too high in the medium, it may kill all cells. In contrast, if it is too low, it may allow non-transgenic cells to escape from the selection.

7. The common particles used for plant biolistic transformation are tungsten and gold particles; however, gold particles are more suitable and stable for plant cells than tungsten particles [29]. Furthermore, 0.6 μm gold particles cause less damage to embryogenic callus than 1.0 μm gold particles. Therefore, 0.6 μm gold particles are more favorable to particle bombardment. The tungsten particles are not recommended.
8. Sorghum normally does not show any symptoms of calcium deficiency when grown in field conditions. However, plants could have calcium deficiency when grown in pots, especially in winter season. The symptoms vary on different sorghum genotypes. For example, Tx430 suffers less than SA281 from calcium deficiency. Typical symptoms are displayed as leaf splitting perpendicular to the lateral vein of leaves, and severe symptoms are observed when the leaf apical meristem begins to curl and die (necrosis).
9. The transgenic plantlets, which go through tissue culture and selection visibly from one spot of tissue, are considered as one transgenic event. In some case, if multiple shoots are regenerated and originate from different spots of one tissue, they can be separated and considered different transgenic events. Usually, transgenic plantlets grow vigorously and produce multiple shoots under selection.

Acknowledgments

We are thankful to the National Natural Science Foundation of China (31671753, 31301383). We really appreciate Dr. Yue Sun for her support during the editing process.

References

1. Gaj T, Gersbach CA, Barbas CF (2013) ZFN, TALEN, and CRISPR/Cas-based methods for genome engineering. Trends Biotechnol 31 (7):397–405
2. Xie KB, Yang YN (2013) RNA-guided genome editing in plants using a CRISPR-Cas system. Mol Plant 6(6):1975–1983
3. Liu X, Wu SR, Xu J, Sui C, Wei JH (2017) Application of CRISPR/Cas9 in plant biology. Acta Pharm Sin B 7(3):292–302
4. Xin ZG, Wang ML, Barkley NA, Burow G, Franks C, Pederson G et al (2008) Applying genotyping (TILLING) and phenotyping analyses to elucidate gene function in a chemically induced sorghum mutant population. BMC Plant Biol 8:103
5. Till BJ, Cooper J, Tai TH, Colowit P, Greene EA, Henikoff S et al (2007) Discovery of chemically induced mutations in rice by TILLING. BMC Plant Biol 7:19
6. Puchta H (2008) Broadening the applicability of CRISPR/Cas9 in plants. Sci China-Life Sci 61(1):126–127
7. Waltz E (2018) With a free pass, CRISPR-edited plants reach market in record time. Nat Biotechnol 36(1):6–7
8. Kim D, Alptekin B, Budak H (2018) CRISPR/Cas9 genome editing in wheat. Funct Integr Genomics 18(1):31–41.9
9. Srivastava V, Underwood JL, Zhao S (2017) Dual-targeting by CRISPR/Cas9 for precise

excision of transgenes from rice genome. Plant Cell Tissue Organ Cult 129(1):153–160
10. Jiang WZ, Zhou HB, Bi HH, Fromm M, Yang B, Weeks DP (2013) Demonstration of CRISPR/Cas9/sgRNA-mediated targeted gene modification in Arabidopsis, tobacco, sorghum and rice. Nucleic Acids Res 41(20):e188
11. Che P, Anand A, Wu E, Sander JD, Simon MK, Zhu W et al (2018) Developing a flexible, high-efficiency agrobacterium-mediated sorghum transformation system with broad application. Plant Biotechnol J 16(7):1388–1395. https://doi.org/10.1111/pbi.12879
12. Shan QW, Wang YP, Li J, Gao CX (2014) Genome editing in rice and wheat using the CRISPR/Cas system. Nat Protoc 9 (10):2395–2410
13. Agarwal A, Yadava P, Kumar K, Singh I, Kaul T, Pattanayak A et al (2018) Insights into maize genome editing via CRISPR/Cas9. Physiol Mol Biol Plants 24(2):175–183
14. LeBlanc C, Zhang F, Mendez J, Lozano Y, Chatpar K, Irish VF, Jacob Y (2018) Increased efficiency of targeted mutagenesis by CRISPR/Cas9 in plants using heat stress. Plant J 93 (2):377–386
15. Liu X, Xie C, Si H, Yang J (2017) CRISPR/Cas9-mediated genome editing in plants. Methods 121:94–102
16. Puchta H (2017) Applying CRISPR/Cas for genome engineering in plants: the best is yet to come. Curr Opin Plant Biol 36:1–8
17. Zhang YL, Ma XL, Xie XR, Liu YG (2017) CRISPR/Cas9-based genome editing in plants. In: Weeks DP, Yang B (eds) Gene editing in plants. Progress in molecular biology and translational science, vol 149. Elsevier Academic Press Inc, San Diego, pp 133–150
18. Zhang Y, Liang Z, Zong Y, Wang YP, Liu JX, Chen KL et al (2016) Efficient and transgene-free genome editing in wheat through transient expression of CRISPR/Cas9 DNA or RNA. Nat Commun 7:12617
19. Woo JW, Kim J, Il Kwon S, Corvalan C, Cho SW, Kim H et al (2015) DNA-free genome editing in plants with preassembled CRISPR-Cas9 ribonucleoproteins. Nat Biotechnol 33 (11):1162–1164
20. Ma X, Zhu Q, Chen Y, Liu Y-G (2016) CRISPR/Cas9 platforms for genome editing in plants: developments and applications. Mol Plant 9(7):961–974
21. Khan AA, El-Sayed A, Akbar A, Mangravita-Novo A, Bibi S, Afzal Z et al (2017) A highly efficient ligation-independent cloning system for CRISPR/Cas9 based genome editing in plants. Plant Methods 13:86
22. Liu GQ, Campbell BC, Godwin ID (2014) Sorghum genetic transformation by particle bombardment. In: Henry RJ, Furtado A (eds) Cereal genomics: methods and protocols. Methods in molecular biology, vol 1099. Springer, New York, Heidelberg Dordrecht London, pp 219–234
23. Liu GQ, Godwin ID (2012) Highly efficient sorghum transformation. Plant Cell Rep 1 (6):999–1007
24. Liu G, Gilding EK, Godwin ID (2015) A robust tissue culture system for sorghum [*Sorghum bicolor* (L.) Moench]. S Afr J Bot 98:157–160
25. Murashige T, Skoog F (1962) A revised medium for rapid growth and bioassays with tobacco tissue cultures. Physiol Plant 15 (3):473–497
26. Jiang W, Zhou H, Bi H, Fromm M, Yang B, Weeks DP (2013) Demonstration of CRISPR/Cas9/sgRNA-mediated targeted gene modification in Arabidopsis, tobacco, sorghum and rice. Nucleic Acids Res 41(20):e188
27. Liu H, Ding Y, Zhou Y, Jin W, Xie K, Chen LL (2017) CRISPR-P 2.0: an improved CRISPR-Cas9 tool for genome editing in plants. Mol Plant 10(3):530–532
28. Haeussler M, Schonig K, Eckert H, Eschstruth A, Mianne J, Renaud JB et al (2016) Evaluation of off-target and on-target scoring algorithms and integration into the guide RNA selection tool CRISPOR. Genome Biol 17(1):148
29. Russell JA, Roy MK, Sanford JC (1992) Physical trauma and tungsten toxicity reduce the efficiency of biolistic transformation. Plant Physiol 98(3):1050–1056

Chapter 13

Novel Ternary Vectors for Efficient Sorghum Transformation

Ajith Anand, Ping Che, Emily Wu, and Todd J. Jones

Abstract

Sorghum has been considered a recalcitrant crop for tissue culture and genetic transformation. A breakthrough in *Agrobacterium*-mediated sorghum transformation was achieved with the use of super-binary cointegrate vectors based on plasmid pSB1. However, even with pSB1, transformation capability was restricted to certain sorghum genotypes, excluding most of the important African sorghum varieties. We recently developed a ternary vector system incorporating the pVIR accessory plasmid. The ternary vector system not only doubled the transformation frequency (TF) in Tx430, but also extended the transformation capability into an important African sorghum elite variety.

Key words *Agrobacterium*, Vector, Ternary vector, Sorghum, Transformation

1 Introduction

Agricultural biotechnology has made important contributions to increased food production needed by the growing global population. However, a limited food supply and malnutrition are still a big challenge in sub-Saharan Africa and other developing countries. Sorghum is the fifth most important cereal crop in the world and the second largest grain crop produced on the African continent (http://www.faostat.fao.org). Enhancing sorghum grain yield would benefit the smallholder farmers and improve the food security in those areas.

Previously considered to be recalcitrant to transformation, many grain crops are now amenable to transformation using the preferred technology of *Agrobacterium*-mediated plant transformation (AMT) [1–4]. Sorghum has been one of the more recalcitrant crop for tissue culture and *Agrobacterium* infection for a long time until the implementation of the super-binary co-integrate vector (CIV), pSB1 [5, 6], for transformation [7]. Application of the pSB1 CIV (Fig. 1) broadened the host range of plants amenable to transformation with *Agrobacterium tumefaciens* making it the primary choice for cereal transformation [8, 9].

Zuo-Yu Zhao and Jeff Dahlberg (eds.), *Sorghum: Methods and Protocols*, Methods in Molecular Biology, vol. 1931,
https://doi.org/10.1007/978-1-4939-9039-9_13,

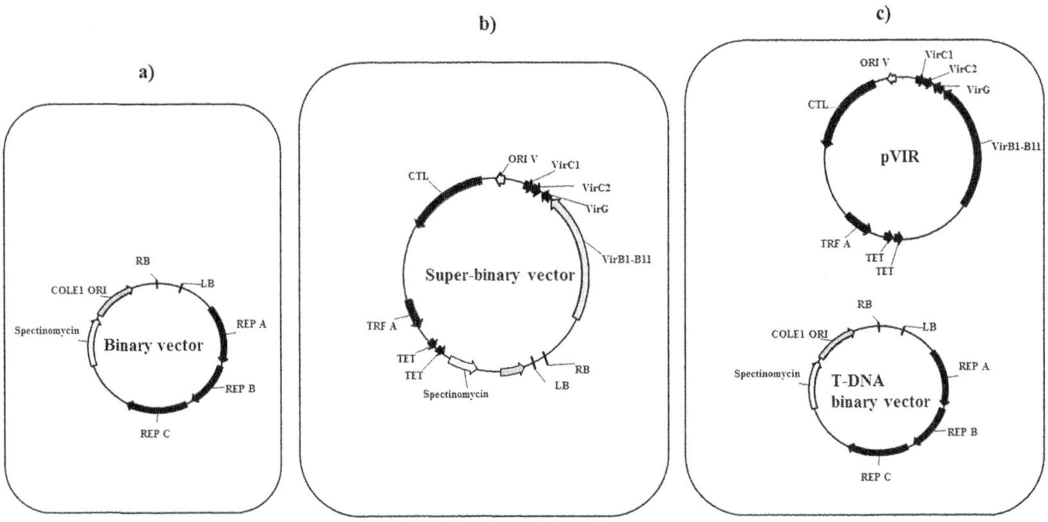

Fig. 1 Schematics of different vectors used for plant transformation, the cartoon depicts an *Agrobacterial* cell with the resident disarmed Ti plasmid. (**a**) A binary vector, (**b**) a super-binary vector, (**c**) a ternary vector

However, the large size of the pSB1 plasmid (~36.9 kb, Fig. 2a) and the co-integration steps necessary for introducing a gene-of-interest can be challenging, making plasmid structural validation difficult and limiting vector construction throughput. In addition, pSB1 relied on the less desirable selectable marker, *tetA* gene conferring resistance to tetracycline, that made the plasmid less stable since spontaneous mutations can arise in the tetracycline gene in certain C58-based *Agrobacterium* strains [10]. To simplify and streamline vector construction for AMT, a series of small and versatile "pVIR" plasmids [11] were designed that could be used as accessory plasmids as a replacement for the pSB1 plasmid. In addition, to overcome the limitations of the cumbersome co-integration steps and the large size of the super-binary vector pSB1, a ternary vector system has been devised. In this chapter, we describe the development and application of novel ternary vectors for sorghum transformation. This vector system has also been successfully used to transform an elite African sorghum variety Macia that could not be transformed with the pSB1 super-binary vector.

2 Materials and Methods

2.1 Vectors Used in Sorghum Transformation

Two ternary vectors containing one of the accessory plasmids (pPHP71539) and two different T-DNA plasmids and one super-binary vector were constructed. Various combinations were evaluated in sorghum transformation in this study. The rational, design, and construction of the ternary vectors are described in the

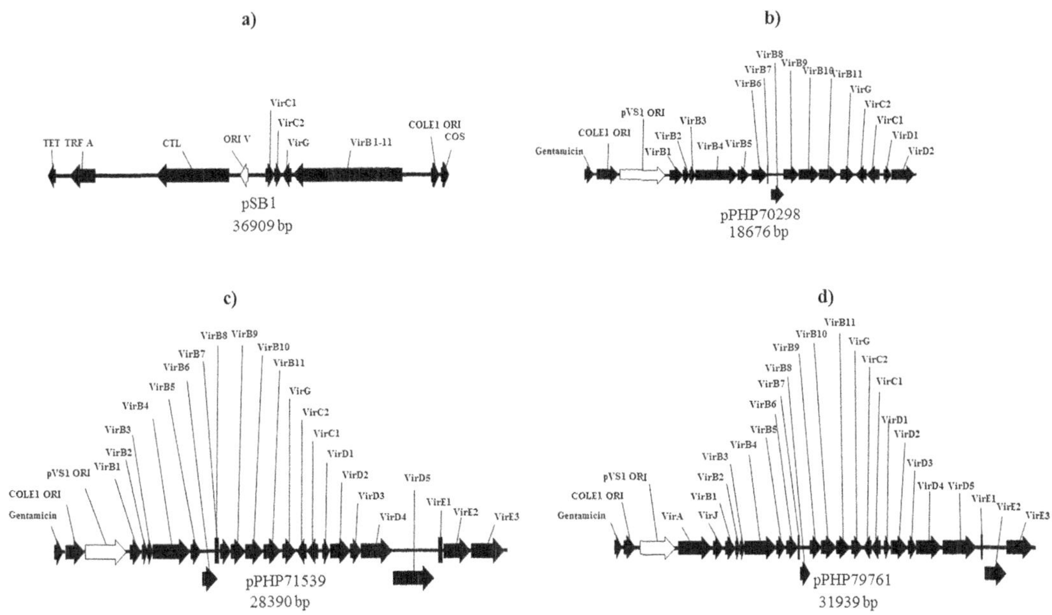

Fig. 2 Schematics of a super-binary vector (pSB1) and three accessory pVIR plasmids built to support the ternary vectors. (**a**) The super-binary vecotr pSB1 (36.9 kb), (**b**) pPHP70298 (18.7 kb), (**c**) pPHP71539 (28.4 kb), and (**d**) pPHP79761 (31.9 kb)

Subheading 3 below. The protocol for vector construction was reported in detail previously [11, 12].

1. Three improved accessory pVIR plasmids, pPHP70298, pPHP71539, and pPHP79761 (Fig. 2), were constructed as a substitute for pSB1.
2. A ternary vector containing accessory plasmid pPHP71539 and T-DNA plasmid pPHP45981 (7.7 kb) (Fig. 3) was constructed in *Agrobacterium* strain LBA4404 THY- (an auxotrophic strain) [1] and used in trasient gene expression studies to evaluate T-DNA delivery. The T-DNA region contained the *Ubiquitin*-driven phosphomannose isomerase (*PMI*) gene as a selectable marker and yellow fluorescent protein (*YFP*) as a reporter marker.
3. Two ternary vectors containing accessory plasmid pPHP71539 and either T-DNA binary vector pPHP78152 or pPHP78233 were constructed, introduced into *Agrobacterium* strain LBA4404 Thy-, and used in sorghum stable transformation. The T-DNA region (17 kb) in pPHP78152 harbors *PMI* and three proprietary genes and the T-DNA region (18 kb) in pPHP78233 harbors *PMI* and four proprietary genes. Both T-DNA plasmids were the derivatives of pHPH82637 (Fig. 3c).

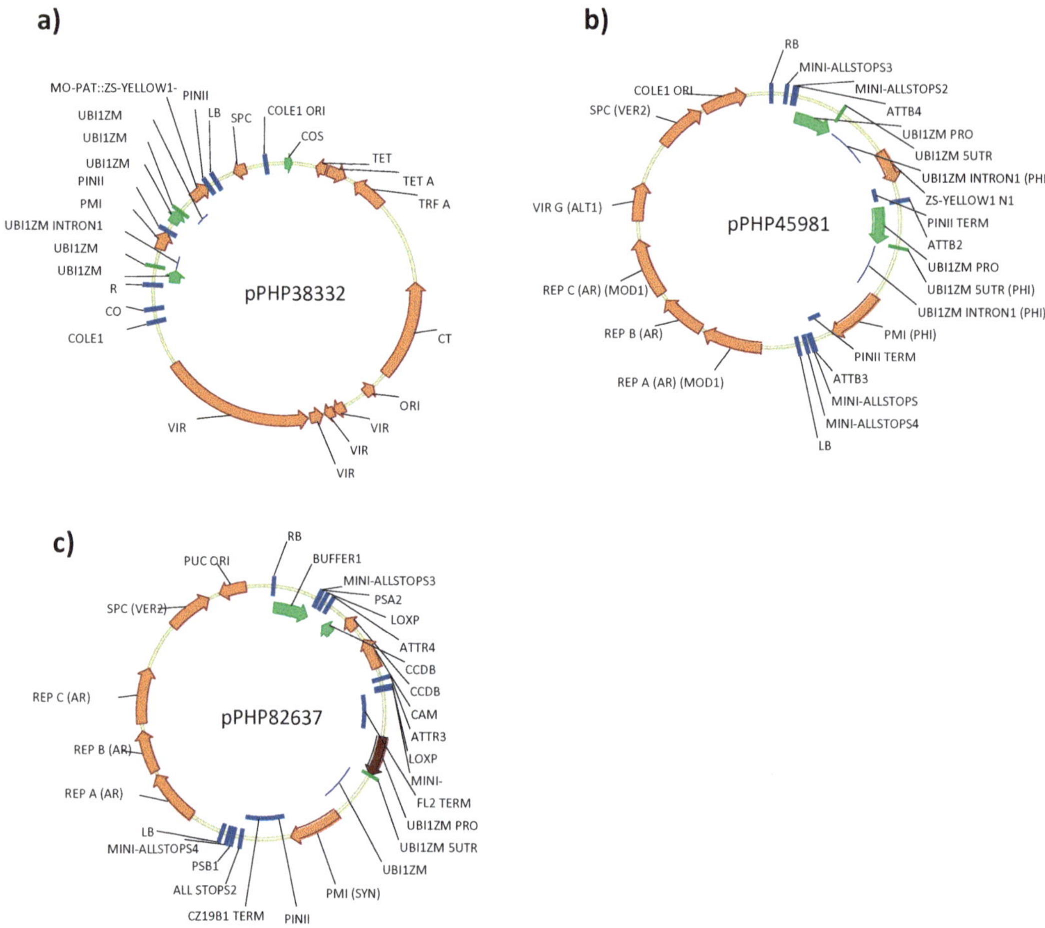

Fig. 3 Plasmid maps of the super-binary vector and the ternary vectors used for sorghum transformation. (**a**) The super-binary vector pPHP38332, (**b**) the T-DNA binary plasmid pPHP45981, and (**c**) the T-DNA binary plasmid pPHP82637

4. A super-binary vector pPHP38332 with 4.6 kb T-DNA carrying *PMI* and *YFP* was also constructed into *Agrobacterium* strain LBA4404 THY- (Fig. 3a) as the control vector in this study.

2.2 *Agrobacterium*-Mediated Sorghum Transformation

Immature embryos (IE) of sorghum inbred line Tx430 and an African elite sorghum variety, Macia, were used for transformation. The transformation protocol used was described previously in detail [2, 13]. Recently, we confirmed that the application of surfactant Silwet-70 during *Agrobacterium* infection increased TF significantly in sorghum [1]. Therefore, 0.005% Silwet-70 was applied to all transformation experiments in this study.

2.3 Molecular Analysis of Transgenic Plants

1. Next-generation sequencing [14, 15] was used to sequence all the vectors prior to using in sorghum transformation.

2. All the transgenic plants were molecularly characterized using a combination of qPCR and PCR assays as reported previously [2, 8]. The assays included estimating the copy number of all transgenes, verifying intact or truncated T-DNA integration, and identifying the presence or absence of any vector backbone sequences in the sorghum genome.

3 Results and Discussion

3.1 Design of the Ternary Vectors

A ternary vector is consisted of three plasmids; (1) a disarmed Ti plasmid, (2) a T-DNA binary plasmid, and (3) an accessory pVIR plasmid [11] (Fig. 1). The incorporation of the accessory plasmid with additional *vir* genes that function in trans with the T-DNA plasmid (Fig. 1c) simplified vector design and construction.

In the early developed vectors for plant transformation, the disarmed Ti plasmid was used to introduce transgenes into the T-DNA region through homologous recombination [16–18]. The complexity of plasmid construction and the structural characterization of the large CIV was simplified through the development of a binary system, where the gene of interest flanked by the T-DNA borders resides on a smaller episome separated from the virulence genes-encoded a disarmed mutant Ti plasmid (Fig. 1a) [19, 20]. The super-binary CIV (pSB1) (Fig. 1b) is an improved version of the binary vector with additional *vir* genes from the Ti plasmid pTiBo542 (Fig. 2a) [5]. However, the large size of the pSB1 plasmid (~37 kb) and the co-integration step has limited the range of strains used for plant transformation, complicated plasmid structural validation, and constrained vector construction throughput. A ternary vector system consists of the disarmed Ti plasmid, a binary T-DNA plasmid, and an accessory plasmid with additional *vir* genes (Fig. 1c) [12]. Ternary vectors harboring a plasmid containing a constitutive *virG* mutant (*vir*GN54D) have previously been described that efficiently transfer T-DNA in dicot plant species [21–23].

The plasmid pSB1 has been used successfully as an accessory plasmid for maize and sorghum transformation, but it worked relatively inefficiently [1, 11, 12]. The work described here improves upon the ternary vector concept for cereal crops with the creation of new pVIR accessory plasmids.

Our accessory pVIR plasmids have many desirable features, such as additional, functional *vir* genes (operons *vir*C, *vir*D, and *vir*E), an improved bacterial selectable marker, enhanced vector stability, and compatibility with Gateway™ cloning technology (Invitrogen) [11, 12]. Three accessory pVIR plasmids, named pPHP70298, pPHP71539, and pPHP79761 (Fig. 2), were constructed. These three accessory plasmids contained a selected set of *vir* genes including *virA*, *virB*, *virC*, *virD*, *virE*, *virG*, and *virJ* from the hypervirulent pTiBo542 plasmid. Design of the pVIR plasmids was based on the open source super-binary vector pCAMBIA5105 (http://www.cambia.org/daisy/cambia/585). The respective sequences of the plasmids can be found under GenBank accession numbers MF788072, MF788073, and MF788074. These plasmids were designed to be smaller (~2.6 kb), more stable [23], and, with the pVS1 replicon, high copy (~20 copies) [24]. A more desirable selectable marker gentamicin (GmR) was introduced and the plasmids were designed with compatibility to support Gateway™ cloning [12].

Plasmid pPHP70298 (Fig. 2b) is an improved version of the super-binary pSB1 and the improved features include fixing functionality in some of the virulence genes, replacing the large size replicon (RK2 ORI) with a smaller replicon (pVS1), removing unnecessary DNA sequences (noncore DNA elements between operons), and excluding the T-DNA region. With all these modifications, plasmid pPHP70298 is only half the size (18.7 kb) of the original pSB1 plasmid (36.9 kb).

Plasmid pPHP71539 (28.4 kb, Fig. 2c) is based on pPHP70298 but contains additional *vir* genes including the complete *vir*D operon (*vir*D1-D5) and *vir*E operon (*vir*E1–3), which are both lacking in pPHP70298.

Plasmid pPHP79761 (31.9 kb, Fig. 2d) is also based on pPHP70298 but contains an expanded set of optimal *vir* operons including *vir*A, *vir*B, *vir*C, *vir*D, *vir*E, *vir*G, and *vir*J, which are suggested to be important for AMT [25]. The arrangement of the *vir* genes in the pVIR plasmids contains only the *vir* coding sequences (with intervening noncoding sequences omitted) from a 31.5 kb fragment present in pTiBo542 plasmid (GenBank Accession No. NC_010929.1). Even with these added features, these three pVIR plasmids are all smaller than pSB1 (Fig. 2). Furthermore, these pVIR plasmids were found to be stably maintained in multiple bacterial backgrounds including different *Agrobacterium* strains [11, 12] which is an important criterion for plasmids used in plant transformation.

A ternary vector was assembled by first mobilizing the accessory plasmid, such as pPHP71539, into an *Agrobacterium* auxotrophic (LBA4404 Thy-) strain and selected on media containing 25 mg/l gentamicin. Subsequently, a T-DNA plasmid with spectinomycin bacterial selectable marker was electroporated into the *Agrobacterium* strain containing the accessory plasmid pPHP71539. Finally, recombinant colonies were selected on media supplemented with both gentamicin and spectinomycin. All constructs used in transformation were first sequenced by next-generation sequencing to confirm accurate vector construction [14, 15]. The ternary design offers simple and versatile vector assembly requiring no co-integration step, which can occasionally result in loss of vector instability, as well as complicating vector assembly and mobilization into multiple *Agrobacterium* strains.

The choice of an accessory plasmid is critical to crop transformation. To elucidate the advantage of the ternary vector for T-DNA delivery in sorghum transformation, we constructed a ternary vector containing the accessory plasmid pPHP71539 and T-DNA plasmid pPHP45981 (7.7 kb) (Fig. 3b) and compared it with a super-binary vector pPHP38332 (T-DNA 4.6 kb) (Fig. 3a). These two plasmids carry a T-DNA containing the phosphomannose isomerase (*PMI*) gene as the selectable marker and yellow fluorescent protein (*YFP*) as the reporter marker gene. They were electroporated into the *Agrobacterium* strain LBA4404 THY- [1]. The two vectors were used to evaluate transient T-DNA delivery into sorghum IE by monitoring *YFP* expression. In addition, two ternary vectors containing T-DNA plasmids either pPHP78152 carrying *PMI* and three proprietary genes (17 kb) or pPHP78233 carrying *PMI* and four proprietary genes (18 kb) plus the accessory pPHP71539 were constructed in *Agrobacterium* strain LBA4404 THY- for stable transformation and comparison to the super-binary vector pPHP38332. Both of these T-DNA plasmids in the ternary vectors were derivatives of pHPH82637 (Fig. 3c).

3.2 Ternary Vector Increased T-DNA Delivery in Sorghum Embryos

The ternary vector carrying pPHP71539 and pPHP45981 and the super-binary vector pPHP38332 were used to infect IEs of sorghum line Tx430. After 3-day of *Agrobacterium* co-cultivation, the embryos were visually evaluated under a fluorescent microscope for YFP expression. The IEs infected with the ternary vector showed both greater number of YFP spots and much brighter YFP expressions compared to IEs infected with the super-binary vector (Fig. 4). This result demonstrated that the ternary vector is superior for transient T-DNA delivery compared to the pPHP38332 super-binary vector in sorghum.

This observation was consistent with data in maize (Fig. 5) [12]. The results of the ternary vector system-mediated T-DNA delivery in both maize and sorghum suggested the *vir* genes located on the accessory plasmid were functional and could be useful to transform recalcitrant crop species or varieties that were not transformable with super-binary vectors.

3.3 Ternary Vector Enhanced Stable Transformation Frequency in Tx430

Based on the above findings, we investigated whether improved T-DNA delivery with the ternary vector could result in enhanced stable transformation in sorghum Tx430. To test this hypothesis, two ternary vectors pPHP78152 and pPHP78233 with T-DNA sizes of 17 kb and 18 kb respectively were used in transformation experiments and compared to super-binary vector pPHP38332 (T-DNA 4.6 kb) as the control. Following molecular assays of the transgenic events, the TF of these two ternary constructs were 29.4% and 24.8%, respectively while the control super-binary vector

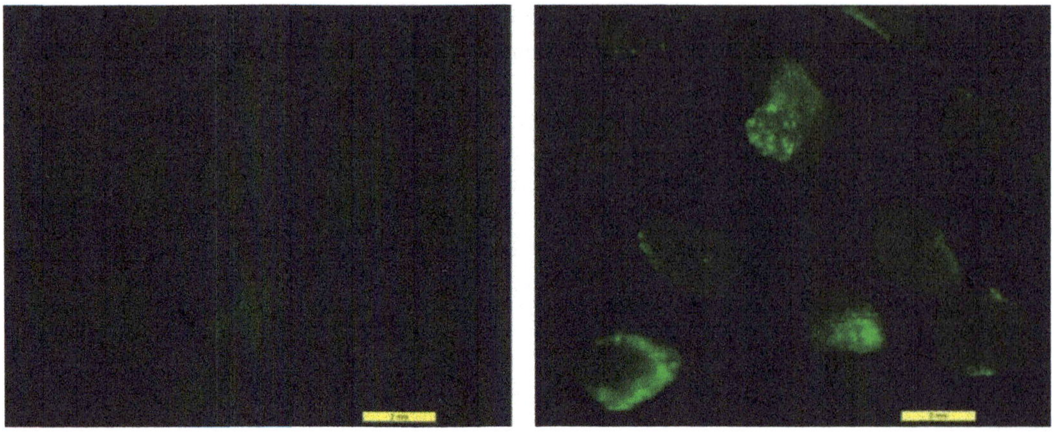

Fig. 4 Transient expression of YFP in Tx430 IE infected by the super-binary vector (pPHP38332) on the left and by the ternary vector (pPHP45981 with pPHP71539) on the right

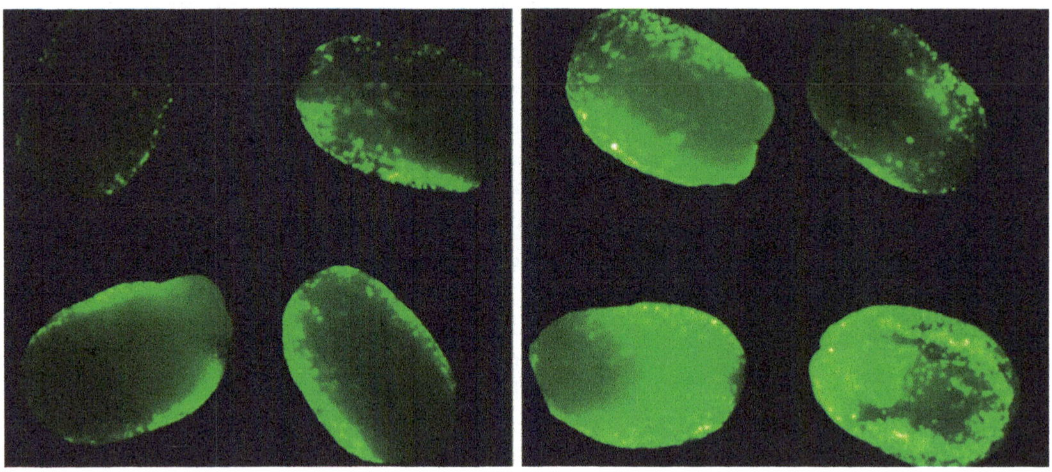

Fig. 5 Transient expression of YFP in maize IE infected by the super-binary vector (pPHP38332) on the left and by the ternary vector (pPHP45981 with pPHP71539) on the right

was 15.1% (Table 1). Even though the T-DNA size of these two ternary vectors is about 3–4 times larger than the T-DNA size of the super-binary vector, the TF of the ternary vectors were still increased 64–95% compared to the super-binary vector.

In addition to the TF, the event quality of the transgenic plants is another important factor to consider. Quality events (QE) have been defined as being free of plasmid vector backbone sequence outside the T-DNA and intact single copy integration of all the transgenes within the T-DNA region into plant genome [2, 8, 9]. Finally, the overall transformation efficiency, called the Usable Event (UE) frequency, was calculated as TF% times QE%. UE% represents a fair measurement of evaluating the final transformation efficiency when comparing *Agrobacterium* strains, vector systems,

Table 1
Transformation efficiencies of the ternary vectors and the super-binary vector in Tx430

Construct (T-DNA size)	# IEs infected	# IEs transformed[a]	TF[b]	QE%	UE%
pPHP78152 (17 kb)	350	103	29.4%	54%	15.9%
pPHP78233 (18 kb)	400	99	24.8%	65%	16.1%
pPHP38332 (4.6 kb)	298	45	15.1%	66%	10.0%

[a]The IEs producing T0 transgenic plants confirmed molecularly are considered as IEs transformed
[b]Stable transformation% was calculated after molecular conformation of T-DNA integration into sorghum genome at T0 plant generation

or other technology related to plant transformation. To assess event quality, all T0 plants generated using super-binary vector pPHP38332 and ternary vector pPHP78152 or pPHP78233 were subjected to detailed molecular characterization to determine the proportion of the QE in each group. Through the analyses of events generated with these two ternary vectors, the QE percentage was 54% and 65%, respectively, while the final UE were 15.9% and 16.1%. This is compared to a QE percentage of 66% and a UE of 10% for events derived from the super-binary vector, pPHP38332 (Table 1). These data demonstrated that the transformation efficiency of these two ternary vectors was approximately 60% better than the super-binary vector in Tx430 transformation, even though the T-DNA sizes in these ternary vectors were 3–4 times in the size of the T-DNA in the super-binary vector.

Previously, we have observed a negative correlation between TF and event quality in sorghum [2] and maize [8] with higher transformation frequencies producing more complex transgene integration events. Complex transgene integrations can contribute to variability in gene expression between sister events [25, 26]. On the contrary, consistent transgene expression has been highly correlated with transformation events of high molecular quality [27]. The two ternary vectors showed similar QE% when compared to the super-binary vector.

3.4 Ternary Vector Extended Transformation Capability to an African Sorghum Elite Variety

Macia is an early-maturing sorghum variety that was released in the 1990s in several African countries. It has been very popular for smallholder farmers in Africa due to its many useful characteristics—short plant height that is both convenient and lodges less, large panicle size, high yield, low dehulling losses, and non-tannin grains without a pigmented testa for good food quality. Macia also has a staygreen characteristic and therefore the residues are suitable for feeding farm livestock. In past years, attempts to transform Macia were unsuccessful. Recently, successful transformation of

Macia has been reported with the use of the pVIR ternary vectors [1].

The above two ternary vectors—pPHP78152 and pPHP78233 as well as the super-binary vector pPHP38332 were used in Macia transformation. As expected, no stable transgenic events were derived from pPHP38332, while three and seven transgenic events were generated from 440 pPHP78152-infected IEs and 521 pPHP78233-infected IEs, respectively. The Macia TF of 0.7% and 1.3% with these two ternary vectors were still low, but this was the first time genetically transformed plants and seeds have been produced in this very recalcitrant African sorghum variety.

The ternary vectors with accessory plasmid pPHP70298 and pPHP79761 have been evaluated in maize [11, 12], and currently are under evaluation in sorghum.

From the data in this study, it is confirmed that the *vir* genes located on the accessory plasmid in the ternary system function as efficiently in trans as in cis in sorghum transformation and with high probability of success in other crops too. In conclusion, through the data presented here and data published from other related research [1, 11, 12], the pVIR ternary vector system has several benefits for crop transformation: (1) simplification of the vector construction process, (2) better plasmid stability in bacterium cells, (3) a smaller plasmid size and stable replicon allowing the introduction of large size T-DNA (17–18 kb), (4) enhanced T-DNA delivery, (5) increase in stable transgenic efficiency, and (6) a broadened transformation capability to more recalcitrant genotypes. This vector system ultimately improved vector construction throughput and resulted in improved transformation efficiencies in both corn [11] and sorghum [1] as well as other crops.

Acknowledgments

The work described here was supported by the Applied Science and Technology department at Corteva Agriscience™, Agriculture Division of DowDuPont™. We are grateful to the assistance received from multiple teams including Vector Construction, Crop Genome Engineering, Controlled Environments, Genomics and Nucleic Acid Analysis. The leadership support received from Scott Betts and Doane Chilcoat is acknowledged.

Reference

1. Che P, Anand A, Wu E, Sander JD, Simon MK, Zhu W et al (2018) Developing a flexible, high-efficiency *Agrobacterium*-mediated sorghum transformation system with broad application. Plant Biotechnol J 16(7):1388–1395 https://doi.org/10.1111/pbi.128792
2. Wu E, Lenderts B, Glassman K, Berezowska-Kaniewska M, Christensen H, Asmus T et al (2014) Optimized *Agrobacterium*-mediated sorghum transformation protocol and molecular data of transgenic sorghum plants. In Vitro Cell Dev Biol Plant 50:9–18
3. Que Q, Elumalai S, Li X, Zhong H, Nalapalli S, Schweiner M et al (2014) Maize transformation technology development for

commercial event generation. Front Plant Sci 5:379

4. Jackson MA, Anderson DJ, Birch RG (2013) Comparison of *Agrobacterium* and particle bombardment using whole plasmid or minimal cassette for production of high-expressing, low-copy transgenic plants. Transgenic Res 22:143–151
5. Komari T (1990) Transformation of cultured cells of *Chenopodium quinoa* by binary vectors that carry a fragment of DNA from the virulence region of pTiBo542. Plant Cell Rep 9:303–306
6. Komari T, Hiei Y, Saito Y, Murai N, Kumashiro T (1996) Vectors carrying two separate T-DNAs for co-transformation of higher plants mediated by agrobacterium tumefaciens and segregation of transformants free from selection markers. Plant J 10:165–174
7. Zhao ZY, Cai T, Tagliani L, Miller M, Wang N, Pang H et al (2000) *Agrobacterium* mediated sorghum transformation. Plant Mol Biol 44:789–798
8. Zhi L, TeRonde S, Meyer S, Arling ML, Register JC III, Zhao ZY et al (2015) Effect of *Agrobacterium* strain and plasmid copy number on transformation frequency, event quality and usable event quality in an elite maize cultivar. Plant Cell Rep 34:745–754
9. Cho M-J, Wu E, Kwan J, Yu M, Banh J, Linn W et al (2014) *Agrobacterium*-mediated high-frequency transformation of an elite commercial maize (*Zea mays* L.) inbred line. Plant Cell Rep 33:1767–1777
10. Luo Z-Q, Farrand SK (1999) Cloning and characterization of a tetracycline resistance determinant present in *Agrobacterium tumefaciens* C58. J Bacteriol 181:618–626
11. Anand A, Bass SH, Cho HJ, Klein TM, Lassner M, McBride KE (2017) Methods and compositions of improved plant transformation. WO2017078836 A1: Patent publication
12. Anand A, Bass SH, Wu E, Wang N, McBride KE, Annaluru N et al (2018) An improved ternary vector system for *Agrobacterium*-mediated rapid maize transformation. Plant Mol Biol. https://doi.org/10.1007/s11103-018-0732-y
13. Wu E, Zhao ZY (2017) *Agrobacterium*-mediated sorghum transformation. In: Schmidt A (ed) Plant Germline development, methods and protocols, methods in molecular biology, vol 1669. Humana Press, New York, pp 355–364
14. Ozsolak F (2012) Third-generation sequencing techniques and applications to drug discovery. Expert Opin Drug Discov 7:231–243
15. Schhuster SC (2008) Next generation sequencing transforms today's biology. Nat Methods 5:16–18. https://doi.org/10.1038/nmeth1156
16. Fraley RT, Rogers SG, Horsch RB, Eichholtz DA, Flick JS, Fink CL et al (1985) The SEV system: a new disarmed Ti plasmid vector system for plant transformation. Bio/Technol 3:629–635
17. Fraley RT, Rogers SG, Horsch RB, Sanders PR, Flick JS, Adams SP et al (1983) Expression of bacterial genes in plant cells. Proc Natl Acad Sci U S A 80:4803–4807
18. Zambryski P, Joos PH, Genetello C, Leemans J, Van Montagu M, Schell J (1983) Ti plasmid vector for the introduction of DNA into plant cells without alteration of their normal regeneration capacity. EMBO J 2:2143–2150
19. Bevan M (1984) Binary agrobacterium vectors for plant transformation. Nuc Acids Res 12:8711–8721
20. Hoekema A, Hirsh PR, Hooykaas PJJ, Schilperoort RA (1983) A binary plant vector strategy based on separation of vir-and T-region of the *Agrobacterium tumefaciens* Ti-plasmid. Nature 303:179–180
21. Kikuchi R, Sage-Ono K, Kamada H, Ono M (2005) Efficient transformation mediated by agrobacterium tumefaciens with a ternary plasmid in *Pharbitis* nil. Plant Biotech J 22:295–302
22. van der Fits L, Deakin EA, Hoge JHC, Memelink J (2000) The ternary transformation system: constitutive virG on a compatible plasmid dramatically increases agrobacterium-mediated plant transformation. Plant MolBiol 43:495–502
23. Nester EW (2014) *Agrobacterium*: nature's genetic engineer. Front Plant Sci 5:730
24. Stanisich VA, Bennett PM, Richmond MH (1997) Characterization of a translocation unit encoding resistance to mercuric ions that occurs on a nonconjugative plasmid in *Pseudomonas aeruginosa*. J Bacteriol 129:1227–1233
25. Depicker A, Sanders M, Meyer P (2005) Transgene silencing. In: Meyer P (ed) Plant

epigenetics. Blackwell Publishing Ltd, Oxford, pp 1–31

26. Srivastava V, Ariza-Nieto M, Wilson AJ (2004) Cre-mediated site-specific gene integration for consistent transgene expression in rice. Plant Biotechnol J 2:169–179

27. De Buck S, Windels P, De Loose M, Depicker A (2004) Single-copy T-DNAs integrated at different positions in the Arabidopsis genome display uniform and comparable β-glucuronidase accumulation levels. Cell Mol Life Sci 61:2632–2645

Chapter 14

Nutritionally Enhanced Sorghum for the Arid and Semiarid Tropical Areas of Africa

Zuo-Yu Zhao, Ping Che, Kimberly Glassman, and Marc Albertsen

Abstract

To help alleviate malnutrition in Africa, nutritionally enhanced sorghum was developed through genetic transformation to increase pro-vitamin A (β-carotene) accumulation and stability, to improve iron and zinc bioavailability, and to improve protein digestibility. Through many years of efforts, significant achievements have been made for these goals. We generated nutritionally enhanced sorghum lines with enhanced and stabilized pro-vitamin A that provide 20–90% of the Estimated Average Requirement (EAR) for children under age 3, lines with a 90% reduction in phytate that increase iron and zinc bioavailability and provide 40–80% of the EAR for iron and zinc, and lines that show no reduction in protein digestibility after cooking compared with normal levels. Once these nutritionally enhanced sorghum lines have undergone biosafety examination and have been deregulated, they will be ready for incorporation into sorghum varieties that will benefit Africa and other areas that rely upon sorghum as a staple food.

Key words Sorghum, Nutrition, Nutrition enhancement, Transgenic sorghum, Enhancing pre-vitamin A, Fe & Zn bioavailability, Phytate reduction, Protein digestibility

1 Introduction

Numerous studies report that millions of people in sub-Saharan Africa suffer from health problems associated with poor nutrition, including impaired immune systems, blindness, and impaired neuropsychological development [1–5]. The most common and clinically significant micronutrient deficiencies in children and childbearing women throughout the world include deficiencies of iron, iodine, zinc, and vitamin A that are estimated to affect as many as two billion people. These four micronutrients are particularly vital for good nutrition and human development, especially among children. Vitamin A Deficiency is a leading cause of preventable blindness in children and increases the risk of severe infections. It can lead to the impairment of the immune system and contributes to the deaths of more than half a million African children annually. Iron deficiency contributes to the deaths of young women during

Zuo-Yu Zhao and Jeff Dahlberg (eds.), *Sorghum: Methods and Protocols*, Methods in Molecular Biology, vol. 1931, https://doi.org/10.1007/978-1-4939-9039-9_14, © Springer Science+Business Media, LLC, part of Springer Nature 2019

pregnancy and childbirth and is a leading cause of anemia. Sub-Saharan Africa has the highest prevalence of anemia among preschool-age children and among both pregnant and non-pregnant women. Although zinc deficiency has proven difficult to quantify and the data are incomplete, sub-Saharan African countries are considered to be among the highest for risk of zinc deficiency [5]. It is estimated that in Africa, 50% of children have a calcium, iron, and zinc deficiency [1]. In addition to micronutrient deficiency, Protein Energy Malnutrition (PEM) in Africa is also an issue [1].

The link between micronutrient deficiency and food security illustrates the challenges in addressing food issues in sub-Saharan Africa. In many African countries, the common diets lack diversity. Traditional foods consist mainly of cereal or root staples with very little micronutrient-rich or animal-source proteins, vegetables, or fruits [6]. Sorghum (*Sorghum bicolor* L) is the sixth most planted crop in the world and it is one of the most important staple foods for an estimated 500 million people, primarily those living in arid and semiarid areas. In Africa, it is the second most important cereal; about 300 million people rely on it as their daily staple food. Although sorghum provides energy in the form of starch, sorghum is considered a nutrient-poor crop [7, 8]. Nutritionally, (1) it is virtually devoid of pro-vitamin A (β-carotene) [9], (2) although it contains about 26–61 mg iron and 21–55 mg zinc per kg grains, most of those micronutrients are biologically unavailable because of chelation with phytate in the sorghum seed, (3) although sorghum seeds containing 8.1–16.8% protein [10], most (~80%) are kafirins, which are lacking in lysine, tryptophan and threonine, and (4) cooked sorghum flour increases protein indigestibility because of disulfide-bridges formed in the kafirins after cooking.

The Grand Challenges in Global Health Initiative was launched in 2003 by the Bill & Melinda Gates Foundation, in partnership with the National Institutes of Health, to harness the power of science and technology to dramatically improve health in the world's poorest countries. The initiative supported groundbreaking research projects to discover and develop scientific breakthroughs for preventing, treating, and curing diseases that kill millions of people each year in developing countries. On June 28, 2005, the Grand Challenges announced funding 43 groundbreaking research projects and the project of "*Nutritionally Enhanced Sorghum for the Arid and Semi-Arid Tropical Areas of Africa*" was one of these 43 funded projects. The African-based organization Africa Harvest Biotech Foundation International was the recipient of this funding and DuPont Pioneer was the scientific lead on this project. DuPont Pioneer donated the initial technology at the beginning of this project.

2 The Targets of Nutrition Improvements in the Nutritionally Enhanced Sorghum Project

The nutrition improvements of the Nutritionally Enhanced Sorghum project have been focused on: (1) enhancing pre-vitamin A to provide at least 20–50% EAR for children under three, (2) increasing Fe (50%) and Zn (35%) bioavailability by reducing phytate up to 80% without disruption of seed germination, and (3) improving protein digestibility—with no digestibility reduction after cooking.

3 Materials and Methods

3.1 Vector Construction and Sorghum Transformation

Sorghum inbred Tx430 was used in this study. Vector construction and *Agrobacterium*-mediated sorghum transformation have been described in detail previously [11, 12]. In addition to wild type (WT) sorghum, null seeds also were used as another control. The null is a non-transgenic line that is derived from the segregation of hemizygous transgenic plants. Null plants and/or seeds have been widely used as controls to evaluate the possible impact of the transgene(s) on the transgenic plants and/or seeds [13].

3.2 Enhancement and Stabilization of Pro-vitamin A in Sorghum Grains

To reach the target goal of pro-vitamin A enhancement, several genes related to carotenoid biosynthesis and stabilization were tested in transgenic sorghum. These genes included maize *PSY1*, *Pantoea ananatis CRTI*, *Arabidopsis At-DXS*, and barley *HGGT*. The methods related to evaluating all-trans-β-carotene level and stability during room temperature storage in the transgenic lines were described by Che et al. in 2016 [14]. Leading events selection through agronomic performance analysis of ABS203 transgenic lines in the confined fields trials is described in Chapter 15 of this book [15].

3.3 Phytate Reduction and Increase of Iron and Zinc Bioavailability in Sorghum Grains

To prove the concept of increasing Fe and Zn bioavailability through phytate reduction in sorghum, we conducted rat pup absorption experiments with in vitro phytate-reduced sorghum flour. Sorghum grain from Tx430 were ground into flour (~1 mm particles). A portion of this flour was incubated with the phytase enzyme to reduce its phytate contents in vitro. Another portion of the flour followed the treatment process but without adding phytase and became the treatment process control. A third portion of the flour did not follow the treatment process and became the unprocessed control sample. The levels of phytate reduction in the treated samples were estimated through measuring phytate and free phosphorus contents in the flour by chromatography [16]. 100% phytate reduction was determined when nearly all the phytate was degraded in the samples. At this point, the phytase enzyme was deactivated by heating at 95°C for 2 h. This 100% phytate-reduced sample then was used to mix with different proportions of the untreated flour (0% phytate reduction sample) to create 80%, 40%, and 20% phytate-reduced sorghum flour samples. These samples were assayed by chromatography again to determine their true phytate and phosphorus contents. These samples subsequently were used in rat pup experiments to estimate Fe and Zn absorption.

The protocol for rat pup absorption experiment was described previously [17–19]. Following depriving of food for 6 h, groups of 14-day-old rat pups ($n = 6$ in each group) were gastric intubated with these sorghum samples extrinsically labeled with either Zn^{65} or Fe^{59}. The positive controls were either 12.5 μM Zn or 12.5 μM Fe solution and the negative controls were 0.5 mM phytate solution. Because there was a cooking process involved in preparing the test meals, there were three other diet controls. These included an uncooked sorghum sample with 100% phytate reduction, an uncooked sorghum sample with 0% phytate reduction, and an unprocessed sorghum sample (no phytase treatment process described above) with 0% phytate reduction. Calculations of radioactivity to estimate Fe or Zn absorption were described previously [19].

3.4 Protein Digestibility Improvement

To improve sorghum protein digestibility, the gene expression of γ-kafirin-1 and -2 and σ-kafirin in sorghum were suppressed using RNAi technology in transgenic lines. The methods of in vitro protein digestibility (IVPD) assay, grain texture observation, and protein nutritional quality analysis were described by da Silva et al. [20].

The protein nutritional quality was evaluated by both the Amino Acid Score (AAS) and Protein Digestibility Corrected Amino Acid Score (PDCAAS). The AAS was calculated as g lysine (limiting essential amino acid)/100 g protein of the sorghum sample/4.8; where 4.8 g lysine/100 g protein is the recommendation for quality protein for 4–18 year olds [21]. The PDCAAS was determined by correcting the AAS by multiplying the wet cooked IVPD values obtained [20].

4 Results and Discussion

The target traits of our nutritionally enhanced sorghum have been focused on the three areas, namely: pro-vitamin A (β-carotene), Fe and Zn bioavailability and protein digestibility. The research strategy was primarily focused on individual traits to identify the molecular elements and to develop the technologies for each trait.

Agrobacterium-mediated sorghum transformation was used to generate genetically-modified sorghum varieties. It has been the essential tool for developing nutritionally enhanced sorghum. The transformation efficiency and the quality of transgenic events played a critical role. Sorghum was first successfully transformed by *Agrobacterium* in 2000 [11] at a relative low frequency (1–2%). Since then transformation technology has been improved continuously, and the transformation efficiency has dramatically jumped to 33% [12] in the sorghum inbred Tx430. The frequency of usable events has improved from 10% up to 16% when a ternary vector transformation system was adopted in sorghum transformation [22, 23]. The significant improvement of sorghum transformation technology provided a solid tool for the success of the biofortified sorghum.

4.1 Pro-vitamin A Enhanced and Stabilized in Sorghum Seeds

The success of pro-vitamin A (beta-carotene) enhancement in cereals began with Golden Rice [24, 25]. We have developed β-carotene-enhanced sorghum that will alleviate vitamin A deficiency among people who rely upon sorghum as their staple diet. Subsequent β-carotene instability during storage, however, negatively impacts full utilization of this essential micronutrient. We determined that oxidation is the main factor causing β-carotene degradation under ambient conditions. We further demonstrated that co-expression of *HGGT* (*homogentisate geranylgeranyl transferase*), stacked with carotenoid biosynthesis genes (*PHY1* and *CRTI*), can mitigate β-carotene oxidative degradation resulting in increased β-carotene accumulation and stability. A kinetic study of β-carotene degradation showed that the half-life of β-carotene is extended from less than 4 weeks to 10 weeks on average with *HGGT* co-expression. Based on the β-carotene bioconversion rate (4.3 μg β-carotene to 1 μg retinol) determined using Mongolian gerbils as an animal model and based on the stability of β-carotene determined in this research, we estimated that fresh harvest β-carotene levels from one of our nutritionally enhanced sorghum lines, ABS203, (Fig. 1) would provide 90% of EAR of vitamin A for children under age 3 and would still provide 20% of EAR after 6 months of seed storage. The detailed description of this nutritionally enhanced sorghum was published in 2016 by Che et al. [14]. Through a series of evaluations of the grain yield and other agronomic performance of the ABS203 in the greenhouse and the confined fields in Iowa and Hawaii, three leading events have been identified for potential product development in Africa [15].

4.2 Phytate Reduction and Bioavailability Increase of Fe and Zin

Iron and zinc are essential micronutrients for human development, and their deficiencies are major public health threats worldwide, especially in Africa. Sorghum contains good levels of Fe and Zn in seeds, ranging from 26 to 61 mg for Fe, and from 21 to 55 mg for Zn per kg seeds. However, the bioavailability of iron and zinc in sorghum grains is low due to the presence of phytate. Phytate or phytic acid, *myo*-inositol 1,2,3,4,5,6-hexakisphosphate, is an abundant component in plants, particularly in cereal grains and legumes.

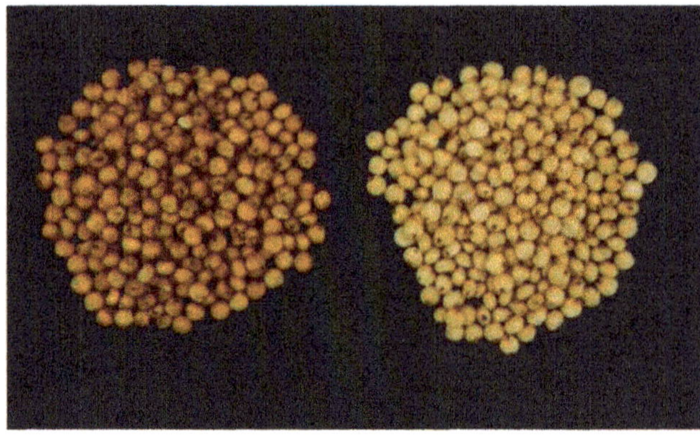

Fig. 1 ABS203 (left) and WT (right) sorghum seeds

Phytate chelates with minerals and inhibits their absorption by monogastric animals, including humans, because their digestive system lacks phytase. Therefore, a reduction of phytate has been widely used as the strategy of increasing Fe and Zn bioavailability in food grains [26]. On the other hand, in plants, phytate plays an important role for seed germination [27]. During seed germination, phytase activity increased three- to five-fold and phytate is degraded to provide energy for seed germination.

Our goal is to increase Iron (50%) and Zinc (30%) bioavailability. Before we generated transgenic sorghum with phytate reduction, we validated the concept of increase of iron and zinc bioavailability through reducing phytate contents in sorghum seeds.

Through in vitro treatment of sorghum flour with enzyme phytase, the sorghum flour samples with 100%, 80%, 40%, 20%, and 0% phytate reduction were obtained in Tx430 and used in rat gastric intubation experiments to estimate the Zn^{65} and Fe^{59} absorptions of 14-day rat pups [17–19].

The results listed in Table 1 showed the Zn and Fe absorptions in rat pups. Sample-1 with 100% phytate reduction showed 89.9% Zn absorption and 90.8% Fe absorption for rat pups respectively and these absorptions were very close to the positive controls (12.5 μM Zn or Fe solution) of 92.4% Zn absorption and 93.3% Fe absorption. Sample-5 with 0% phytate reduction showed 53.8% Zn absorption and 59.5% Fe absorption for rat pups and very close to the negative control (0.5 mM phytate solution) of 61.2% Zn absorption and 59.3% Fe absorption. Samples-2 to 4 with 79%, 39% to 20% phytate reductions, showed similar results to Sample-1, that is, significantly improved Zn absorption for rat pups compared to the Sample-5 (0% phytate reduction). For Fe absorption, significantly increased absorptions in rat pups were observed with 100% and 79% phytate reduced samples, but not with 39% and 20% phytate reduced samples. The cooking process to prepare the

Table 1
Zn^{65} and Fe^{59} absorption in rat pups

Sample	Phytate reduction% (mg/g)	Zn^{65} absorption% ± SD	Fe^{59} absorption% ± SD
Positive control	Notes§	92.4% ± 0.030	93.3% ± 0.005
Negative control	Notes§	61.2% ± 0.153	59.3% ± 0.144
Sample-1	100%, (0.002)	89.9%* ± 0.035	90.8%* ± 0.008
Sample-2	79%, (0.312)	82.8%* ± 0.076	77.9%^±0.091
Sample-3	39%, (0.909)	85.6%*±0.163	61.6%±0.116
Sample-4	20%, (1.193)	94.5%*±0.052	65.0%±0.105
Sample-5	0%, (1.490)	53.8%±0.034	59.5%±0.062
Uncooked-1	100%, (0.002)	92.1*%±0.033	67.1% ±0.158
Uncooked-5	0%, (1.490)	55.8%±0.075	69.8% ±0.097
Unprocessed-5	0%, (1.983)	48.5%±0.120	64.3%

*$p<0.001$, ^$p<0.05$, §positive control:12.5 μM Zn or Fe solution, and §negative control:0.5 mM phytate solution

meal samples for rat feeding did not impact the Zn absorption (Zn absorptions in Sample-1 vs. Uncooked-1 and in Sample-5- vs. Uncooked-5 were not statistically different). However, the cooking process may be helpful for Fe absorption (Sample-1: 90.8% vs. Uncooked-1: 67.1%, statistically significant difference).

Data from the rat pup gastric intubation experiments confirmed that a significant enhancement of iron and zinc bioavailability could be achieved through phytate reduction in sorghum seeds. The data also indicated there was no significant difference of Zn absorptions among 100% to 20% phytate reduced samples, although perhaps this suggests that this intubation experiment was better suited for qualitative bioavailability estimates rather than for quantitatively measurements.

4.3 Protein Digestibility Improvement

Sorghum protein digestibility has been studied extensively both in vivo and in vitro [7, 28, 29]. These studies show that the proteins of wet cooked sorghum are significantly less digestible in comparison with that of other cereals such as wheat and maize. It was found that sorghum contains 8.1–16.8% protein [10] in which the majority are kafirins, such as α-kafirins, β-kafirins, and γ-kafirins etc. [30, 31]. In sorghum, the reduction of protein solubility and digestibility after cooking is due to the formation of disulfide cross-linking of kafirin proteins, thus preventing enzymatic access to the protein bodies for further digestion. Among these three major classes of kafirins, γ-kafirins is reported to contain the highest proportion (7 mol%) of cysteine which provides the main source of the disulfide bonds [32].

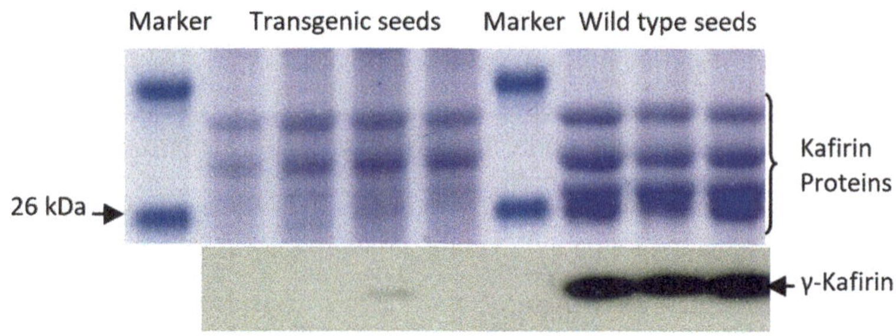

Fig. 2 The protein electrophoresis (top portion) and the Western (bottom portion) gel demonstrated the 26 kDa γ-kafirins in transgenic sorghum seeds (lane 2–5) were suppressed compared to WT seeds (lane 7–9)

Table 2
Protein digestibility of raw and cooked transgenic and control flour

Sorghum flour	Transgenic	Null	WT
Raw	84% (±1.2)	63% (±1.3)	65% (±0.8)
Cooked	72% (±0.7)	49% (±0.1)	48% (±0.3)

Our goal for sorghum protein digestibility improvement was to achieve sorghum with no protein digestibility loss after cooking relative to protein levels found uncooked, unmodified sorghum. We approached this goal through suppression of γ-kafirin accumulation in sorghum grain. Using RNAi technology, we successfully generated transgenic sorghum lines with suppressed γ-kafirins accumulation as demonstrated in Fig. 2. The protein electrophoresis and the Western assay demonstrated that compared to WT, the 26 kDa γ-kafirin proteins were significantly reduced in the transgenic seeds (Fig. 2), potentially improving protein digestibility. To confirm that suppression of γ-kafirins accumulation improves protein digestibility, protein digestibility assays were conducted for raw and cooked sorghum flour prepared from transgenic, null and WT sorghum seeds. As shown in Table 2, protein digestibility of WT and null dropped significantly after cooking from 65% and 63% to 48% and 49%, respectively. Alternately, protein digestibility of the transgenic seed only slightly dropped after cooking, from 84% to 72%, significantly higher than that of either WT and null before or after cooking.

Grain texture or hardness is another factor that needs to be considered when achieving protein digestibility improvement. Grain hardness impacts various aspects of seed development and food processing from fungal resistance to cooking quality. Like maize, sorghum varieties with soft endosperm are considered to have low grain quality and will not be accepted by farmers and consumers [33]. Therefore, endosperm hardness is one of the

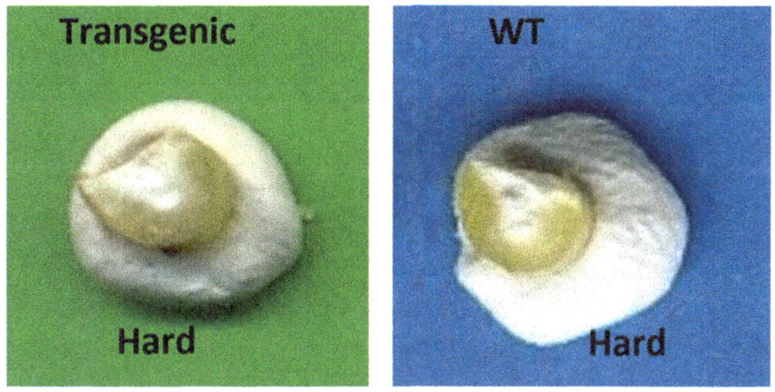

Fig. 3 Hard sorghum grains of the transgenic (left) and WT (right) seed

Table 3
Protein nutritional quality of transgenic and null seeds

Line	Protein content[a]	AAS	PDCAAS
Null	13.6%	0.45	0.16
Transgenic	13.7%	0.51	0.21

[a]Protein content calculated based on g protein/100 g flour

important criteria to measure the acceptability of digestibility-improved sorghum. Seed hardness can be analyzed by both visual and stereo-microscopic observation of the sectioned transgenic and WT sorghum kernels [20]. As shown in Fig. 3, no obvious visual seed hardness changes were observed for the transgenic seeds when compared to that of the WT.

After suppression of γ-kafirins accumulation in sorghum seeds, it was important to verify the total protein content in the transgenic seeds. The comparison of protein content in transgenic seeds with their null seeds confirmed that total protein in the γ-kafirins suppressed transgenic seeds was almost identical to its null seeds and that both were in the normal range of sorghum protein content [10] (Table 3). These results show a substantial increase in the proportion of non-kafirin storage proteins and non-storage proteins in the seeds where the γ-kafirins were suppressed. In addition, both AAS and PDCAAS of our transgenic line were better than the null control (Table 3) [20].

These results confirmed that the co-suppression of γ-kafirins in sorghum grains could be used to improve protein digestibility in sorghum.

Through many years' efforts, we generated transgenic lines with nutritionally enhanced sorghums meeting our initial target goals. This is not only a successful research accomplishment, but also a potential for delivering nutritionally enhanced sorghum products to alleviate critical micronutrient deficiencies in people who rely upon sorghum as their staple diet.

Acknowledgments

This research was supported by Bill and Melinda Gates Foundation Grand Challenges in Global Health Grant ID-37877 and by The Howard G. Buffett Foundation funding of the Africa Biofortified Sorghum Project. DowDuPont™ provided funding and in-kind donations. Dr. Paul Anderson and Dr. Rudolf Jung made contributions to this project in early stage. Analyses of sorghum protein digestibility and endosperm texture were carried out in John Taylor's lab of University Pretoria, South African. Rat pup absorption experiments for Fe/Zn bioavailability analysis were conducted in Bo Lonnerdal's lab of University of California, Davis; financially supported by HarvestPlus International Food Policy Research Institute, Washington DC. This work was done in close collaboration with Dr. Florence Wambugu and her staff at Africa Harvest Biotech Foundation International headquartered in Nairobi, Kenya.

References

1. Bain LE, Awah PK, Geraldine N, Kindong NP, Sigal Y, Bernard N et al (2013) Malnutrition in sub-Saharan Africa: burden, causes and prospects. Pan Afr Med J 15(120):2535
2. Global Nutrition Report (2016) From promise to impact: ending malnutrition by 2030. International Food Policy Research Institute, Washington, DC
3. Food insecurity and malnutrition in Africa: current trends, causes and consequences (September 19, 2012) polity, other opinions. Accessed 1 May 2018
4. Akombi BJ, Agho KE, Merom D, Renzaho AM, Hall JJ (2017) Child malnutrition in sub-Saharan Africa: a meta-analysis of demographic and health surveys (2006-2016). PLoS One 12(5):e0177338. https://doi.org/10.1371/journal.pone.0177338
5. UNICEF and MI (2004) Vitamin and mineral deficiency: a global damage assessment report. United Nations Children's Fund and the Micronutrient Initiative, New York and Ottawa
6. Food and Agriculture Organization of the United Nations (2012) Executive brief of FAO on Sahel crisis. http://www.fao.org. Accessed 1 May 2018
7. Duodu KG, Taylor JRN, Belton PS, Hamaker BR (2003) Factors affecting sorghum protein digestibility. J Cereal Sci 38(2):117–131
8. Shewry PR (2007) Improving the protein content and composition of cereal grain. J Cereal Sci 46(3):239–250
9. Kean EG, Ejeta G, Hamaker BR, Ferruzzi MG (2007) Characterization of carotenoid pigments in mature and developing kernels of selected yellow-endosperm sorghum varieties. J Agric Food Chem 55(7):2619–2626
10. Rooney LW, Serna-Saldivar SO (1990) Sorghum. Chapter V. In: Lorenz KJ, Kulp K (eds) Handbook of cereal science and technology. Markel Dekker, New York, pp 233–270
11. Zhao ZY, Cai T, Tagliani L, Miller M, Wang N, Pang H et al (2000) *Agrobacterium* mediated sorghum transformation. Plant Mol Biol 44:789–798
12. Wu E, Lenderts B, Glassman K, Berezowska-Kaniewska M, Christensen H, Asmus T et al (2014) Optimized agrobacterium-mediated sorghum transformation protocol and molecular data of transgenic sorghum plants. In Vitro Cell Dev Biol Plant 50(1):9–18
13. Freeman J, Sparks CA, West J, Shewry PR, Jones HD (2011) Temporal and spatial control of transgene expression using a heat-inducible promoter in transgenic wheat. Plant Biotechnol J 9:788–796
14. Che P, Zhao ZY, Glassman K, Dolde D, Hu TX, Jones TJ et al (2016) Elevated vitamin E content improves all-trans beta-carotene accumulation and stability in biofortified sorghum. Proc Natl Acad Sci U S A 113:11040–11045
15. Che P, Zhao ZY, Hinds M, Rinehart K, Glassman K, Albertsen M (2018) Evaluation of agronomic performance of β-carotene

elevated sorghum in confined field conditions. Chapter 15 In: Zhao Z-Y, Dahlberg J (eds) Sorghum: methods and protocols. Springer, New York

16. Shi J, Wang H, Wu Y, Hazebroek J, Meeley RB, Ertl DS (2003) The maize low-Phytic acid mutant *lpa2* is caused by mutation in an inositol phosphate kinase gene. Plant Physiol 131:507–515
17. Sandstrom B, Keen CL, Lonnerdal B (1983) An experimental model for studies of zinc bioavailability from milk and infant formulas using extrinsic labeling. Am J Clin Nutr 38:420–428
18. Stuart SMA, Johnson PE, Hamaker B, Kirleis A (1987) Absorption of zinc and iron by rats fed meals containing sorghum food products. J Cereal Sci 6:81–90
19. Kruger J, Taylor JRN, Du X, De Moura FF, Lönnerdal B, Oelofse A (2013) Effect of phytate reduction of sorghum, through genetic modification, on iron and zinc availability as assessed by an in vitro dialysability bioaccessibility assay, Caco-2 cell uptake assay, and suckling rat pup absorption model. Food Chem 141:1019–1025
20. da Silva LS, Jung R, Zhao ZY, Glassman K, Taylor J, Taylor JRN (2011) Effect of suppressing the synthesis of different kafirin sub-classes on grain endosperm texture, protein body structure and protein nutritional quality in improved sorghum lines. J Cereal Sci 54:160–167
21. WHO/FAO/UNU Expert Consultation (2007) Protein and amino acid requirements in Human Nutrition, Report of a Joint WHO/FAO/UNU Expert Consultation. World Health Organization Technical Report No. 935. World Health Organization, Geneva
22. Che P, Anand A, Wu E, Sander JD, Simon MK, Zhu W et al (2018) Developing high efficient *Agrobacterium*-mediated sorghum genetic transformation system with broad applications. Plant Biotechnol J 16(7):1388–1395. https://doi.org/10.1111/pbi.12879
23. Anand A, Che P, Wu E, Jones TJ (2018) Novel ternary vectors for efficient sorghum transformation. Chapter 13 In: Zhao Z-Y, Dahlberg J (eds) Sorghum: methods and protocols. Springer, New York
24. Ye X, Al-Babili S, Kloti A, Zhang J, Lucca P, Beyer P et al (2000) Engineering the provitamin A (b-carotene) biosynthetic pathway into (carotenoid-free) rice endosperm. Science 287:303–305
25. Paine JA, Shipton CA, Chaggar S, Howells RM, Kennedy MJ, Vernon G et al (2005) Improving the nutritional value of Golden Rice through increased pro-vitamin A content. Nat Biotechnol 23(4):482–487
26. Gupta RK, Gangoliya SS, Singh NK (2015) Reduction of phytic acid and enhancement of bioavailable micronutrients in food grains. J Food Sci Technol 52(2):676–684
27. Strother S (1980) Homeostasis in germinating seeds. Ann Bot 45:217–218
28. Maclean WC, De Romana GL, Placko RP, Graham GG (1981) Protein quality and digestibility of sorghum in pre-school children: balance studies and plasma free amino acids. J Nutr 111:1928–1936
29. Axtell JD, Kirleis A, Hassen MM, D'Croz-Mason N, Mertz ET, Munck L (1981) Digestibility of sorghum proteins. Proc Natl Acad Sci U S A 78:1333–1335
30. Hamaker BR, Schull JM (1995) In vitro protein digestibility of developing and mature sorghum grain in relation to α-, β-, and γ-kafirin disulfide crosslinking. J Cereal Sci 22(1):85–93
31. Shull JM, Watterson JJ, Kirleis AW (1991) Proposed nomenclature for the alcohol-soluble proteins (kafirins) of Sorghum bicolor (L. Moench) based on molecular weight, solubility and structure. J Agr Food Chem 39:83–87
32. Shull JM, Watterson JJ, Kirleis AW (1992) Purification and immunocytochemical localisation of kafirins in Sorghum bicolor (L. Moench) endosperm. Protoplasma 171:64–74
33. Chandrashekar A, Mazhar H (1999) The biochemical basis and implications of grain strength in sorghum and maize. J Cereal Sci 30(3):193–207

Chapter 15

Evaluation of Agronomic Performance of β-Carotene Elevated Sorghum in Confined Field Conditions

Ping Che, Zuo-Yu Zhao, Mark Hinds, Kristen Rinehart, Kimberly Glassman, and Marc Albertsen

Abstract

To help alleviate vitamin A deficiency in Africa, we have developed nutritionally enhanced sorghum with stabilized high all-trans-β-carotene accumulation. Toward the finalization of this nutritionally enhanced sorghum for food production, confined field trials were conducted to determine the agronomic performance of thirteen independent transgenic events in Iowa and Hawaii. Through these trials, three leading events with no negative impact on agronomic performance were identified. The studies described in this chapter have laid the groundwork for development of the next generation of β-carotene elevated sorghum as a food product.

Key words Sorghum, Agronomic performance, Field trial, β-carotene elevated, β-carotene stability, Transgenic, Nutritionally enhanced sorghum

1 Introduction

Alleviating vitamin A deficiency is a serious challenge. The WHO [1] has stated that "Vitamin A deficiency (VAD) is the leading cause of preventable blindness in children and increases the risk of disease and death from severe infections. In pregnant women VAD causes night blindness and may increase the risk of maternal mortality. Vitamin A deficiency is a public health problem in more than half of all countries, especially in Africa and South-East Asia, hitting hardest young children and pregnant women in low-income countries. Crucial for maternal and child survival, supplying adequate vitamin A in high-risk areas can significantly reduce mortality. Conversely, its absence causes a needlessly high risk of disease and death." Considerable effort across the past decade [2, 3] has been focused to improve VAD in a variety of food crops.

VAD is one of the most serious malnutrition issues in Africa [4]. Worldwide, the five countries with highest proportions of preschool age children with VAD ranged from 68.8% to 95.6%, all

Zuo-Yu Zhao and Jeff Dahlberg (eds.), *Sorghum: Methods and Protocols*, Methods in Molecular Biology, vol. 1931, https://doi.org/10.1007/978-1-4939-9039-9_15,

of which were all located in Africa [4]. Sorghum (*Sorghum bicolor* L.) is the second most important cereal in Africa; about 300 million people rely on it as their daily staple food. It is, however, considered a nutrient-poor crop [5, 6] with very low amounts of β-carotene. To address the VAD challenge in Africa, we developed nutritionally enhanced sorghum lines through genetic transformation. In these sorghum lines, all-trans-β-carotene accumulation in grain has increased 20-fold from 0.5 μg/g in the wild-type (WT) to about 10 μg/g. The stability of all-trans-β-carotene during room temperature storage also increased. Its half-life has been extended from less than 4 weeks to 10 weeks [7]. A Mongolian gerbil feeding study [8] quantified the bioefficacy of these β-carotene elevated sorghum lines as 4.3:1, that is, 4.3 μg β-carotene required for conversion to 1 μg retinol. We estimated that when freshly harvested, our β-carotene elevated sorghum could provide 90% of the Estimated Average Requirement (EAR) of vitamin A for children under age 3 and still provide 20% of EAR after 6 months of grain storage.

Field evaluation is a critical step in the development and release of any crop. This includes evaluating characteristics such as agronomics, seed germination rates, greenhouse grain yield, and open field trials. The transgenic nature of these materials required that the trials be grown in a confined field trial setting, according to the Biosafety Regulations established by Federal and State governments. Thirteen ABS203 homozygous transgenic lines (designated ABS203) were grown under confined field trials conditions in Johnston, Iowa, USA. Through these evaluations, three candidate events were selected and further evaluated under Hawaiian tropical confined field trial condition. Following the Hawaii field trial, these three events were molecularly evaluated with Southern blotting to confirm their single copy and intact insertion of the transformed T-DNA in sorghum genome. These three events met the criteria for practical sorghum production and validated as the first transgenic events that could alleviate VAD.

2 Materials

Thirteen ABS203 independent transgenic sorghum events [7] were generated in a Tx430 [9] background. T_0 plants directly derived from the transformation and regeneration process were self-pollinated to produce T_1 seeds in the greenhouse. The T_1 plants of these thirteen events were analyzed by qPCR [9] to determine the homozygosity of the transgenes (homozygous, hemizygous, and null). T_2 seeds with homozygous transgenes derived from self-pollinated T_1 homozygous plants were used in this study. In addition to these homozygous transgenic seeds, WT and the null seeds corresponding to the thirteen transgenic events were used as the controls for the field trials. Nulls have been widely used as controls to evaluate the possible impact of the transgene(s) on the transgenic plants or seeds [10]. The all-trans-β-carotene content and the β-carotene half-life in these transgenic seeds were determined by Che et al. [7].

3 Methods

3.1 Field Experimental Design and Data Collections in Iowa Field Trial

The homozygous transgenic seeds of thirteen ABS203 events, with their corresponding nulls and the WT as the controls, were planted in a confined field trial site located in Johnston, Iowa, USA. For each transgenic event and its controls, two replications (reps) with a 2-row plot per rep were distributed randomly in the field. At least ten rows of maize plants surrounding this field were planted as the borders to avoid edge affects (Fig. 1). Twenty sorghum seeds were sowed in a 4-meter row. Seed germination data were collected after 4-week sowing. Sorghum plant phenotypes that were collected included plant height, plant shape, tiller development, plant uniformity, flowering time, maturation time, panicle size and disease susceptibility. They were recorded during plant growth at three stages: pre-panicle emergence, flowering, and pre-harvesting. All panicles were covered with bags starting from pre-flowering until maturation. To estimate yield, panicles were harvested from the plants in a 0.9-m section in the middle of the row to avoid edge effect variation. One to five sections were harvested in each plot based on the spacing of plants in the plot. After drying and threshing, seeds from each 0.9-m section were pooled and weighed. The total seed weight of the threshed panicles harvested in each 0.9-m section was used to represent the yield.

3.2 Field Experimental Design and Data Collections in Hawaii Field Trial

The homozygous transgenic and their corresponding null seeds of event -3, -8, and -11 were planted in Kauai, Hawaii, USA to test the adaptability of the transgenic events to tropical growth condition. For each transgenic event, two reps with a 1-row plot per rep were randomly distributed in the field along with its corresponding null. Twenty-five seeds were sowed in a 4-m row. Five rows of WT sorghum plants were grown surrounding this field as the borders (Fig. 2). The same observations of plant phenotypes as in Iowa field trial were conducted. The total seed weight harvested from all panicles after drying in each row was used to represent the yield.

3.3 Statistical Analysis

Yield was analyzed as a two-way treatment structure in a one-way ANOVA [11]. The data were analyzed in a two-step process. The first step was to analyze all possible treatment combinations (e.g., in Iowa field trial: thirteen events, thirteen nulls, one WT and two reps of each plot; in Hawaii trial: three events, three nulls and two reps of each plot) in a one-way ANOVA. This enabled the testing of any possible combination to be directly compared with any other combination. The second step was to use single degree of freedom

Fig. 1 The confined sorghum field located in Johnston, Iowa, USA. The taller plants in the back were corn plants as boarders and the lower plants were sorghum plants at panicle pre-merging stage. The arrow-pointed two rows in the center were dwarf plants with short plant status and narrow leaves

Fig. 2 The confined sorghum field located in Kauai, Hawaii, USA. The un-bagged sorghum plants were WT as the boarder rows and the bagged sorghum plants in the center were the sorghum plants for trial

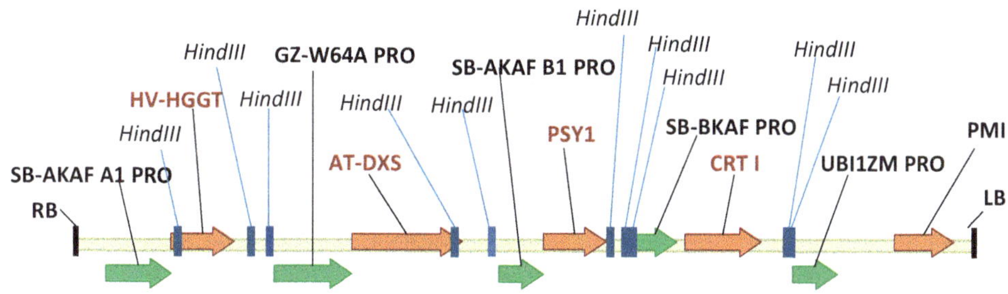

Fig. 3 The map of the T-DNA fragment used to generate transgenic sorghum ABS203. A number of *HindIII* sites and the *SB-AKAF A1* promoter, *CRTI* gene and *AT-DXS* gene were illustrated

contrast statements to estimate the levels of either main effect (lines or repeats) or the differences between levels within a main effect. This enabled testing of a level of an effect against any other level or the WT. Significant differences were deemed when the probability of the difference was less than 0.05.

3.4 Sothern Blotting and Q-PCR Analysis

Standard Southern blotting protocol [12] was used to verify the copy number and the intactness of T-DNA integrations. Transgenic sorghum genomic DNA was digested with *HindIII* and probed with the *SB-AKAF A1* promoter for the identification of the copy number of inserted T-DNA (Fig. 3) and with *CRTI* and *AT-DXS* for the intactness of the inserted T-DNA.

Q-PCR was used to verify the integration of the vector backbone. The detailed Q-PCR procedure was as described previously [9].

4 Results and Discussion

4.1 Events Selected for Iowa Field Trials

Based on the observations of plant growth in greenhouse and the level of all-trans-β-carotene and vitamin E in the T_1 seeds [7], thirteen transgenic sorghum events were selected to evaluate their agronomic performance under field conditions. These thirteen events had (1) single copy insertion of all transgenes with no vector backbone integration based on qPCR analysis, (2) good seed production with normal seed germination rate and normal plant phenotype compared to WT sorghum grown under the same greenhouse conditions, (3) a range of all-trans-β-carotene contents from 5.0 to 10.6 μg/g, and (4) a range of total tocopherols from 6 to 18 μg/g [7].

4.2 Seed Germination Rate in Iowa Field Trial

The seed germination rates of these thirteen events, their corresponding nulls, and the WT were recorded 4 weeks after sowing. The seed germination rates ranged from 60% to 82% for the transgenic events, 60%–90% for the nulls and 72% for the WT (Fig. 4). Most importantly, we tested whether the elevated

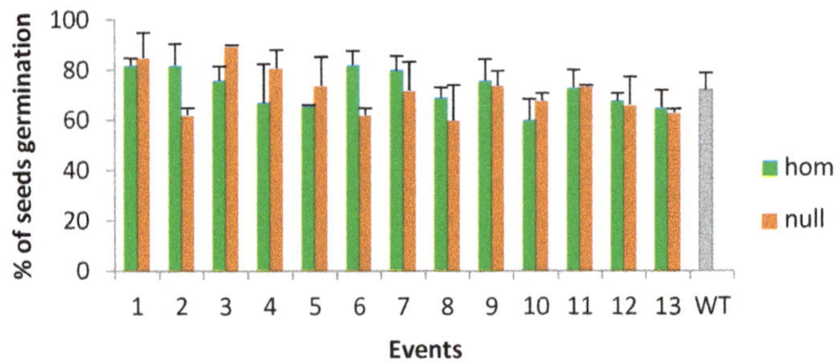

Fig. 4 Seed germination rates of the thirteen transgenic events, their corresponding nulls and the WT control in Iowa field (hom = homozygous transgenic event)

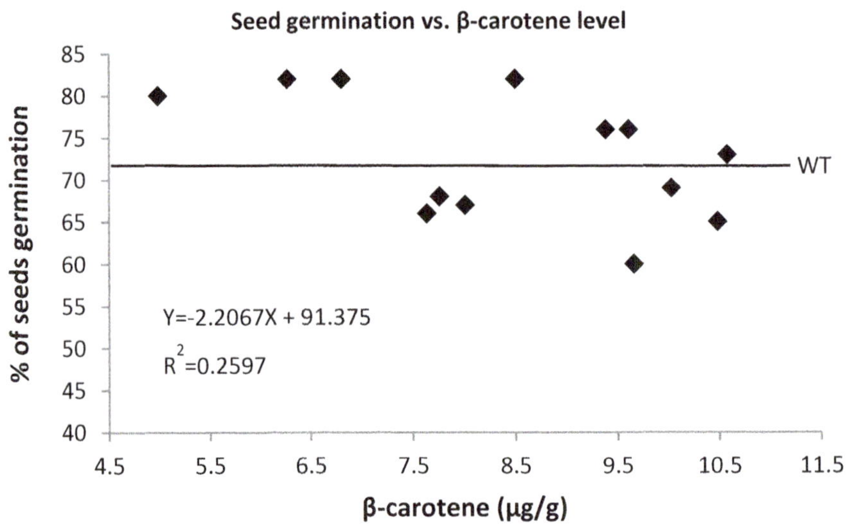

Fig. 5 The correlation of seed germination rate vs. all-trans-β-carotene content in these thirteen events (no correlation if $R^2 < 0.5$)

β-carotene content in these transgenic seeds have a negative impact on seed germination. The correlation coefficient analysis showed no significant correlation ($R^2 = 0.2597, < 0.5$) between the seed germination rates and the all-trans-β-carotene contents of these thirteen transgenic events as shown in Fig. 5. Based on these results, we concluded that these levels of elevated all-trans-β-carotene do not impact sorghum seed germination.

4.3 Plant Phenotypes in Iowa Field Trial

Plant phenotypes including plant height, plant shape, tiller development, plant uniformity, flowering time, maturation time, panicle size, and disease susceptibility were evaluated at three different growth stages (pre-panicle merging, flowering, and pre-harvesting) of the selected transgenic events under field conditions. Among these thirteen transgenic events, ten events showed normal phenotypes during the entire growth season and three, event-6, -9 and -10, showed some off-type phenotypes. A few tall plants were found in the rows of event-6. These tall plants could have been the result of outcrosses of its parental T_1 plant in greenhouse. A dwarf plant phenotype (Fig. 1) with narrow leaves and small panicles were found for all plants of event-9 and -10. In contrast, plants from null-9 were of normal stature and phenotype, whereas the plants from null-10 were all dwarfs. These results suggest that, although the dwarf phenotype in event-9 could be due to a mutation caused by the insertion of the T-DNA into the sorghum genome, the dwarf phenotype in event-10 was a mutation that may not be related to T-DNA integration. In addition, a few of plants with albino leaves were found in null-2, a few tall plants also were found in null-12 and null-13.

4.4 Grain Yield in Iowa Field Trial

Sorghum yield was evaluated for all thirteen events, the corresponding nulls, and the WT after the grain was dried and threshed. The yield (represented by the total seed weight in a 0.9-m section within a row) ranged from 246.7 to 480 g for these thirteen transgenic events, from 201.7 to 542.1 g for their nulls (except null-4, 45 g only) and 447 g for the WT control (Fig. 6). Statistical analysis showed that three events (1, 2, and 6) had lower yield than their nulls and four events (1, 2, 10, and 12) had lower yield than the WT. The remainder of the events had similar or higher yields to both controls (Fig. 6). In addition, the potential impact of the all-trans β-carotene level on yield also was analyzed. As indicated in Fig. 7, there was no significant correlation ($R^2 = 0.069$, <0.5) between yield and the all-trans β-carotene level. This is a critical result as it showed that grain yield was not affected by the elevated all-trans-β-carotene in these transgenic sorghums. The lower yields of event-1 and -2 could be due to the

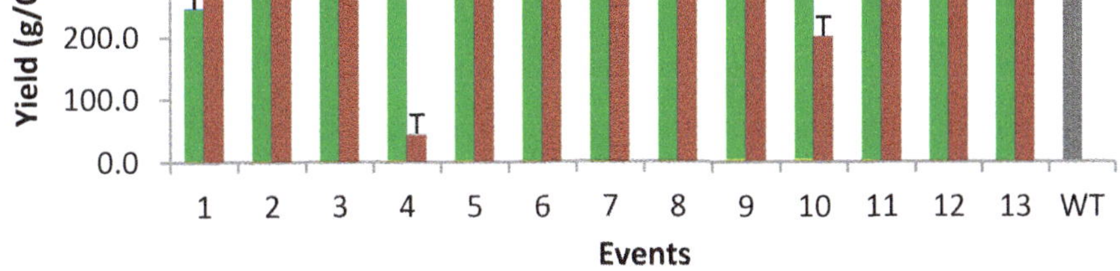

Fig. 6 The grain yields of these thirteen transgenic events, their corresponding nulls and the WT control in Iowa field trial. The yield data were collected from 0.9-meter sections (hom = homozygous transgenic event)

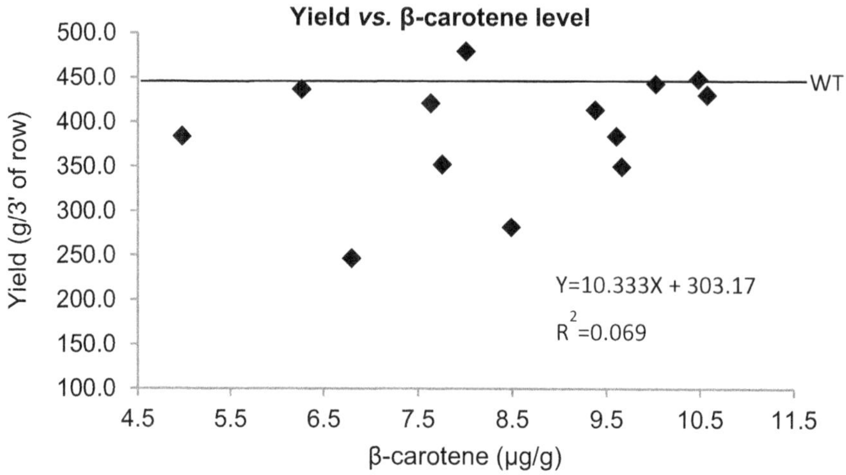

Fig. 7 The grain yields vs. their β-carotene contents of these thirteen transgenic events and the grain yield of the WT control in Iowa field trial (no correlation if $R^2 < 0.5$)

specific transgene integration sites as the yields of these two transgenic events were lower than both their related nulls and the WT. The lower yield in event-10 is likely the result of its dwarf plant phenotype caused by a genetic mutation rather than transgene insertion. The dwarf mutation is likely to have occurred during the transformation stage of the tissue culture process. The explanation of the lower yield in both event-12 and null-12 compared to the WT control is not clear.

Based on the data generated through the Iowa confined field trials, we selected and advanced three transgenic events, event-3, event-8, and event-11 for further evaluation under the tropical field conditions of Hawaii.

4.5 Yield Evaluation in Hawaii Field Trial

The objective of this project is to develop transgenic sorghum with enhanced and stabilized β-carotene to alleviate VAD in Africa. To mimic some of the tropical growth conditions in African countries, we selected three events for evaluation in Hawaii. Based upon the Iowa field data, event-3, -8, and -11 were selected for Hawaiian field trials (Fig. 2). Plant phenotypes observed in Hawaii confirmed the phenotypes observed for all three events in Iowa. The yields of these three transgenic events and their corresponding nulls are shown in Fig. 8. The yields of the transgenic events were slightly higher than their corresponding nulls, although statistics analysis showed no significant differences of the yields between the transgenic events and their nulls, confirming no significant differences between Iowa and Hawaii field growth and performance for the tested events.

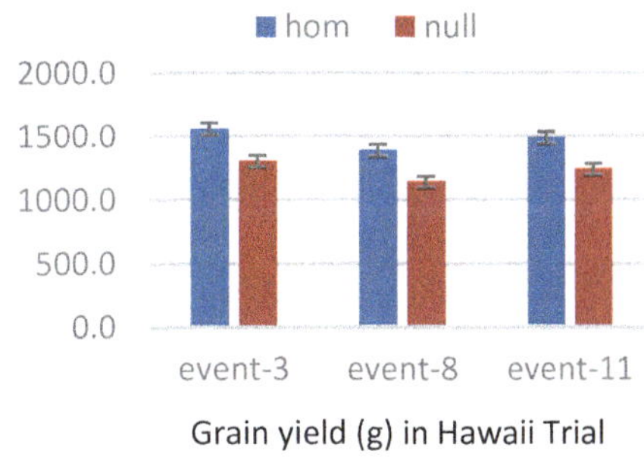

Fig. 8 The grain yields of the 3 transgenic events, their corresponding nulls in Hawaii field trial. The yield data were collected from 4-meter rows (hom = homozygous transgenic event)

4.6 Confirmation of the Molecular Quality by Southern Blotting

To confirm a single copy and intact integration of the T-DNA into the sorghum genome in these three ABS203 events, Southern blotting analyses were conducted. The genomic DNA isolated from six individual plants in each of these three transgenic events were digested with restriction enzyme *HindIII* (Fig. 3) and probed with *SB-AKAF A1* promoter (Fig. 9) for copy number identification and with *CRTI* (Fig. 10) and *AT-DXS* (Fig. 11) for intact integration of the T-DNA in these three events. A plasmid DNA containing the T-DNA fragment used in ABS203 transformation and the WT sorghum genomic DNA spiked with or without the plasmid DNA were used as controls. The following three Southern blots illustrate that an intact single copy of the T-DNA was integrated into three different locations in the sorghum genome, defining these three events.

4.7 The Three Leading Events

Based on all the data obtained from greenhouse, lab analysis [7], two field trials, and the Southern blots, three transgenic sorghum ABS203 events were identified as the first lead events that could alleviate VAD among people who consume sorghum as their staple diet. The major parameters of these three events are listed in Table 1. The all-trans-β-carotene contents were 9.4, 10.0, and 10.6 μg/g, which is about 20-fold higher than WT. The total tocols (including α-tocotrienol, α- and γ-tocopherol [7]) were 14.3, 16.8 and 18.4 μg/g, which is about three-fold higher than WT. The β-carotene half-life were 9.3, 9.5, and 10.1 weeks respectively in these three events. They did not have a yield penalty, off-type phenotypes or seed germination issue and they have single copy integration of intact transgenes without presence of the vector backbone in their genome [9]. These three events were selected as potential leading events for product development.

We conducted a Mongolian gerbil feeding study [8] to determine the β-carotene bioconversion rate of ABS203 transgenic, revealing it as 4.3:1 (4.3 μg β-carotene converting to 1 μg retinol) which is in the similar ranges of other pro-vitamin A enhanced transgenic crops [13–17]. Based on this bioefficacy, we estimated

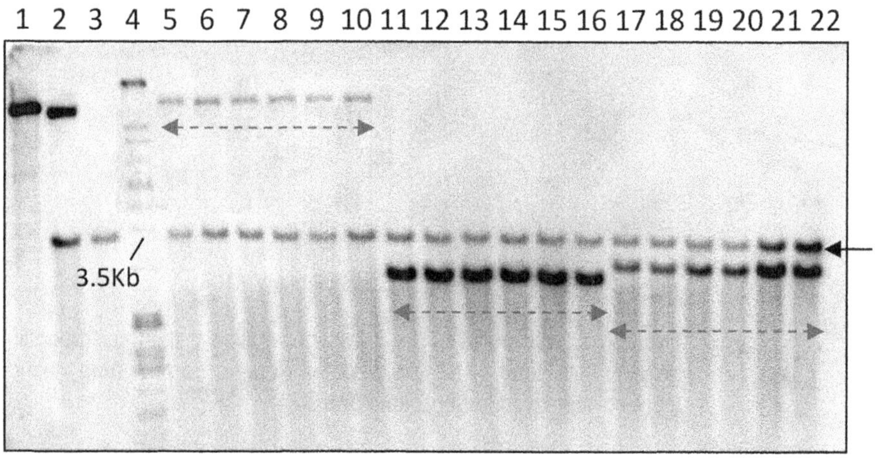

Fig. 9 A Southern blot of 3 selected transgenic ABS203 events to confirm a single copy of the T-DNA integration. The genomic DNA digested with Hind III and probed with *SB-AKAF A1* promoter showing a single copy of the T-DNA integration in three different locations in there three events. Lanes: 1: plasmid DNA, 2: WT sorghum genomic DNA spiked with the plasmid DAN, 3: WT sorghum genomic DNA, 4: DNA size marker, a 3.5Kb band marked; 5–10: sorghum genomic DNA from 6 transgenic plants of ABS203 event-3, 11–16: sorghum genomic DNA from 6 transgenic plants of ABS203 event-8, 17–22: sorghum genomic DNA from 6 transgenic plants of ABS203 event-11. The common band (arrow pointed) across all lanes including these two WT samples (lane 2 and 3) but excluding the plasmid DNA (lane 1) was the sorghum endogenous band

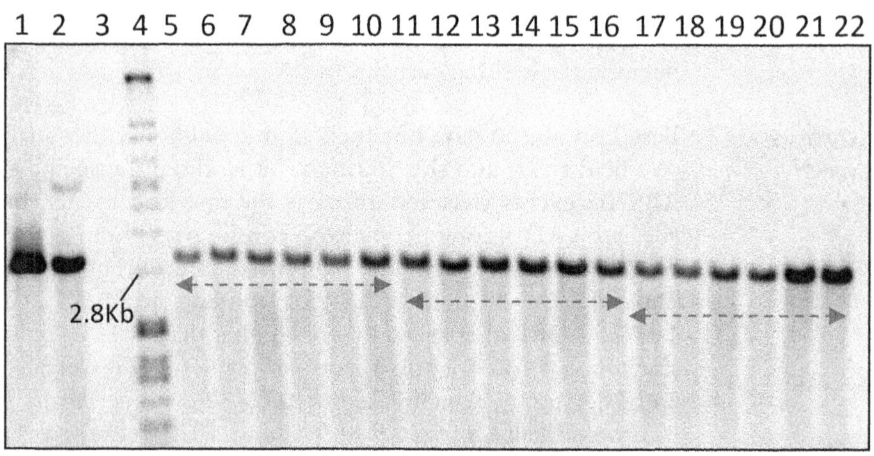

Fig. 10 A Southern blot of 3 selected transgenic ABS203 events to confirm intactness of the T-DNA integration. The genomic DNA digested with Hind III and probed with *CRTI* showing intactness of the T-DNA integration in these 3 events. Lanes: 1: plasmid DNA, 2: WT sorghum genomic DNA spiked with the plasmid DAN, 3: WT sorghum genomic DNA, 4: DNA size marker, a 2.8Kb band marked; 5–10: sorghum genomic DNA from 6 transgenic plants of ABS203 event-3; 11–16: sorghum genomic DNA from 6 transgenic plants of ABS203 event-8; 17–22: sorghum genomic DNA from 6 transgenic plants of ABS203 event-11

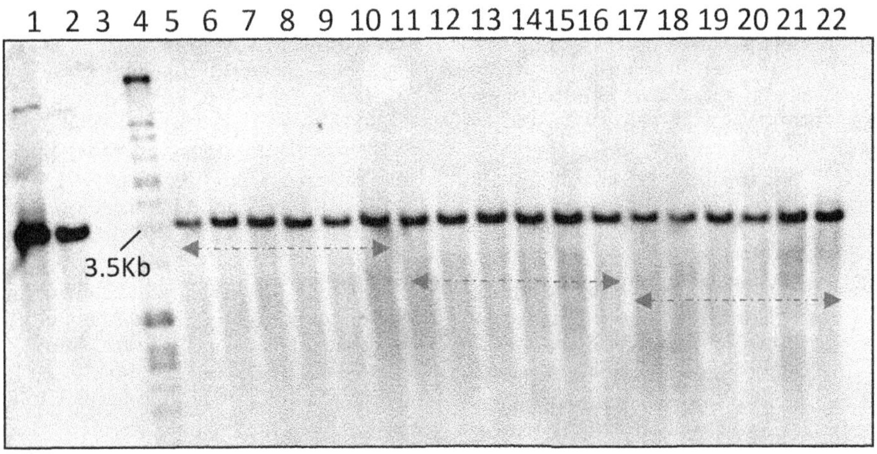

Fig. 11 A Southern blot of 3 selected transgenic ABS203 events to confirm intactness of the T-DNA integration. The genomic DNA digested with Hind III and probed with *AT-DXS* showing intactness of the T-DNA integration in these 3 events. Lanes: 1: plasmid DNA, 2: WT sorghum genomic DNA spiked with the plasmid DAN, 3: WT sorghum genomic DNA, 4: DNA size marker, a 3.5Kb band marked; 5–10: sorghum genomic DNA from 6 transgenic plants of ABS203 event-3; 11–16: sorghum genomic DNA from 6 transgenic plants of ABS203 event-8; 17–22: sorghum genomic DNA from 6 transgenic plants of ABS203 event-11

Table 1
The major characters of three leading events compared with WT

Event	β-Carotene (μg/g)	Total tocols (μg/g)	$T_{1/2}$ (weeks)	Yield in Iowa trial (g)/0.9-meter section	Yield in Hawaii trial (g)/4-meter row
3	9.4	14.3	10.1 ± 0.9	414	1557
8	10.0	16.8	9.5 ± 0.9	443	1387
11	10.6	18.4	9.3 ± 0.7	431	1485
Control	0.5	5.0	NA	447[a]	1226[b]

[a]WT yield
[b]Average yield of the three nulls

that our β-carotene enhanced sorghum could provide 90% of the estimated average requirement (EAR) of vitamin A for children under age three when freshly harvested and still would provide 20% of EAR after 6 months of grain storage.

Acknowledgment

This research was supported by Bill and Melinda Gates Foundation Grand Challenges in Global Health Grant ID-37877 and by The Howard G. Buffett Foundation funding of the Africa Biofortified Sorghum Project. DowDuPont™ provided funding and in-kind donations. This work was done in close collaboration with Dr. Florence Wambugu and her staff at Africa Harvest Biotech Foundation International headquartered in Nairobi, Kenya.

References

1. WHO (2017) Micronutrient deficiencies. World Health Organization. http://www.who.int/nutrition/topics/vad/en/. Accessed 1 May 2018
2. Bouis HE, Saltzman A (2017) Improving nutrition through biofortification: a review of evidence from HarvestPlus, 2003 through 2016. Glob Food Sec 12:49–58
3. Giuliano G (2017) Provitamin A biofortification of crop plants: a gold rush with many miners. Curr Opin Biotechnol 44:169–180
4. Report G (2009) Investing in the future—a united call to action on vitamin and mineral deficiencies. www.unitedcalltoaction.org/index.asp. Accessed 1 May 2018
5. Duodu KG, Taylor JRN, Belton PS, Hamaker BR (2003) Factors affecting sorghum protein digestibility. J Cereal Sci 38(2):117–131
6. Shewry PR (2007) Improving the protein content and composition of cereal grain. J Cereal Sci 46(3):239–250
7. Che P, Zhao ZY, Glassman K, Dolde D, Hu TX, Jones TJ et al (2016) Elevated vitamin E content improves all-trans beta-carotene accumulation and stability in transgenic sorghum. Proc Natl Acad Sci U S A 113:11040–11045
8. You H, Zhang Y, Zhao ZY, Che P, Albertsen M, Glassman K et al (2015) Quantifying the bioefficacy of β-carotene-biofortified sorghum using a Mongolian gerbil model. FASEB J 29(1). Supplement 605.3
9. Wu E, Lenderts B, Glassman K, Berezowska-Kaniewska M, Christensen H, Asmus T et al (2014) Optimized agrobacterium-mediated sorghum transformation protocol and molecular data of transgenic sorghum plants. In Vitro Cell Dev Biol Plant 50(1):9–18
10. Shen B, Allen WB, Zheng P, Li C, Glassman K, Ranch J et al (2010) Expression of ZmLEC1 and ZmWRI1 increases seed oil production in maize. Plant Physiol 153:980–987
11. Ostertagová E, Ostertag O (2013) Methodology and application of one-way ANOVA. Am J Mech Eng 1(7):256–261. https://doi.org/10.12691/ajme-1-7-21
12. Brown T (2003) Current protocols in molecular biology, Unit 2.9 A southern blotting, published online: 1 May 2001. https://doi.org/10.1002/0471142727.mb0209as21, Copyright © 2003 by John Wiley and Sons
13. Tang G, Qin J, Dolnikowski GG, Russell RM, Grusak MA (2009) Golden rice is an effective source of vitamin A. Am J Clin Nutr 89:1–8
14. Tang G, Hu Y, Yin SA, Wang Y, Dallal GE, Grusak MA et al (2012) Β-carotene in Golden Rice is as good as b-carotene in oil at providing vitamin A to children. Am J Clin Nutr 96:658–664
15. Howe JA, Maziya-Dixon B, Tanumihardjo SA (2009) Cassava with enhanced b-carotene maintains adequate vitamin A status in Mongolian gerbils (Meriones unguiculatus) despite substantial cis-isomer content. Br J Nutr 102:342–349
16. La Frano MR, Woodhouse LR, Burnett DJ, Burri BJ (2013) Transgenic cassava increases b-carotene and vitamin A concentrations in the TAG-rich plasma layer of American women. Br J Nutr 110:310–320
17. Li S, Nugroho A, Rocheford T, White WS (2010) Vitamin A equivalence of the b-carotene in b-carotene–transgenic maize porridge consumed by women. Am J Clin Nutr 92:1105–1112

Part IV

Economics and Broad Application

Chapter 16

Economics of Increasing Sorghum Productivity in Sub-Saharan Africa: The Mali Case

John H. Sanders, Botorou Ouendeba, Ababacar Ndoye, Niaba Témé, and Soungalo Traore

Abstract

Sorghum retains a crucial role in Sub-Saharan Africa for food and in the future feed. Unfortunately, the movement of sorghum technology onto farmers' fields in Sub-Saharan Africa has been slow in spite of substantial research since the great African drought of 1968–1973. What is necessary to get African sorghum yields and profitability up?

After reviewing the situation of sorghum in the world, the USA, and Sub-Saharan Africa from 2007 to 2017 the results and the lessons of a twelve-year program in the Sahel of West Africa to introduce new sorghum technology and marketing strategies are the focus of the rest of the paper. In Mali, the program identified new technologies that were extended into a large number of farmers' associations. The Mali program then collaborated with two other agencies to scale up this pilot program. The pilot project demonstrated that yields with moderate fertilization, new varieties, and improved agronomic practices could be increased 50 to 100% and prices increased 30 to 50%. The 2012 military coup and then invasion of Al Queda from the north shut down both the pilot and the scaling up activities as the US government banned collaboration with Malian government agencies after the coup. The pilots were continued in Niger and Burkina Faso through 2014 and then with a Gates Foundation grant from 2014 to 2016. The pilot program in Mali responded to two of the three Second Generation problems identified. But more significantly the pilot project identified the lack of funds for responding to Second Generation problems as a major constraint for implementing a technology-marketing program in a low-income country.

Key words Semi-arid, Sahel, Ethanol, Yield gap, Lodging, Farmers'association, "Soudure", Processors, Kiosks, Grinkan, "Tô", Tannin, Anaerobic

1 Introduction

Sorghum [*Sorghum bicolor* (L.) Moench] and maize [*Zea* mays (L.)] production and consumption trends in the world, the United States (USA), and Sub-Saharan Africa during the last decade are reviewed. We then review 12 years of fieldwork in one of the

Botorou Ouendeba was deceased at the time of publication.

Zuo-Yu Zhao and Jeff Dahlberg (eds.), *Sorghum: Methods and Protocols*, Methods in Molecular Biology, vol. 1931, https://doi.org/10.1007/978-1-4939-9039-9_16,

poorest areas of the world, the semiarid, Sub-Saharan Africa country of Mali. The program developed a pilot program and then advised on scaling up activities of the new sorghum technologies and marketing strategies. Yields and marketing practices were substantially improved but continuing investments will be necessary to maintain these gains.

Maize has progressed faster than sorghum in the last decade in all three regions. Both have become important for feed and ethanol in the USA. Compared with the USA, both maize and sorghum lag far behind in yields in Sub-Saharan Africa. The predominant use of both cereals is for food in Sub-Saharan Africa. A substantial increase in feed grain use has occurred in maize and is beginning for sorghum in Sub-Saharan Africa. The large yield gap with the USA for both coarse grains indicates the potential for substantial productivity improvements in Sub-Saharan Africa.

2 Sorghum in the World

Over one billion metric tons of maize were produced in 2017/18 worldwide with 36% of that production taking place in the USA. Sorghum production in the world and the USA were only 5.5% and 2.5% of maize. In Sub-Saharan Africa sorghum was 41% of the production of maize (Table 1). Nevertheless, nearly fifty-seven million metric tons and 40 million harvested ha still qualifies sorghum as a major cereal of the world (data for the 2017/18 crop year; Table 2). Although maize increases over the last decade, there is a decline in sorghum production (Figs. 1 and 2).

Among the coarse grains sorghum is dwarfed by the production and productivity of maize in the world and the USA. However, the production and area differences between maize and sorghum are much less in Sub-Saharan Africa indicating sorghum's greater importance within semiarid regions (Appendix to the chapter, Tables 2, 3, 4). Sorghum tolerates water shortages much better than maize and its production is concentrated in the lower rainfall regions of the world on primarily alluvial and heavier soils. The importance of sorghum to Sub-Saharan Africa can be illustrated by the yield potential witnessed in the USA. Though planted areas are much lower than those in Africa, production in the USA nearly tops that of all of Africa demonstrating the potential of improving yields for the welfare and nutritional benefits of many low-income farmers and consumers in Sub-Saharan Africa.

The USA and Sub-Saharan African harvested areas in maize are approximately equal whereas Sub-Saharan African harvested sorghum area remained about ten times that of the USA during the past decade (Figs. 3 and 4).

United States maize yields are over twice world yields and almost six times those of Sub-Saharan Africa in spite of the

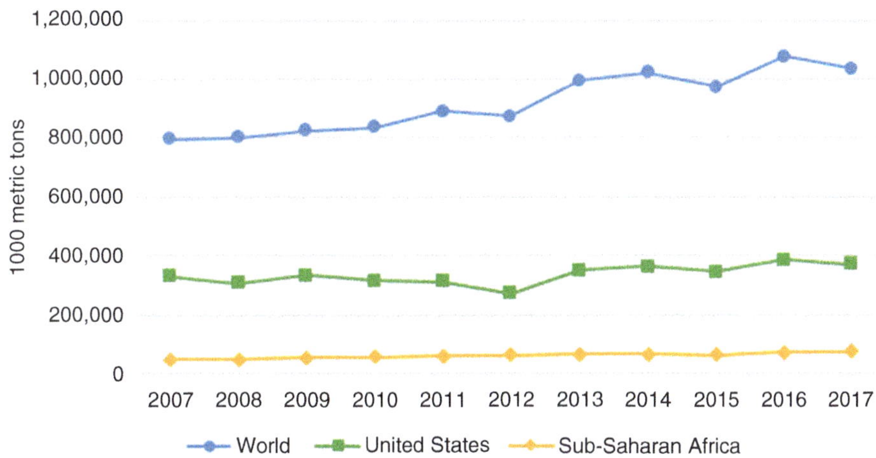

Fig. 1 Maize production. World, USA, and Sub-Saharan Africa, 2007–2017. Source: Appendix Table 2

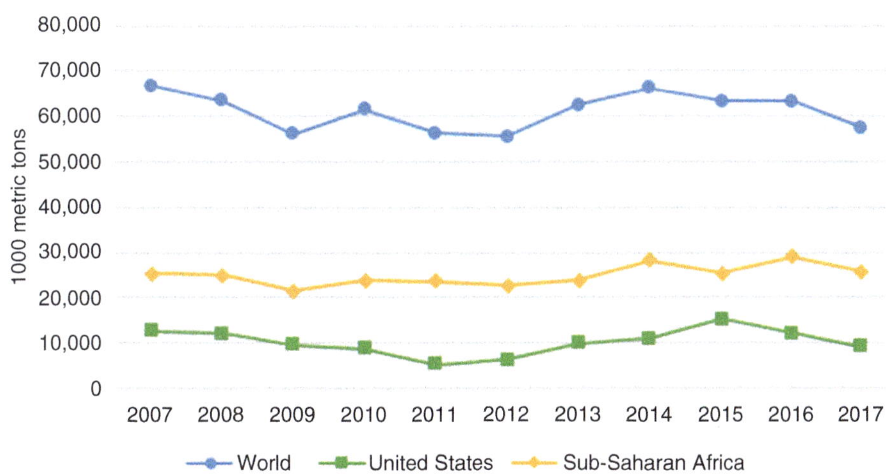

Fig. 2 Sorghum production. World, USA, Sub-Saharan Africa, 2007–2017. Source: Appendix Table 2

introduction of new varieties and higher fertilization levels during the last decade in Sub-Saharan Africa. For sorghum the world yields are only 50% higher than the Sub-Saharan one ton level. However, USA sorghum yields were almost five times those of Sub-Saharan Africa by the end of the decade. This yield gap represents an opportunity to attain higher yields by using a known combination of inorganic fertilizer, new cultivars, and improved agronomic practices. The process of getting the technology to farmers and selling the output to the final consumers is considered in the next section of this paper.

In the USA, maize generally has the advantage of higher and better distribution of rainfall than for sorghum, higher use of inputs, and higher prices, especially during the period of rapidly rising demand for maize to produce ethanol in the USA. United

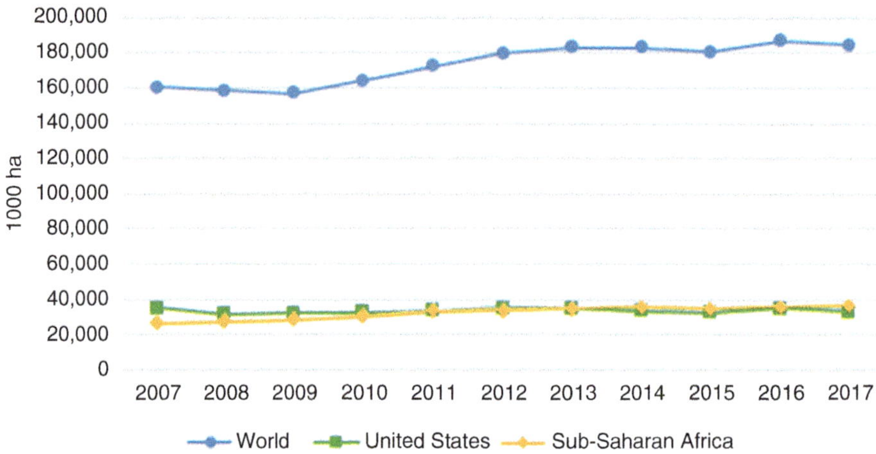

Fig. 3 Maize area harvested. World, USA, Sub-Saharan Africa, 2007–2017. Source: Appendix Table 3

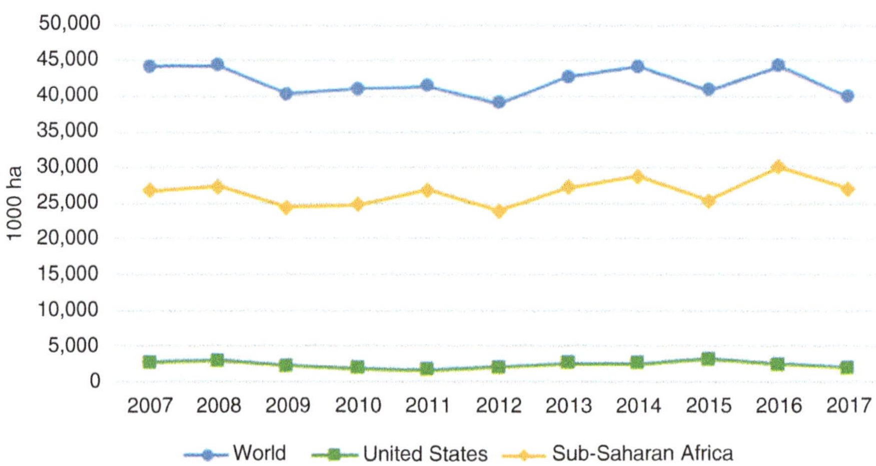

Fig. 4 Sorghum area harvested. World, USA, Sub-Saharan Africa, 2007–2017. Source: Appendix Table 3

States maize yields are more than double those of sorghum but both are impressive internationally with yields of maize reaching 11 mt per ha and sorghum 4.5 mt per ha in 2017–2018 (Table 4). For the three regions maize yields are not only substantially higher than sorghum but also increasing whereas sorghum yields are moderately increasing in the USA and stagnant in the world and Sub-Saharan Africa (Figs. 5 and 6).

In Sub-Saharan Africa the predominate use of sorghum is as a primary staple food or a reserve cereal when maize or millet yields falter. Feed use is beginning in Sub-Saharan Africa which is using roughly 6% of consumption for animal feed compared to 53% in the USA (Table 1). As incomes increase, higher quality diets are demanded, hence cereal (and tubers in higher rainfall regions)

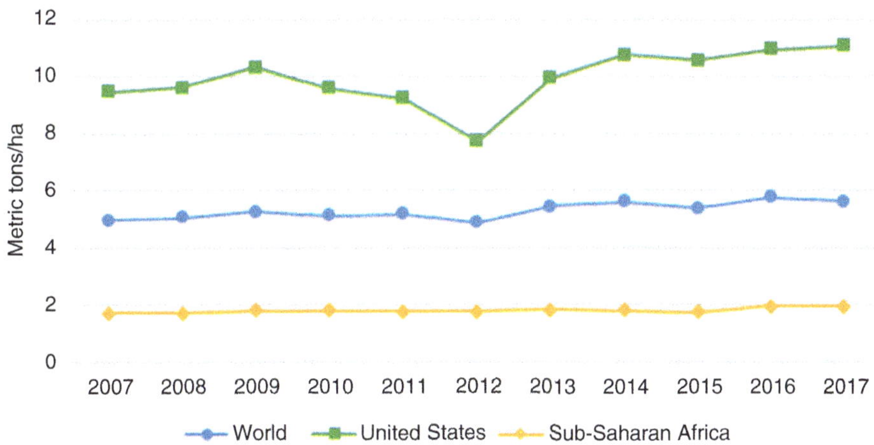

Fig. 5 Maize yields. World, USA, Sub-Saharan Africa, 2007–2017. Source. Appendix Table 4

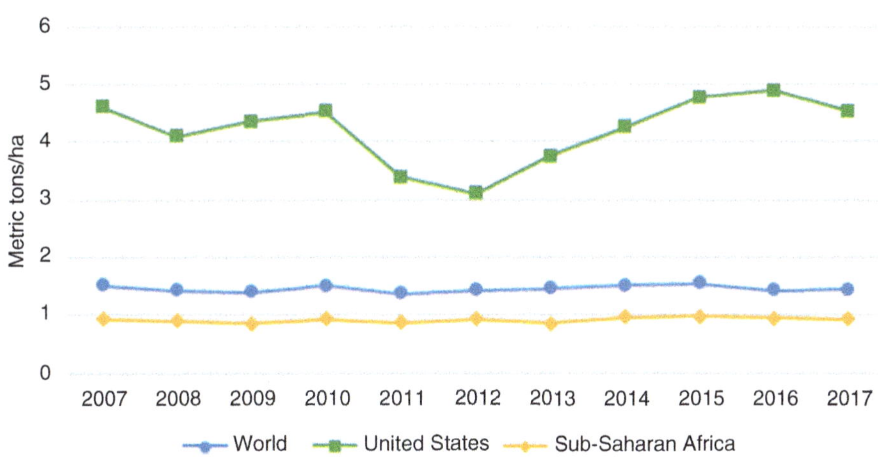

Fig. 6 Sorghum yields. World, USA, Sub-Saharan Africa. Source: Appendix Table 4

consumption for food is reduced but fruit, vegetable, meat, milk, and cheese consumption rapidly increase.

It is often difficult for rapidly developing countries to expand maize or sorghum production fast enough to keep up with increasing demand. This is illustrated by the recent international trade in sorghum by the USA with China, a country that has experienced rapid economic growth over the past 40 years. Over the last 5 years China has averaged 71% of the world sorghum imports at an average demand of 6.46 million mt [1]. During this period the USA dominated the world export market for sorghum with 78% of the market with an average of 7.05 million exported mt. In China this imported sorghum is primarily utilized for feed.

Breaking maize and sorghum consumption into their component parts in Table 1 maize dominance is clear. Sorghum is only 6% of maize consumption in the world. However, in Sub-Saharan

Table 1
Domestic consumption of food, feed, and ethanol from maize and sorghum, 2016–2017

Crop	Region	Domestic cons. (1000 metric tons)	Food, seed, industrial[a] (1000 mt)	Feed (1000 mt)	Ethanol (1000 mt)
Maize					
	World	1,036,799	263,877	632,911	140,011
			25%	61%	14%
	USA	313,854	34,836	139,007	140,011
			11%	44%	45%
	Sub-Saharan Africa	70,785	55,975	14,810	–
			79%	21%	
Sorghum					
	World	63,407	36,582	24,268	2557
			58%	38%	4%
	USA	6212	353	3302	2557
			6%	53%	41%
	Sub-Saharan Africa	29,416	27,718	1698	–
			94%	6%	

Source: Sorghum consumption data were from Foreign Agricultural Service, USDA, FAS Home (http://www.fas.uada.gov/) Market and Trade data. Ethanol data of bushels of maize and sorghum used for ethanol were from: www.ethanolrfa.org/resources/industry/statistics/#1461260019819-00afed-a339

This is the ethanol data for 2017 combined with the other consumption data for the 2016–17 crop year. The conversion factors were 56 lbs per bushel of corn or sorghum and 2.2 kg per lb

[a]There are other industrial uses besides ethanol in this USDA category of food, seed and industrial use but ethanol production was separated using the RFA data collected from industry. *See* the above URL site

Africa sorghum is 46% of world sorghum consumption. Since world sorghum yields are not much higher than Sub-Saharan sorghum yields, demonstrating yield increases in Sub-Saharan Africa would be expected to be a good model for many of the developing world sorghum producers. There are three basic uses of sorghum in the table; first human food, seed and industrial use (but excluding ethanol production), then animal feed and finally ethanol production. The production data are from 2016 to 2017 with ethanol production figures from 2017 [2].

United States maize use for feed and ethanol were almost equal at 45% or 140 million mt with 35 million mt used in food systems. In Sub-Saharan Africa, feed consumption for maize is 14.8 million mt (21% of consumption) with sorghum just beginning to be used for feed at 1.7 million mt (6%). Both maize and sorghum food consumption are very important regionally in Sub-Saharan Africa, at 56 (79% of consumption) and 28 (94%) million mt.

In Sub-Saharan Africa, sorghum remains a major crop but its production has remained constant over the past decade while maize has made substantial yield progress. Nevertheless, 29.4 million mt of sorghum remains a key element in increasing the nutrition and welfare levels in Sub-Saharan Africa. Moreover, the same productivity increasing innovations seen with maize are being introduced with sorghum as will be discussed in the next section.

3 Overcoming the Poverty Trap in Low-Income Countries

With "subsistence" crops, farmers use few inputs and consequently have low yields. Then they are pressed to sell at the lowest annual prices right after the harvest by their need for cash. These farmers also sell a low quality product to the lowest level of the marketing chain.

To change all of this requires coordinated action on finance, production, and marketing. Farmers generally need credit to finance higher input levels especially inorganic fertilizers. At least moderate fertilization is required to respond to the low soil fertility levels generally found in Sub-Saharan Africa. A frequent response to the high costs of inputs is to reduce or eliminate expenditures on fertilizers. However, savings on costs for inorganic fertilizer will not function on these low fertility soils. The other alternative is to increase profitability rather than to reduce costs by utilizing marketing or processing improvements to obtain higher prices. More on marketing strategies will be presented in the next section.

Methods of increasing yields in cereals are well known, with the use of inorganic fertilizer, new cultivars responsive to fertilization, and improved agronomics being the primary drivers for yield improvements. In semiarid regions, this improved management includes improved water retention activities such as tied ridges and organic fertilizers.

Risk is often raised as a limiting factor in farmers' management decisions especially low-income farmers. Farmers in Sub-Saharan Africa practice a series of measures to reduce risks. Small farmers diversify their crops, their varieties, and land use. Early and late cultivars are often planted in the same fields. Farmers use the topography differences to reduce the risks of drought and flooding, planting sorghum on the lowlands and millet on the slopes for example. Households send off family members either permanently or seasonally to urban areas and mines. Then the migrants help the households in poor rainfall years and receive some of the harvest in good years. Water retention measures, both ridges and manure, reduce risks. However, farmers also understand the risk-return concept of needing to take more risk to earn more income. They mainly prefer not to put their subsistence at risk.

After dealing with risk the next step is to identify improved varieties, fertilizer levels, and obtaining financing for fertilizer purchases. In the Sahel of West Africa, it took several years to find a

Fig. 7 (Picture 1): Niatichama in 2007 near Bamako with farmers

Fig. 8 (Picture 2): Grinkan in 2008 in Garasso with farmers, extension agents and researchers

high yielding sorghum variety that farmers would adopt with adequate food grain and animal fodder qualities. The initial varieties from the experiment station had been selected for adverse low input conditions. They were too tall and lodged with even moderate fertilization (Fig. 7).

Farmers selected an intermediate height variety from a regional variety trial and pushed to introduce this into the pilot project trials. Farmers gave it the name "Grinkan" which roughly translates as "Look at This" (Fig. 8).

Fig. 9 (Picture 3): Garasso with farmer and extension agent

By contrast, the local cultivars principally produced stalks and little grain even with good production conditions (Fig. 9). Moderate fertilization in the pilot project was initially two 50 kg sacks of NPK (15-15-15) and one sack of Urea (46-0-0). Then there was a switch to the higher nutrient level DAP (18-48-0) with one sack of DAP and one sack of Urea. The first treatment has 38 kg ha^{-1} of N, and 15 of both P and K. The second treatment with only two bags but more nutrient concentration had 32 kg ha^{-1} of N, 24 kg of P and 0 of K [3]. Nitrogen and phosphorous are generally the most limiting nutrients. Potassium can become limiting over time as sorghum extracts substantial K but the more immediate problems are the deficiencies in N and P.

Farmers generally need financing to purchase higher inorganic fertilizers. This can be handled with farmers' associations in pilot projects. Our pilot project provided this credit to be reimbursed in grain at harvest time back to the farmers' association. Then the association was expected to store and hold the cereal prior to selling before the next harvest to provide a revolving fund to pay for input credits at planting time. These farmers' associations are useful for getting members to adopt, for providing storage of the grain to avoid forced sales at the period after the harvest when prices can collapse, and to handle the marketing of the cereals and the group purchases of inputs.

In 2009 there were six farmers' associations in southern Mali producing 606 ha of new sorghum varieties with the associated technologies mentioned above. Another three associations with 270 ha in central and northern Mali were producing millet with the same moderate fertilization and a well-known but improved millet cultivar, Toroniou [4].

The flagship village for sorghum was Garasso. The village enthusiastically joined the program and farmers followed the pilot program recommendations very closely. In 2008, the first year of the program, the 50 participating farmers averaged grain yields of 1.56 t ha^{-1} with the improved cultivar Grinkan. In 2009, participation increased to 138 members and sorghum grain yields were 1.94 t ha^{-1}. In her PhD thesis in Garasso, J. Coulibaly estimated grain yields of 1.64 t ha^{-1} in 2008. Regional yields of sorghum rotated after highly fertilized cottons averaged 800 kg to 1 ton ha^{-1}. Utilizing the improved cultivar Grinkan, yield increases compared to regional yields were 50–100% greater in 2008–2009 [5].

In 2008, the Garasso farmers' association with some financial help from a NGO had constructed storage facilities. In 2009, they accumulated 200 tons to sell in only the second year of the program. As they approached the start of the rainy season there was considerable excitement as the price had already passed 150 FCFA per kg. The farmers' association had been advised to wait for the best price and to sell in large quantities. However, an important logistic factor had been forgotten. During the rainy season, it becomes very difficult for large trucks to travel to and from Garasso on their dirt roads. This realization took place when the rainy season had begun and unfortunately, the buyer discounted the price offered to the farmers' association to 105 FCFA per kg because of these transportation issues. There was a feeling of disillusionment with the price and anger with the leadership of the farmers' association especially given the earlier price.

In the USAID-Mali office in Bamako there was the opposite reaction. There were pictures of Grinkan yields in farmers' fields on the walls of the USAID office. The reaction of USAID was to begin scaling up the program with new varieties, associated technology, and marketing strategies as articulated in the pilot. Before reporting on this scaling up, the marketing components that were used to support the production innovations are considered in the next section.

4 The Marketing Strategy

This is a fundamental aspect of the introduction of new technology. Input levels have to be increased to raise yields yet costs of imported fertilizers are high. However, five basic strategies can raise prices especially for farmers moving to more market involvement through farmers' associations: (1) Enable farmers to avoid the post-harvest price collapse through storage and group sales; (2) Add value to the cereal and obtain a premium price; (3) Sell to the highest bidder within the marketing chain and obtain some of the margins previously obtained by collectors and regional merchants; (4) Sell larger quantities and do more systematic marketing searches for better prices; and, (5) Buy inputs in quantity (cost reducing) and disseminate and monitor technology and marketing techniques through functioning farmers' associations. Examples of these strategies follow.

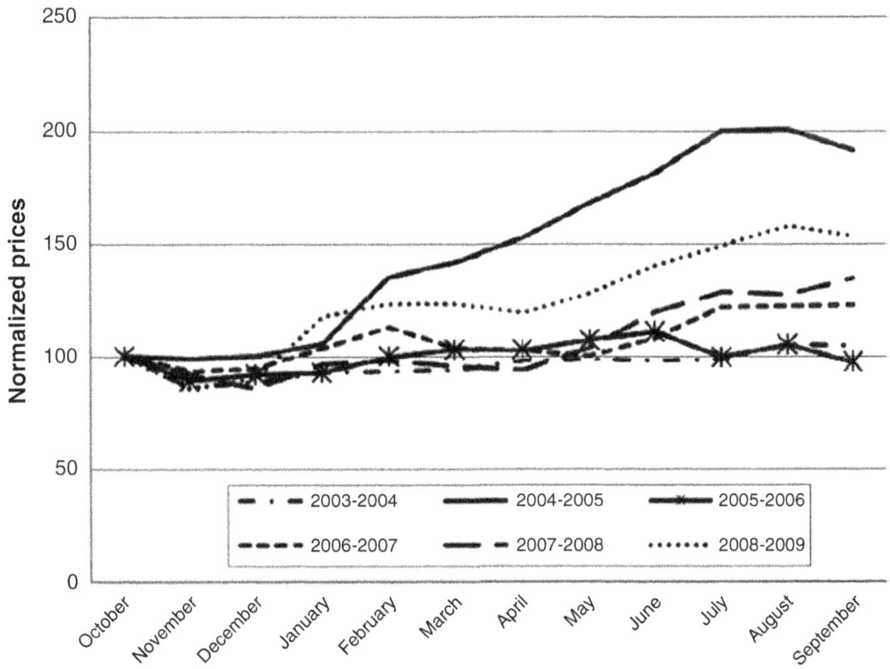

Fig. 10 Sorghum monthly price variation in Koutiala Retail Markets, 2003–2009. Source: J. Coulibaly, J.H. Sanders, P.V. Preckel, and T.G. Baker (2015) "Will Cotton Make a Comeback in Mali?"*Agricultural Economics.* 46:53–67

4.1 Storage and Group Sales

In poor rainfall years, such as 2004–2005, the farmers' association could have doubled the harvest price if they waited for the *soudure* (the hungry season) before the next harvest or waited for a 60% price increase in April before the next planting season (Fig. 10).[1] Since the base price for each year is an index with 100 at the harvest period the figure does not illustrate the well-known phenomenon of prices collapsing at harvest time in good rainfall years.

One difficulty farmers have in selling later is the liquidity pressure at harvest. Farmers need cash at this time to pay school fees, pay their workers including their wives,[2] pay for naming ceremonies (equivalent to baptisms in Islam), marriages, health and housing expenditures that had been deferred, and pay for migration for family members going off for part or full time work after the harvest. However, with increased productivity of 50 to 100% with their sorghum, farmers can respond to this liquidity demand at harvest and still have more cereal to sell later. Therefore, storage in poor rainfall years and in average years can bring increased prices,

[1] In normal and good rainfall years the technology gains are greater but the seasonal price increases are less.

[2] In the Islamic multiple wife system wives in Mali are paid for their fieldwork on the family plot in money, cereal, clothing or reciprocal labor services for their private plots [6].

which facilitates the financing of higher input costs and reduces the risk in adopting new technologies.

4.2 Added Value and Premium Prices

The traditional threshing method with mortar and pestle is breaking down with yield increases. So threshing is often done by running over the grain on the ground with a car or a tractor. Moreover, the heads of the sorghum and millet are traditionally put on the ground after cutting. In a small sample of grain purchased in Bamako in 2007 from retail cereal merchants, impurities were found to be approximately 5 to 25% of the purchased grains [7]. The predominantly female cereal processors (Fig. 11) sell the processed sorghum and millet in small plastic sacks from the kiosks of Bamako. They are reluctantly willing to pay a premium price for clean sorghum and millet so they do not have to clean it. This premium has so far been between 10 and 20 FCFA per kg. Keeping the cereal off the ground means putting the heads on the cut stalks and putting down a tarp where the cereal will be threshed. In addition, screens can be used to separate the grain from dirt, rocks and plant residue and some NGOs handling cereals do this.

4.3 Finding More Profitable Markets

The farmers' association enables the farmer and the association to bypass both the collector and the regional merchant (Fig. 11). The farmers' association then can sell the clean cereal directly to the processor or the wholesaler. There may be a struggle between the wholesaler(s) and the farmers' association for shares of the margins saved by eliminating the margins paid to the collectors and regional merchants. Price fixing by the wholesalers has been observed as they attempt to capture more of the saved margins. Selling larger quantities of clean grain cereal and being prepared to sell outside the region to overcome the price fixing would give the farmers' associations more bargaining power [8].

4.4 Selling Large Quantities of Grain and Understanding Markets

The farmers' association has an advantage in that it can sell larger quantities of grain compared to individual farmers and can also train some members to become informed about various marketing prospects outside the region. This can increase the bargaining power of the association when negotiating prices.

4.5 Buying Power, Dissemination, and Monitoring

Related to the previous points is the function of the farmers' association ability to buy large quantities of inputs and negotiate price discounts because of their larger purchases. The farmers' association can also facilitate the more rapid extension of new production practices and improved marketing techniques, such as the need for clean grain.

The effect of various combinations of these marketing practices was to increase prices received by the associations by 30 to 50% [9]. Over time, the farmers' associations will get better at these

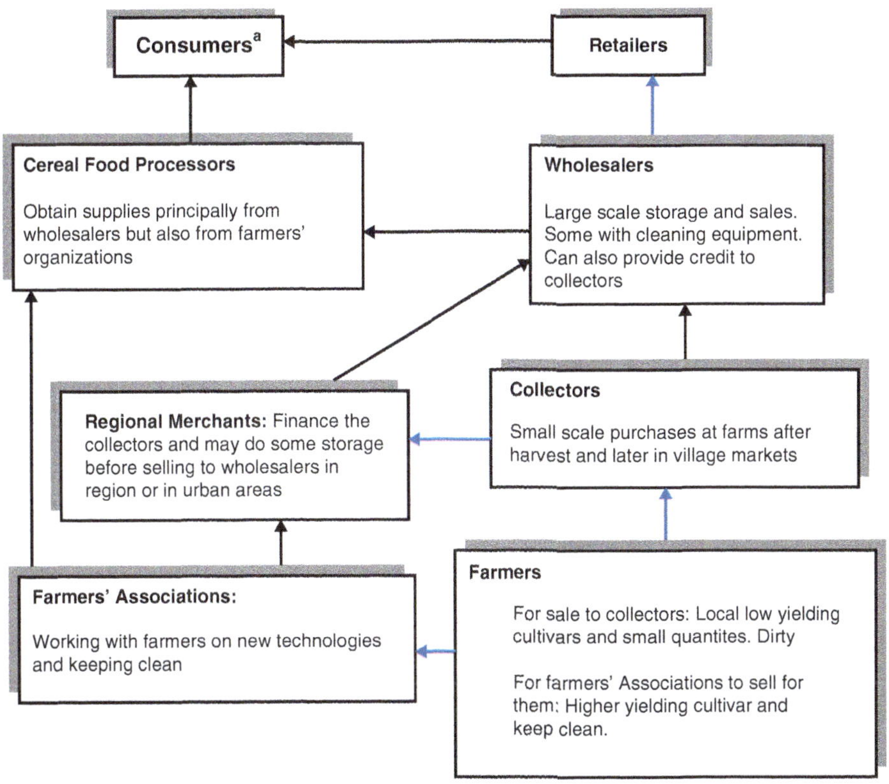

Fig. 11 Marketing structure for Sorghum/Millet in Mali, 2016. Source: J. Sanders et al. 2016

marketing practices and further increase prices as they have only partially adopted these practices. The farmers' associations need to be more transparent and build more confidence among their members. With increased confidence in the farmers' associations, farmers are expected to entrust more of their production to the associations for storage and marketing.

5 Scaling Up

After observing the success of the Grinkan pilot project from 2007 to 2009, Jean Harmon, the Agriculture Director of USAID-Mali, decided to switch activities in IICEM from providing services for six minor export crops to a program to rapidly diffuse Grinkan and Toroniou with the associated inputs and marketing strategies discussed above. IICEM was an AID program with substantial experience in financing, which was supported by Abt consulting. There was initial discussion of a 85,000 ha expansion but the scaling up program began with 2000 ha in the Mopti region for Toroniou and 3000 ha in the Koutiala and Sikasso regions.

The focus of the scaling up was twofold: obtain financing for farmers' associations from the national development bank of Mali (BNDA); and, lining up contracts with wholesalers for the association's sorghum and millet grains. The financing activity was very successful in the 3 years of the program, 2010–2012. The farmers' association and individual farmers had to provide sufficient information so that the BNDA knew they had not defaulted on previous loans for cotton. Second, IICEM had to put in the banks 50% of the value of the loans in the first year as a guarantee of repayment. This guarantee was reduced to 1/3 of the loan value in the second year and zero in the third year as the repayment experience was excellent.

Unfortunately, in 2012, the third year of this program, there was a coup by the Malian army. In response most donors cut off financing to all agencies related to the government until the civilian government was restored. Later in the year, jihadists invaded the north and started toward Bamako before the French Air Force intervened and drove them back into the desert.

In 2012, both USAID and the Dutch government were concerned with the welfare consequences for farmers after the coup and the jihadist attack. The banks had left the Mopti region in 2012. USAID-Mali provided credit to many of the farmers' associations in the north that the BNDA could not. This credit was to be reimbursed to the farmers' association as grain at harvest as in the pilot project. A similar program was implemented by the Dutch in the south, but the Dutch program only required one-half of the credit to be reimbursed.

Buyers had contracted with farmers' associations who would sell grain soon after the harvest and sell one-third of their production to the wholesalers. The price would be the market price plus 10 FCFA for quality. Farmers recognized that the wholesalers were trying to pay lower prices close to harvest and delayed threshing and supplying grain to the association. Farmers also sold sufficient grain to pay off loans for the fertilizer rather than one third of their production. Farmers were beginning to understand their market power associated with their ability to provide larger quantities of cleaner grains to the market.

The management and financing in scaling up and the substantial area expansion of the program were excellent. Unfortunately, problems in scaling up and technology failures occurred. We label this "technology erosion." There is the tendency of implementers of technology programs to move away from the recommendations of the pilot program often with deleterious effects on yields. There were problems getting good quality Grinkan seed so local varieties were often substituted. Little effort was made to obtain DAP and some farmers only utilized one sack of 15-15-15. Inputs are always a problem with the rapid expansion and scaling up of a pilot program. There was also an ownership issue in scaling up when a different organization took over and brought their own ideas about appropriate program components which is a common and recurring problem.

The scaling up program demonstrated the viability of the technology and support for financing from the BNDA. More importantly were the very high repayment rates of 90 to 100% from the various farmers' associations (Conversation with the Director of the IICEM program, winter 2013). In addition, the success of the farmers' associations in dealing directly with wholesalers and in honoring these contracts for loan repayment further showcased the success of the program.

The events of 2012 were very disruptive and a new consulting firm obtained the IICEM contract. The pilot project activities were shifted to Niger and Burkina Faso as the prohibitions on working with Malian governmental agencies continued. It became critical to respond to one Second-Generation problem that farmers complained about with Grinkan.

6 Second Generation Problems

Breeders typically have one of or several of the following objectives; pest and disease resistance, higher yields, and acceptable food functionality. Having too many objectives can limit progress in developing new cultivars that are acceptable to farmers. Many objectives can be difficult to achieve, such as drought tolerance and tolerance to low soil fertility. Most of the time the biologic nature of agriculture results in a continuous evolution of pests and diseases that need to be responded to.[3] In other cases some elements of farmers or consumer preferences slow, depress, or reverse introduction of new varieties. In Grinkan's case major issue arose that started with storage insect problems, then seed deterioration in the anaerobic storages sacks for insect control, and finally a taste preference for sticky "tô", a principal staple food dish in Mali that reversed the expansion of the use of this cultivar.

Grinkan is a soft seeded cultivar compared to the local hard sorghums in Mali. This means that insect damage in storage can be more serious if the grain is not properly stored. To handle this storage insect problem, triple overlapping polyethylene sacks (PIC sacks)[4] were employed to insure that it was anaerobic. However, if the sorghum is not sufficiently dry, seed decay can result in a substantial decrease in the viability of the seeds. Rains occurred late in both 2010 and 2011 and no drying equipment was available.

[3] With the introduction of IR-8 continuous cultivation of rice was enabled which substantially exacerbated the insect problems. The International Rice Research Institute needed to respond to this problem.

[4] Purdue had developed this technique to handle the extremely damaging storage insects with cowpeas. These bags are increasingly used across the developing world now with various crops.

Substantial decreases in germination rates of Grinkan from stored seeds were observed on the farm. Professional seed producers did not have this problem. Moreover, women stopping buying Grinkan in the market because the "tô" was not sticky enough. Even though both people and animals (forage) appreciated the taste, the stickiness problem and the poor germination rates of farmer stored seeds quickly shut down the market for and hence the production interest in Grinkan.

Fortunately, a food scientist in the team, Dr. Ababacar Ndoye, quickly identified the solution to the stickiness problem [10]. One of the villages had also independently discovered and adjusted for this problem. With the soft cultivars, it is not necessary to soak overnight as part of the processing. Eliminating this operation resolves the problem. A YouTube type movie was made of this process with participation from the village adopting it and shown to representatives of 19 of the farmers' associations, a seed producer, and a wholesaler promising to buy all the Grinkan produced. The wholesaler also offered free transportation to his warehouse for lots over 10 tons. Each farmers' association was given a flash drive with this movie to show in their village.

7 Demand Increases for Sorghum: Broilers

When trying to increase the production of a "subsistence" crop there is often a concern with the price collapsing if technology introduction rapidly occurs on a large scale. The very low-income sector may not be able to afford the commodity in processed form and the rest of the society may already have as much as they want of the commodity. Fortunately, the demand for high-quality foods increases as incomes grow. The demand shifts from cereals (in higher rainfall zones tubers) to fruits, vegetables, meat, milk, and cheese. Predominant beneficiaries of this process are broilers and dairy products. Broilers are generally confined and fed a prepared mix of 40 to 60% cereal, maize, and/or sorghum (for studies of the substitution of sorghum for maize [11–13]). The variation in percentage depends upon the growth stage of the chickens. However, field sampling and lab testing have shown the availability of a number of low tannin sorghum primarily but not exclusively on the experiment station [14].

High tannins in the sorghum can cause digestion problems if too much high tannin sorghum is in the feed ration. The USA has moved to non-tannin sorghums to address this feed issue, but this is often not widely done in developing countries. One animal nutrition expert has reported that non-tannin sorghum has 95 to 97% of the nutritional quality of maize (Hancock, conversation, fall, 2015) [15]. This implies that if the price of sorghum is below 95% of maize chicken producers and feed mixers should prefer non-tannin

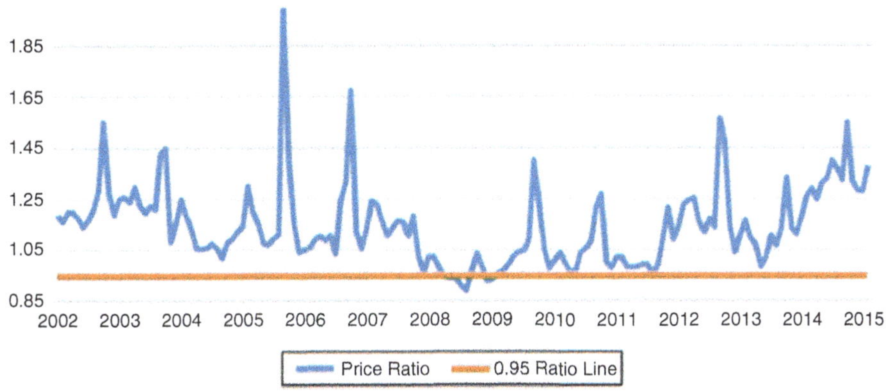

Fig. 12 Price ratios of sorghum to maize in Mali, 2002–2015. Source: Sanders et al. 2018

sorghum as the principal feed base rather than maize. Farmers and feed producers would also need to know that the sorghum was a non-tannin variety and that there was a consistent supply of this high-quality grain.

Therefore, a relevant question to ask is how often it would have been economical to switch from maize to sorghum. Between 2002 and 2015, 2008 was the only year in which the price ratio barely favored sorghum in Mali (Fig. 12). Unfortunately, most veterinarians and animal nutritionist in Mali advising chicken producers on feed composition did not know about the substitution potential of non-tannin sorghum for maize (interviews in Mali, summer 2015).

As demand grows for broilers and feed there will be pressure in many countries to push feed production into semiarid regions, where there are advantages to the production of sorghum over maize. Countries would also help their farmers by not banning maize exports as Mali frequently does, especially in adverse rainfall years. Removing the bans and other selective incentives for maize would benefit sorghum and maize producers with higher prices and therefore more incentive to adopt technological change.

8 Conclusions

A system of sorghum and millet technology introduction, improved institutional support of farmers' associations, and marketing improvements were introduced into several Sahelian countries focusing on Mali from 2008–2012. The program demonstrated that substantial yield gains are possible with moderate fertilization, an improved cultivar, and better agronomic management. This combination can also raise low-income farmers' incomes. Marketing improvements are simple as the markets evolve. Seasonal price variation indicates the potential for profitable returns through proper storage and marketing of the grain. A price premium for quality is an obvious target as more processed products of millet and sorghum become available in urban areas. Using a farmers' association enables farmers to sell at higher prices within marketing chains (wholesale and processors) and also allows them to obtain access to bank credit. Larger volume sales and some specialization in the farmers' associations on marketing alternatives are also taking place.

A problem of the soft seeded Grinkan was increased insect attacks. This required the use of the triple poly- ethylene sacks creating an anaerobic atmosphere killing the insects. But this procedure required thorough drying of the grain when utilized for seed. With late rains in 2010 and 2011 and the lack of drying equipment, there were problems of seed quality from storage in 2011 and 2012.

Moreover, the stickiness problem when making "tô" required changes in the processing methods for this soft seeded cultivar. A simple solution to this was the elimination of overnight soaking and this was demonstrated to 19 farmers' association representatives. However, drying still remains a problem with late rains for seed production of Grinkan. Farmers need to use professional seed producers with adequate equipment or as the value of seed increases invest in drying equipment themselves or in the associations.

With regard to second generation problems there is often a need for continuity in research and extension support as new problems occur in the field. This includes not only natural biological changes requiring further research and extension but also farmer or consumer requirements not foreseen by the breeder. Developing countries tend to use their funding in research and extension agencies to pay for personnel and then depend upon donors and NGOs for operational support. As one donor leaves, the next generally has his/her own program priorities which may not include overcoming the constraints now facing earlier programs, even successful ones. As more efficiency and transparency in the use of resources are attained within national research and extension programs, long term donor funding for continuity would be expected to have a high pay off.

Acknowledgments

Amanda Fuller found and organized the data reviewing sorghum production by region. Jean Harmon helped focus the concern of USAID and the Malian government on the importance of raising productivity of sorghum and millet. INTSORMIL, USAID-Washington, USAID-Mali, the McKnight Foundation and the Gates Foundation supported various phases of the fieldwork and research. John Yohe of INTSORMIL supported us over the entire period.

Appendix: Data for Maize and Sorghum, 2007–2008 to 2017–2018

Table 2
Maize and sorghum production (1000 metric tons). World, USA, and Sub-Saharan Africa, 2007–2017

	Maize				Sorghum		
	Year	World	United States	Sub-Saharan Africa	World	United States	Sub-Saharan Africa
2007/2008	2007	795,739	331,177	47,488	66,683	12,636	25,235
2008/2009	2008	799,985	305,911	48,774	63,498	12,087	24,999
2009/2010	2009	824,495	331,921	53,189	56,028	9693	21,387
2010/2011	2010	835,775	315,618	56,745	61,546	8775	23,786
2011/2012	2011	890,578	312,789	61,205	56,255	5410	23,639
2012/2013	2012	873,893	273,192	61,913	55,564	6293	22,588
2013/2014	2013	995,690	351,272	64,984	62,420	9966	23,729
2014/2015	2014	1,022,400	361,091	66,809	66,164	10,988	28,211
2015/2016	2015	972,813	345,506	62,466	63,365	15,158	25,359
2016/2017	2016	1,078,314	384,778	71,790	63,342	12,199	29,075
2017/2018	2017	1,036,662	370,960	73,285	57,429	9242	25,667

Source: Foreign Agricultural Service, *Grain World Markets and Trade*, Office of Global Analysis, United States Department of Agriculture, Washington, DC. Appendix, pp. 12–20

Table 3
Maize and sorghum harvested area (1000 ha). World, USA, and Sub-Saharan Africa, 2007–2017

	Maize				Sorghum		
	Year	World	United States	Sub-Saharan Africa	World	United States	Sub-Saharan Africa
2007/2008	2007	160,223	35,014	26,928	44,153	2749	26,820
2008/2009	2008	158,010	31,796	27,983	44,318	2959	27,456
2009/2010	2009	156,952	32,169	28,962	40,319	2227	24,482
2010/2011	2010	164,023	32,960	30,863	41,014	1945	24,949
2011/2012	2011	172,403	33,945	33,855	41,357	1596	26,969
2012/2013	2012	179,594	35,356	34,098	38,991	2021	24,019
2013/2014	2013	182,806	35,390	34,971	42,684	2665	27,276
2014/2015	2014	182,889	33,644	36,394	44,182	2590	28,881
2015/2016	2015	180,616	32,680	35,118	40,843	3177	25,500
2016/2017	2016	186,816	35,106	36,342	44,294	2494	30,265
2017/2018	2017	184,317	33,469	37,285	39,930	2042	27,179

Source: Foreign Agricultural Service, *Grain World Markets and Trade,* Office of Global Analysis, United States Department of Agriculture, Washington, DC. Appendix, pp. 12–20

Table 4
Maize and sorghum yields (metric tons per ha) World, USA, and Sub-Saharan Africa, 2007–2017

	Maize				Sorghum		
	Year	World	United States	Sub-Saharan Africa	World	United States	Sub-Saharan Africa
2007/2008	2007	4.97	9.46	1.76	1.51	4.6	0.94
2008/2009	2008	5.06	9.62	1.74	1.43	4.09	0.91
2009/2010	2009	5.25	10.32	1.84	1.39	4.35	0.87
2010/2011	2010	5.1	9.58	1.84	1.5	4.51	0.95
2011/2012	2011	5.17	9.22	1.81	1.36	3.39	0.88
2012/2013	2012	4.87	7.73	1.82	1.43	3.11	0.94
2013/2014	2013	5.45	9.93	1.86	1.46	3.74	0.87
2014/2015	2014	5.59	10.73	1.84	1.5	4.24	0.98
2015/2016	2015	5.39	10.57	1.78	1.55	4.77	0.99
2016/2017	2016	5.77	10.96	1.98	1.43	4.89	0.96
2017/2018	2017	5.62	11.08	1.97	1.44	4.53	0.94

Source: Foreign Agricultural Service, *Grain World Markets and Trade,* Office of Global Analysis, United States Department of Agriculture, Washington, DC. Appendix, pp. 12–20

References

1. Foreign Agricultural Service, USDA (2018) Grain world markets and trade. Foreign Agricultural Service, USDA, Washington, DC
2. Foreign Agricultural Service, USDA (2018) (http://www.fas.uada.gov/) Market and Trade data. Ethanol data for maize and sorghum were from. www.ethanolrfa.org/resources/industry/statistics/#1461260019819-00afed-a339. (This is the data for the 2017 calender year for ethanol from maize and sorghum combined with the other consumption data for the 2016–17 crop year)
3. Ouendeba B, Teme N (2013) Fiche de production et de commericialision de sorgho Grinkan. INTSORMIL, Lincoln, NE https://ag.purdue.edu/agecon/Pages/IPIM_Fran%C3%A7ais.aspx. (Pilot program recommendations for technology introduction and marketing strategy
4. Sanders JH, Ouendeba B, Ndoye A, Teme N (2018) Introducing new technology and marketing strategies for sorghum producers in developing countries: the Sahel case. In: Rooney W (ed) Achieving sustainable cultivation of sorghum, vol 2. Burleigh Dobbs Science Publishing Co, Cambridge England, p 156
5. Coulibaly YJ (2011) Diversification or cotton recovery in the Malian cotton zone: effects on households and women. Dissertation. Purdue University
6. Lilja N, Sanders JH, Duham CA, Groote Hd, Dembele I (1996) Factors influencing the payments of women in agriculture. Am J Agric Econ 78:1342
7. Toure A, Abdoulaye T, Sanders JH, Ouendeba B (2006) Transformacion commerciale du mil et du sorgho au Mali, Bulletin No. 7. INTSORMIL, Lincoln, NE https://ag.purdue.edu/agecon/Pages/IPIM_Fran%C3%A7ais.aspx
8. Sanders JH, Ouendeba B, Ndoye A, Traore S, Teme N (2016) Development of sorghum and millet production systems in Mali and Burkina Faso. Report to the Gates Foundation, Purdue University, West Lafayette, IN, p 183 https://ag.purdue.edu/agecon/Pages/IPIM.aspx
9. Baquedano F, Sanders JH, Vitale J (2010) Increasing incomes of Malian farmers: is elimination of US subsidies the only solution? Agric Econ 103:418–432; Coulibaly J, Kumeraswamy G, Sanders JH (2013) Economic impact of sorghum and millet technologies in Mali, Program results 2010–2011, bulletin no. 11. INTSORMIL, Lincoln, NE https://ag.purdue.edu/agecon/Pages/IPIM.aspx; Coulibaly J (2010) Evaluation des technologie de producion et de commercialisi de sorgho et du mil, Program results 2008–2009 bulletin no. 10. INTSORMIL, Lincoln, NE Also in English on above site. https://ag.purdue.edu/agecon/Pages/IPIM_Fran%C3%A7ais.aspx
10. Ndoye A, Ouendeba B (2016) Demand for millet/sorghum processed products in Mali and Burkina Faso. In: Sanders JH, Ouendeba B, Ndoye A, Traore S, Teme N (eds) Development of sorghum and millet production systems in Mali and Burkina Faso. Report to the Gates Foundation, Purdue University, West Lafayette, IN, p 183 https://ag.purdue.edu/agecon/Pages/IPIM.aspx
11. Carmencita DM, Nelia FC (2006) Feeding and economic evaluation of corn, wheat, and sorghum based diets in broilers. Philipp J Sci 135 (1):49–58
12. Clement IM, Kwari ID, IGwebuike J, Nkama I, Mohammad D, Hamaker B (2010) Performance and economics of production of broiler chickens fed sorghum or millet as replacement for maize in the semiarid zone of Nigeria. Agric Biol J N Am 1(3):321–325
13. Tandiang DM, Diop MT (2014) Effect of corn substitution by sorghum grain with low tannin content on broiler production: animal performance, nutrient digestibility, and carcass characteristics. Int J Poult Sci 13(10):568–574
14. Abdoulaye T, Sanders JH (2006) Quelle Cereal par los aliments de volaille en Afrique de l'Oest: Sorgho ou Mais? Bulletin No. 4. INTSORMIL, Lincoln, NE https://ag.purdue.edu/agecon/Pages/IPIM_Fran%C3%A7ais.aspx (This bulletin identifies the farmer and experiment station sorghum tannin levels based upon collection activities in four Sahelian countries with lab tests of the samples in the countries of the Sahel and at Texas A& M)
15. Hancock JD (2000) Value of sorghum and sorghum co-products in diets for livestock. In: Smith W, Frederickson RA (eds) Sorgum, origin, history, technology and production. Wiley Series in Crop Science, Hoboken, NJ, pp 731–751

Chapter 17

Background on Grain Sorghum Usage

John Duff, Austin Vincent, Doug Bice, and Ian Hoeffner

Abstract

Sorghum acreage increased significantly as a result of hybridization of the crop in the 1950s. This increase brought the cattle feeding industry to the High Plains; however, demand for the crop in this industry waned, and acres fell significantly due to unfavorable shifts in federal policy. The rise of the modern ethanol industry led to a resurgence in demand for sorghum and a subsequent increase in acres, and the interest generated by these occurrences led to greater interest in other end-uses such as food manufacturing. Sustainability is important to both these industries, so sorghum offers both significant benefits through its low water requirements and small carbon footprint.

Key words Ethanol, Renewable fuels, Cereal grains, Sustainability, Carbon intensity, Greenhouse gas

1 Review

The history of the U.S. grain sorghum [*Sorghum bicolor* (L.) Moench] industry has been impacted significantly by demand for the crop and interactions between marketplace participants. In 1955, grain sorghum was hybridized, and the resulting increase in yields and yield stability significantly boosted production in just a few years. Virtually all U.S. grain sorghum farmers adopted hybrids by 1961, and by the end of 1965, 5-year average production stood at 548 million bushels or more than triple the 170 million-bushel 5-year average ending in 1955. This increase was driven both by higher yields (the 5-year average yield ending in 1955 was 19 bushels per acre while the 5-year average yield ending in 1965 was 45 bushels per acre) and greater harvestability manifest in a higher harvest ratio (the 5-year average share of planted acres harvested ending in 1955 was 52% while the 5-year average share of planted acres harvested ending in 1965 was 75%) [1].

While this dramatic rise in grain sorghum yields and harvestable acreage greatly increased production, prices fell drastically in many areas. One such area was the High Plains, where proximity to markets in the 1950s was low and growing conditions *were* very

Zuo-Yu Zhao and Jeff Dahlberg (eds.), *Sorghum: Methods and Protocols*, Methods in Molecular Biology, vol. 1931, https://doi.org/10.1007/978-1-4939-9039-9_17,

challenging. The resulting oversupply of grain led farmers to explore value-added opportunities such as beef cattle feeding. In 1955, area farmers formed the Grain Sorghum Producers Association, and over the next 7 years the organization engaged in intensive study of cattle feeding and the feasibility of a large-scale industry on the High Plains. Significant feedyard expansion efforts began in 1962, and by 1973, the number of fed cattle marketed had grown to 4.8 million head. This represented a 1600% increase from the 300,000 head marketed in 1958 before concerted efforts to build a large-scale industry had begun [2].

Although an abundance of grain sorghum helped bring the cattle feeding industry to the High Plains, the crop was eventually supplanted by corn as the feed of choice for area feedyards. There were multiple reasons for this, including a large contraction in grain sorghum acres triggered by a series of negative policy signals in the Food Security Act of 1985 and a shift in grain attributes preferred by feeders. Rail access to low-cost corn produced in the Midwest was also a factor in this shift, and by 2017, feedyards fed just 60 million bushels of grain sorghum. By contrast, at the 1973 peak of grain sorghum use by cattle feeders, 694 million bushels were fed. Figure 1 illustrates this decline.

Fortunately for grain sorghum farmers, feeding beef cattle is often a suboptimum use for the crop, so the decline in this market did not adversely affect the industry. While grain sorghum is an integral part of the history of cattle feeding on the High Plains, the crop's lower energy value and higher protein level make it better suited to applications with greater indifference toward fat content (i.e., ethanol production) and a greater need for protein (e.g., swine production, dairy production). Grain sorghum's status as a gluten free grain rich in protein, fiber, and antioxidants also garners significant value outside traditional commercial livestock feed-based marketplaces such as the human consumption and companion animal markets. Furthermore, grain sorghum offers certain intangible benefits such as sustainability attributes on which only end-users like ethanol producers and food companies can capitalize.

Today, U.S. farmers plant grain sorghum on approximately six million acres from south Texas to central South Dakota and from eastern Colorado to just east of the Mississippi River. Eastern seaboard states from Georgia to Maryland have seen increased grain sorghum acres in recent years, as well. This trend has been driven largely by swine producers' need for the reliable protein source grain sorghum can provide regardless of environmental conditions. Figure 2 depicts the 5.7 million grain sorghum acres planted in the U.S. in 2017.

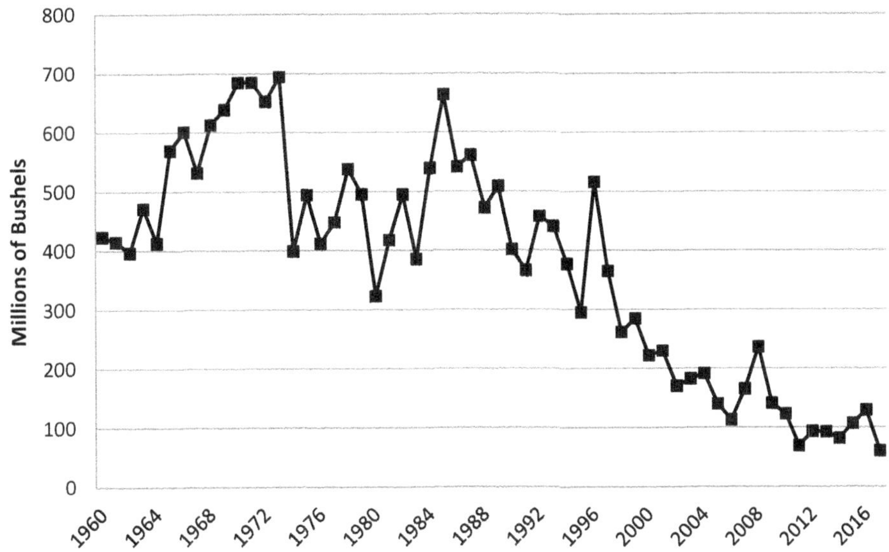

Fig. 1 Feed and residual usage of grain sorghum

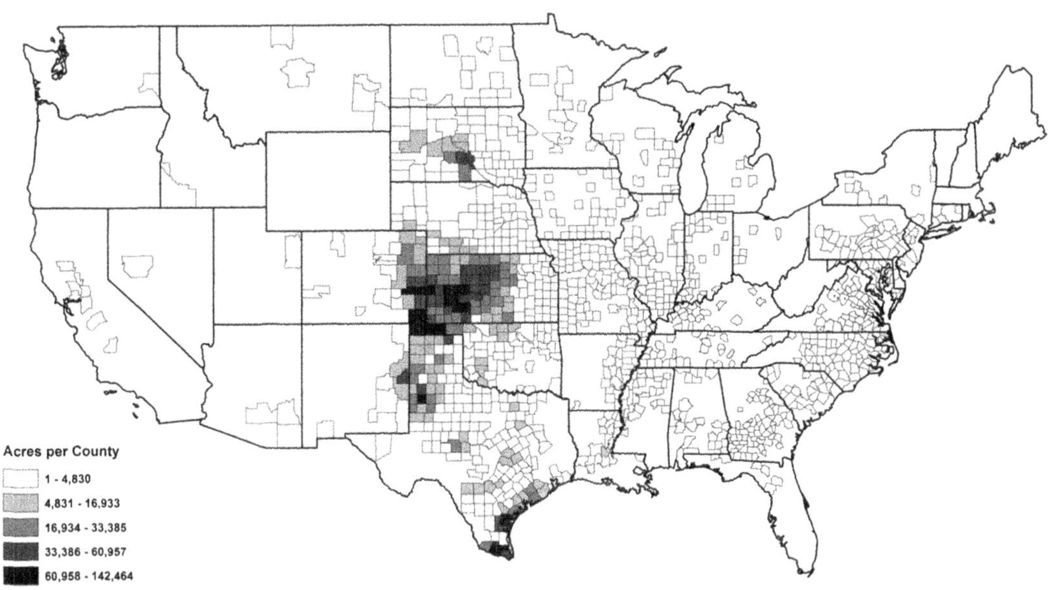

Fig. 2 Planted grain sorghum acres by county

2 Grain Sorghum Usage in Ethanol Applications

In 1896, Henry Ford introduced his first vehicle, the quadricycle. The quadricycle was fully powered by ethanol, and the legendary Model-T that followed it could be powered by ethanol, gasoline, or kerosene. Ford often publically predicted a future in which vehicles were powered by organic biomass, lending significant credence to the burgeoning fuel ethanol industry. Although producers experienced a setback when Prohibition began in 1919 (denatured alcohol, or alcohol mixed with petroleum fuel, was legal during Prohibition, but many on-farm ethanol plants were destroyed by federal authorities anyway), continued interest and investment in ethanol by Henry Ford, oil producers and other automobile manufacturers kept the industry growing. A need for rural economic stimulus also drove increases in production, and by 1945, wartime demand for ethanol had pushed production to 600 million gallons per year.

After nearly a half-century of development, the U.S. ethanol industry effectively ceased to exist with the conclusion of World War II and the end of contract ethanol production. Grain exports soared over the next three decades, and ethanol was no longer offered as a transportation fuel in much of the country. This changed with the energy crisis of the late 1970s and action taken by the U.S. Environmental Protection Agency (EPA) to phase lead out of gasoline. By 1978, the first federal policy promoting ethanol was instituted giving birth to the modern ethanol industry. Over the next decade, federal and state incentives promoted a rapid expansion of ethanol production; however, many facilities built as a result of these incentives did not survive [3].

In the early 1990s, ethanol production began growing again with production eclipsing 1.2 billion gallons per year by 1993 and 2.1 billion gallons by 2002. The specter of man-made climate change coupled with high oil prices soon led to additional federal action, and in 2005, the U.S. Congress instituted the first Renewable Fuel Standard (RFS1). This legislation was followed in 2007 by the second Renewable Fuel Standard (RFS2) which set aggressive renewable fuel blending requirements culminating with a requirement to blend 36 billion gallons of renewable fuel into the U.S. fuel supply in 2022. This action helped facilitate an historic build-out of ethanol production facilities, and today, U.S. production totals 15.4 billion gallons and commands approximately 10% of the domestic transportation fuel market [4]. This has meant significant value not only to corn farmers but to grain sorghum farmers, as well.

After federal incentives were introduced in the late 1970s, many Midwestern farmers again began building on-farm ethanol production facilities to add value to their own corn. Farmers in the state of Minnesota provided many of the most notable examples; however, farmers from southern plains states such as Kansas and New Mexico also built on-farm facilities to add value to their own grain sorghum. These small facilities provided a proof of concept, and when the mid-2000s build-out began, a significant amount of ethanol production infrastructure was added in important grain sorghum growing areas including Kansas and Texas. Figure 3 illustrates the explosive growth of food, seed and industrial usage of

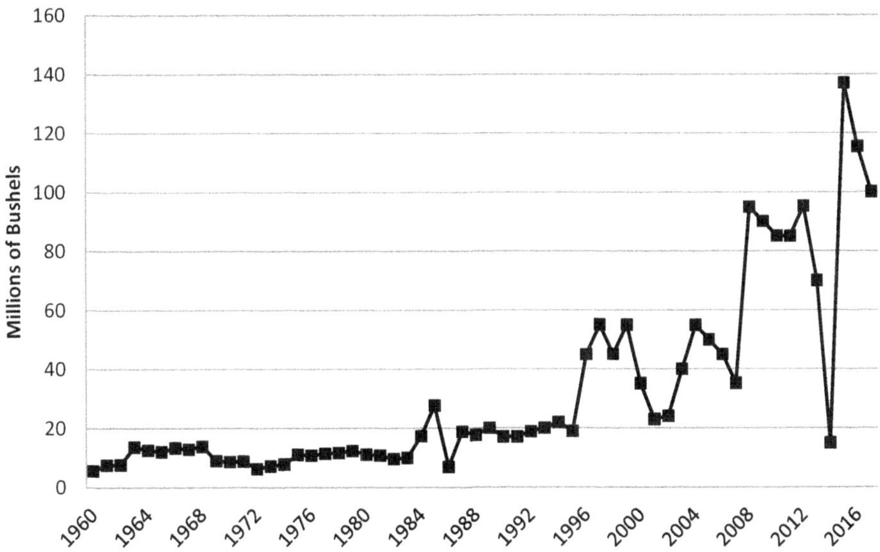

Fig. 3 Food, seed, and industrial usage of grain sorghum

grain sorghum while Fig. 4 maps the facilities that today produce a majority of U.S. grain sorghum ethanol. The size of each circle depicts the represented facility's relative demand for grain sorghum.

Grain sorghum is an important part of these facilities' strategies not only for their current feedstock procurement needs but for their growth strategies over the long term, as well. It is important to note grain sorghum is grown and used to produce ethanol in arid regions. For this reason, relying wholly on locally-produced corn is a significant risk for these ethanol producers. Because of drought and heat stress, locally-produced nonirrigated corn is often of low quality or unavailable. Grain sorghum provides a much more stable supply in terms of both quality and quantity, and due to nearly annual occurrences of drought and heat stress in some portion of the High Plains growing season, the starch levels in grain sorghum are comparable to those found in corn. This fact coupled with reliability and often favorable pricing makes grain sorghum the preferred choice for these ethanol producers.

In addition to ongoing concerns about the stability of supply, concerns about the future of irrigated agriculture on the High Plains persist. With approximately ten million acres of irrigated corn situated above the Ogallala Aquifer in Colorado, Kansas, Nebraska, New Mexico, Oklahoma, and Texas, many farmers, end-users, and other supply chain participants have expressed concern over the economic repercussions of severe and sudden declines in groundwater availability. Given the High Plains is currently a grain-deficit region (more grain is used in the region than can be produced locally), reductions in grain production would have an

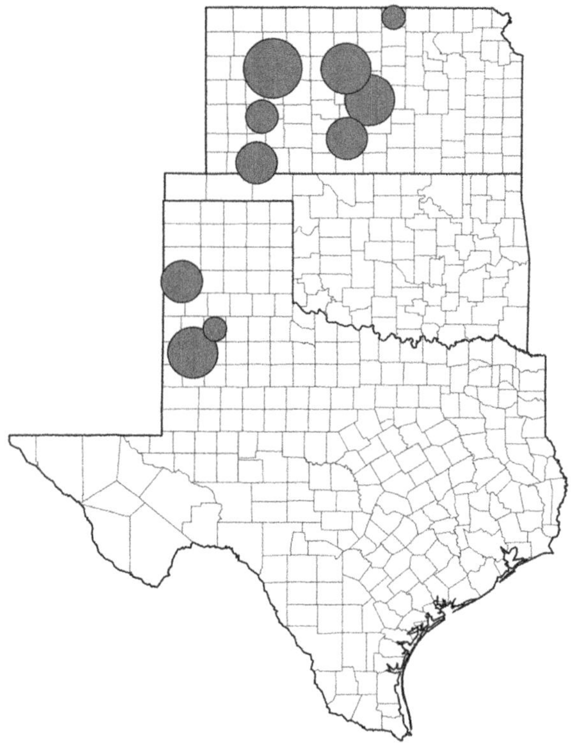

Fig. 4 Grain sorghum ethanol producers by location and relative demand for grain sorghum

outsized impact on supply-demand relationships and compound the effects of farm financial stress on rural communities by threatening these economies' only other job-creating engines. Ethanol producers on the High Plains are keenly aware of this situation and see grain sorghum, which uses one-third less water than corn, as an important part of the solution to this growing problem.

3 Grain Sorghum Usage in Food Applications

By production volume, grain sorghum is the fifth-largest cereal grain in the world. The crop is a dietary staple in many countries, but its use in U.S. human consumption and companion animal food markets is often overlooked compared to that of other grains such as corn and wheat. This relative obscurity is due to the geographic location of the crop's largest consumers (much of the grain sorghum consumed outside the U.S. is produced on a subsistence basis independent of global supply-demand relationships, and thus the crop rarely attracts significant media focus) and its demand in other applications. Outside of these traditional production and demand regions, grain sorghum as a food crop has only recently begun to gain attention. Often referred to as an ancient grain, remnants of grain sorghum plants dating to 8000 BCE have been discovered in archeological sites in southern Egypt, and breeders use these genetic foundations for crop improvement today. Whether as porridge (as originally used in Africa) or finely ground flour (as originally used in India), grain sorghum has an array of food applications due to its nutrient profile and variety of usage and preparation methods [5].

In the U.S., consumer demand for ancient grains is increasing, and with this increase has come higher visibility and demand for food-grade grain sorghum (food-grade grain sorghum is often of higher quality than commodity grain sorghum, or grain sorghum sold for other uses, and the two types are rarely used interchangeably). Comprised of 75% complex carbohydrates, grain sorghum is an energy-dense cereal also high in iron, zinc, B complex vitamins, and numerous phytochemicals. Although grain sorghum is high in protein, it does not contain the storage protein known as gluten. Gluten is found in wheat, barley, and rye, so its absence in grain sorghum has fueled demand from consumers with celiac disease or other types of gluten-intolerance [6]. The crop's water-efficiency is also a key demand driver as U.S. consumers are becoming increasingly conscious of the ways in which their dietary choices impact the environment.

Like demand for food-grade grain sorghum for human consumption, U.S. consumer demand for the crop for companion animal food is increasing, as well. Primarily those owning dogs and cats as well as the food manufacturers that supply them seek high-quality ingredients, and grain sorghum has many of the attributes these groups demand. The same attributes that drive demand for grain sorghum in human food markets make the crop attractive for companion animal formulations. These attributes include a low glycemic index, an absence of gluten as well as high levels of antioxidants, dietary fiber, minerals and protein. Grain sorghum's water-efficiency is also an important factor in demand from companion animal owners. In response to this demand, 15 companion animal food companies are now using sorghum to produce more than 130 products [7]. Figure 5 illustrates the increase in grain sorghum usage in human and companion animal food markets [8]. Combined, these industries have increased grain sorghum usage by 250%.

4 Grain Sorghum Sustainability Benefits in Ethanol and Food Applications

Aside from the direct benefits grain sorghum offers ethanol producers and food companies (e.g., reliability and resource conservation for ethanol producers, nutritional attributes for food companies), the crop brings significant and often intangible benefits only

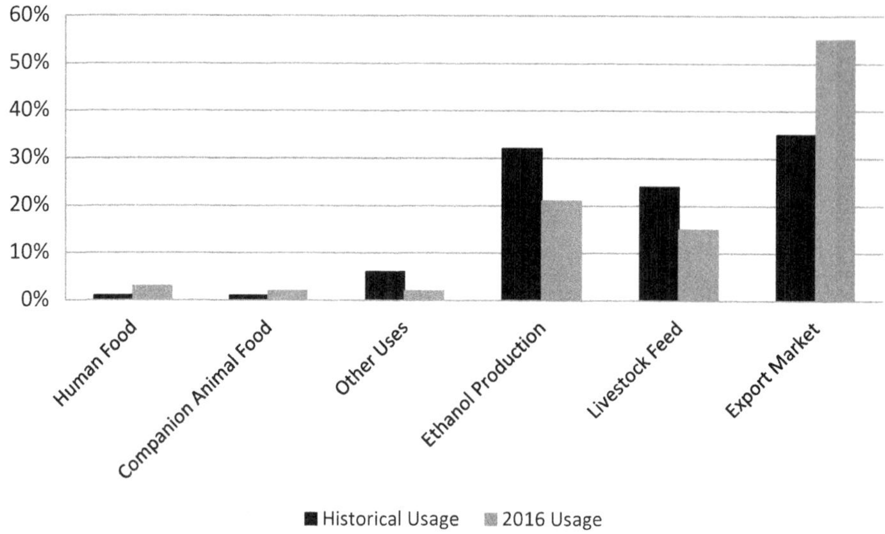

Fig. 5 Historical grain sorghum usage compared to 2016 grain sorghum usage by industry share

valuable in these applications. For example, grain sorghum ethanol's ability to reduce greenhouse gas (GHG) emissions compared to a petroleum fuel baseline makes it eligible to generate additional revenue for its producers under the RFS2's market-based system that rewards progressively higher GHG emissions reductions. In its analysis of grain sorghum ethanol, the EPA concluded by 2022, an average gallon of grain sorghum ethanol would reduce GHG emissions by 32% compared to the petroleum fuel baseline [9] while an average gallon of corn ethanol would reduce GHG emissions by 20% compared to the petroleum fuel baseline [10]. Figure 6 depicts the output of EPA probability models of grain sorghum and corn ethanol GHG emissions.

In addition to the RFS2, California has enacted its own legislation aimed at lowering GHG emissions and curbing man-made climate change. Similar to the RFS2, the landmark California Low Carbon Fuel Standard (LCFS) rewards the production of fuels with smaller carbon footprints. The LCFS uses a California-specific version of the Greenhouse Gases, Regulated Emissions and Energy Use in Transportation Model (CA-GREET) to assign each fuel a carbon intensity (CI) score. Approximately 95% of the GHG emissions associated with farming inputs can be accounted for by the nitrogen fertilizer application rate, and a growing body of evidence supports an improving grain sorghum nitrogen application rate that is comparable to that of corn. Table 1 compares five key farming input values for grain sorghum as found in CA-GREET version 2.0 and based on U.S. Department of Agriculture (USDA) data collected beginning in 1991 [11] and as collected by SGS North America from 2008 through 2011 [12].

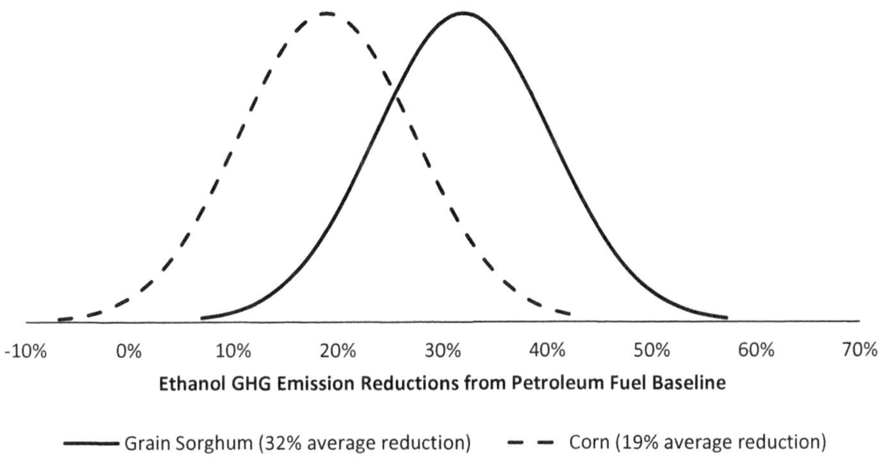

Fig. 6 Probabilities of ethanol GHG emission reductions modeled by the EPA

Table 1
Grain sorghum farming input values

Input	USDA	SGS North America
Nitrogen in grams per bushel	511.53	411.30
Phosphorus in grams per bushel	120.74	99.73
Potassium in grams per bushel	18.64	19.83
Herbicide in grams per bushel	28.10	27.46
Energy in BTUs per bushel	16,405.73	10,096.15

State extension service hybrid trial data also point toward increasing fertilizer use efficiency in grain sorghum. Table 2 summarizes the results of hybrid trials conducted from 2008 through 2016 on the High Plains. This database includes 3216 observations and an average nitrogen application rate of 390.59 grams per bushel [13]. This fact is noteworthy as hybrid trials are often over-fertilized to prevent inefficiencies like leaching and volatilization from limiting yields. Equally noteworthy is the fact this database includes multiple drought years but finds no adverse impact on fertilizer use efficiency. A significant problem with USDA grain sorghum data is most surveys are conducted in only a limited number of years due to budget constraints and the relatively small size of the U.S. grain sorghum crop. However, limited collection periods often result in skewed outcomes due to increased chances of sampling a drought year and finding abnormally high inefficiency. An average of 390.59 g per bushel collected over large areas and significant amounts of time demonstrates grain sorghum fertilizer usage efficiency is much higher than USDA data show.

Table 2
Grain sorghum farming input values observed in state extension hybrid trials

State	Observations	Grams per bushel applied		
		Nitrogen	Phosphorus	Potassium
Colorado	349	485.84	119.85	6.76
Kansas	1331	424.54	61.18	0.00
Nebraska	120	398.08	64.60	0.00
New Mexico	222	408.63	176.60	0.00
Oklahoma	319	365.88	116.02	0.00
South Dakota	462	387.94	19.66	0.00
Texas	413	266.78	70.00	0.00
Average	459	390.59	72.48	0.41

This efficiency is due to a combination of factors including a root system which enables grain sorghum to scavenge for nitrogen and production systems which enable greater nitrogen accumulation in the soil. While the former factor is currently being studied in grain sorghum by the U.S. Department of Energy's $35 million program known as Rhizosphere Observations Optimizing Terrestrial Sequestration or ROOTS [14], a significant amount of literature has been published on the latter factor. For example, one long-term study conducted by the USDA found that in wheat-grain sorghum rotations enough nitrogen is mineralized to produce both wheat and grain sorghum yields in the absence of supplemental fertilizer [15]. Figure 7 maps the Kansas fields on which grain sorghum is produced in rotation with wheat. In Kansas, where a significant majority of grain sorghum ethanol originates, 59% of grain sorghum acres can be found in this type of rotation [16].

Other studies highlight grain sorghum's nitrogen usage efficiency, as well. A Kansas State University paper in publication based on 26 years of grain sorghum yields as well as nitrogen application and uptake rates found that high levels of residual nitrogen carry over from year to year [17]. This implies land on which grain sorghum is produced—particularly when no-till systems are employed as in this study—loses little or no nitrogen between growing seasons. In terms of GHG emissions reductions, this is significant in that it lowers the amount of supplemental nitrogen needed for the following crop and means less nitrogen escapes into the atmosphere. This beneficial carbon footprint coupled with a water requirement one-third less than that of corn has driven increased demand for grain sorghum for food use, as well.

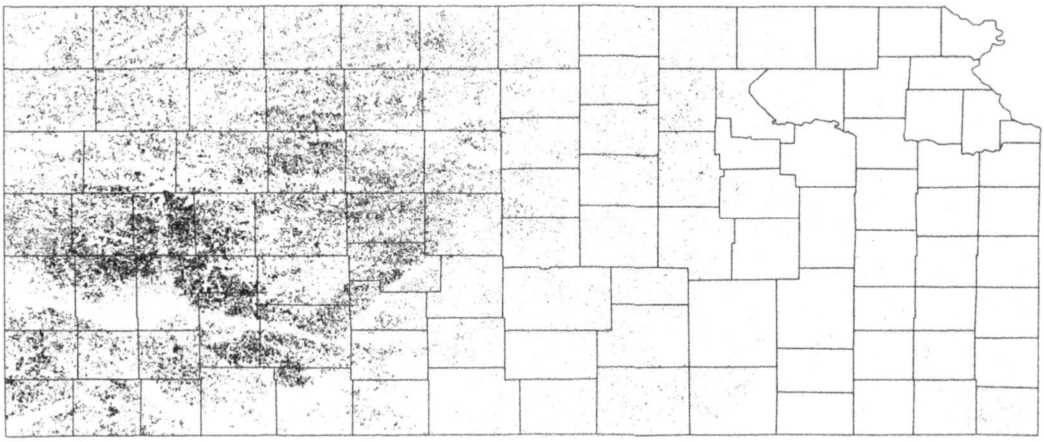

Fig. 7 Fields on which grain sorghum was produced in rotation with wheat in Kansas

5 Conclusions

Demand for grain sorghum has had a significant effect on the crop and the industry that supports it over the past six decades. Recently, grain sorghum has experienced a resurgence in demand driven by new applications such as ethanol and food production. Neither of these uses is truly new; however, they continue to increase in both market share and importance. Due in part to favorable physiological characteristics and in part to positive sustainability attributes, grain sorghum demand from these applications is and will likely continue to be strong.

References

1. USDA-NASS (2018) Quickstats. https://quickstats.nass.usda.gov/. Accessed 14 Sept 2017
2. Ball CE (2010) Cattle feeding. https://tshaonline.org/handbook/online/articles/aucrw. Accessed 14 Sept 2017
3. ICM (2017) It's still (essentially) your great-granfather's ethanol. http://www.icminc.com/innovation/ethanol/ethanol-timeline.html. Accessed 15 Sept 2017
4. U.S. Energy Information Administration (2018) Monthly energy review April 2018. https://www.eia.gov/renewable/data.php#biomass. Accessed 17 Oct 2017
5. Henley EC (2010) The sorghum story: an ancient, healthy and nutritious old world cereal. United Sorghum Checkoff, Lubbock TX
6. United Sorghum Checkoff Program (2018) Consumer food. http://www.sorghumcheckoff.com/market-opportunities/consumer-food. Accessed 1 Nov 2017
7. Aldrich G (2015) Is sorghum a superfood for modern pet food? http://www.sorghumcheckoff.com/market-opportunities/pet-food. Accessed 1 Nov 2017
8. United Sorghum Checkoff Program (2018) All about sorghum. http://www.sorghumcheckoff.com/all-about-sorghum. Accessed 1 Nov 2017
9. US EPA (2012) Federal register Vol. 77, No. 242. https://www.gpo.gov/fdsys/pkg/FR-2012-12-17/pdf/2012-30100.pdf. Monday, 17 Dec 2012
10. US EPA (2010) Federal register Vol 75, No. 58. https://www.gpo.gov/fdsys/pkg/FR-2010-03-26/pdf/2010-3851.pdf. 26 March 2010

11. CA-GREET version 2.0 (2017) https://www.arb.ca.gov/fuels/lcfs/ca-greet/ca-greet2.0-tier1.xlsm. Accessed 18 Oct 2017
12. Oulton M, Seiter M, Whitfield S, Meerburg M, Jan Olthuis H (2015) Agricultural market research: the carbon footprint of sorghum. SGS North America (Société Générale de Surveillance). http://www.sorghumcheckoff.com/assets/media/pdfs/The%20Carbon%20Footprint%20of%20Sorghum.pdf. Accessed 14 Sept 2017
13. United Sorghum Checkoff Program (2018) Links website. http://sorghumgrowers.com/fg5cb812c0l2.html. Accessed 14 Sept 2017
14. US-DOE ARPA-e (2016) ARPA-e roots program overview presentation. https://arpa-e.energy.gov/sites/default/files/documents/files/ROOTS_ProgramOverview.pdf. Accessed 14 Sept 2017
15. Jones OR, Johnson GL (1996) A ten year comparison of cropping and tillage systems for dryland grain production. USDA-ARS Report # 96–04. http://amarillo.tamu.edu/files/2010/11/TenYear_dryland_grain_production.pdf
16. USDA-NASS (2018) CropScape-cropland data layer. https://nassgeodata.gmu.edu/CropScape/. Accessed 14 Sept 2017
17. Mengel D (2018) Personal Communication

Chapter 18

Assaying Sorghum for Fuel Production

Courtney Payne, Justin Sluiter, and Ed Wolfrum

Abstract

The publicly available NREL suite of laboratory analytical procedures (LAPs) provides researchers with the analytical tools to determine the composition of woody and herbaceous feedstocks. Feedstock characterization includes several steps: sample preparation, starch determination, moisture analysis, ashing, extraction, and hydrolysis. These steps provide information on specific compounds and classes of compounds such as ash, protein, moisture, extractives, sucrose, starch, glucan, xylan, galactan, arabinan, acetate and lignin. Here, we describe the use of these procedures to characterize sorghum [*Sorghum bicolor* (L.) Moench] with specific consideration of four main sorghum types: grain, forage, sweet, and biomass, which can vary significantly in composition. Special attention is paid to the extraction and the differentiation of glucose from starch and glucan from cellulose in the hybrids with significant amounts of nonstructural carbohydrates.

Key words Sorghum, Glucan, Xylan, Lignin, Characterization, LAPs, Structural carbohydrates, Starch, Cellulose, Hemicellulose, Lignocellulose, Nonstructural carbohydrates

1 Introduction

The NREL Laboratory Analytical Procedures (LAPs) for biomass compositional analysis were developed more than 20 years ago and are based on the Uppsala extraction method and the Saeman hydrolysis [1]. The LAPs were developed as a suite of procedures for the characterization of woody and herbaceous biomass. The procedures measure a number of individual analytes and compound classes including ash, water extractives, ethanol extractives, sucrose, free glucose, free fructose, protein, lignin, glucan, xylan, arabinan, galactan, and acetate. The LAPs work well in their characterization of woody and most herbaceous materials and when used as a suite of procedures often provide a summative mass closure between 95 and 100%. These procedures are publicly available online and provide considerable details and practical considerations regarding execution of the methods [2]. The history and description of the methods as well as an evaluation of method uncertainties have been published [1, 3].

Zuo-Yu Zhao and Jeff Dahlberg (eds.), *Sorghum: Methods and Protocols*, Methods in Molecular Biology, vol. 1931,
https://doi.org/10.1007/978-1-4939-9039-9_18,

Here, we outline these methods as applied to the characterization of sorghum [*Sorghum bicolor* (L.) Moench]. Because the procedures have been published elsewhere, we will reference the relevant sections of these methods that require a change or would benefit from additional information. Four basic types of sorghum as defined by their end—use are grain, sweet, forage, and biomass. These four types can be significantly different compositionally [4]. For example, grain sorghum is bred for grain and is often high in starch. Sweet sorghum, like sugar cane, is bred for its nonstructural carbohydrates specifically sucrose. Forage sorghum, as the name implies, is used for animal nutrition and is primarily grown with an emphasis on dry tonnage. Biomass sorghum is typically bred to minimize panicle formation and maximize cellulose content and (like forage sorghum) dry tonnage [4–9]. With consideration of these differences the extraction method and starch assay used in the overall analysis can play significant roles, or in the case of starch can be omitted completely if no grain is present.

Figure 1 shows a flow diagram for the individual analyses required to determine the appropriate constituents in sorghum samples. These procedures start with sample preparation to ensure a more uniform material which recommends milling to a 2 mm particle size. Methods are optimized for a specific moisture content and therefore samples should be dried to less than 10% moisture. The resulting percentage of moisture in the sample is determined to calculate all analytes back to a dry weight basis. Samples are extracted using a two-step procedure using water and then ethanol with soxhlet or automated solvent extractor (ASE). Water extractives may include inorganic material, nonstructural carbohydrates, and extractable protein. This step is particularly important for sorghum as some types, specifically sweet sorghum, can contain large amount of nonstructural carbohydrates. These methods were originally optimized with corn stover in mind which is typically low in nonstructural sugars. If these sugars are not removed prior to hydrolysis, they would interfere with the quantification of glucan from cellulose. The ethanol extractives may include waxes, chlorophyll, and other minor compounds. The extractives free sorghum is then subjected to a two-step acid hydrolysis, 72% sulfuric acid followed by 4% under autoclave conditions. The initial hydrolysis serves to hydrolyze cellulose and hemicellulose into oligomers and the hydrolysis at temperature to further break the oligomers into their monomeric sugars. The lignin fractionates into both soluble (low molecular weight) and insoluble (high molecular weight) lignin during hydrolysis. The insoluble lignin portion is referred to as Klason lignin. The liberated acetate is also measured. The inorganic content of the sorghum is measured on the whole material through ashing at an elevated temperature. Ash can also be quantified on the extractives free material to correct the Klason lignin for any inorganic contributions. Protein content can also be

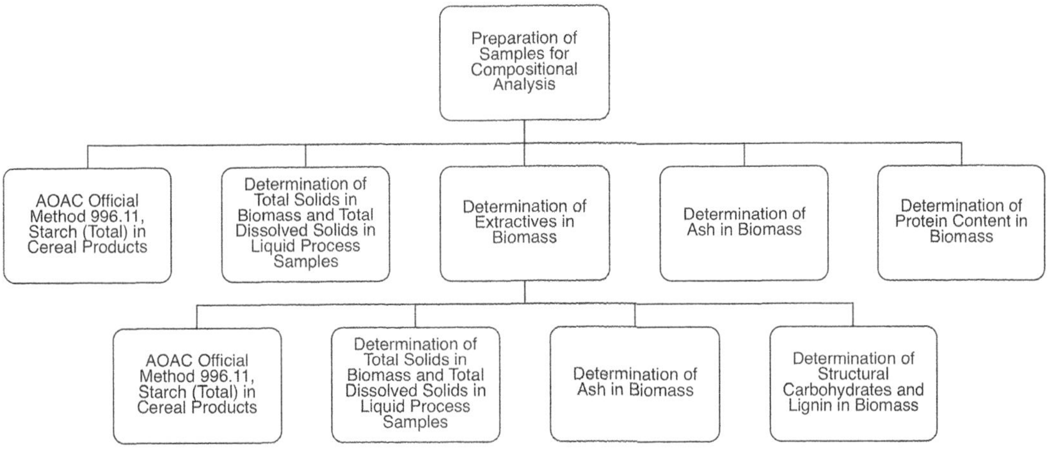

Fig. 1 Flow diagram of individual LAPs and AOAC procedures for analyzing a variety of sorghum hybrid samples. This diagram is not intended for use in analyzing pure grain or seed head only samples

determined like ash on the whole and extractives free biomass. Protein is determined through calculation from the Kjeldahl nitrogen value.

Starch is an important polymer for consideration in the compositional analysis of sorghum. It may play little to no role in forage and biomass sorghums or a large role in grain sorghums. Quantification of starch is necessary as it can overestimate the cellulose content if not properly accounted for prior to hydrolysis. NREL employs the Megazyme or AOAC method for quantification of resistant starch [10]. This assay is performed on the whole sorghum material to determine an overall starch content as well as on the extractives free material. This is necessary because neither the water nor ethanol extraction completely removes starch from the biomass. Therefore, to ensure proper quantification of glucan from cellulose and not starch, the extractives free material should be assayed. From the previously described NREL methods and independent starch assay, we can obtain mass closures of 100 ± 3% dry weight for most sorghum feedstocks.

2 Materials

Each publicly available LAP has an extensive materials section which is not reproduced here but can be accessed at: https://www.nrel.gov/bioenergy/biomass-compositional-analysis.html. Please access the below listed documents online under the *Procedures* subheading and refer to the "Apparatus" and "Reagents and materials" sections of each technical report. Numbered references from these reports (e.g., 7.1.1) are indicated below where *Notes* have been provided or changes to the method have been made.

2.1 Sample Preparation

"Preparation of Samples for Compositional Analysis" (*see* **Note 1**).

2.2 Starch Assay

We follow the AOAC recommended method (996.11) or the Megazymes Total Starch assay (AOAC Method 996.11, AACC Method 76.13, and ICC Standard method No. 168) for, "Determination of total starch content of samples containing resistant starch, but no D-glucose and/or maltodextrins" [10]. This method provides detailed information on reagents and equipment for proper execution (*see* **Note 2**). NREL deviates from this with the incorporation of a high-purity starch standard in place of flour and the exclusion of the GOPD reagent and use of UV/Vis for quantification. NREL analyzes the hydrolyzed solution using high-pressure liquid chromatography (HPLC) with additional details provided below.

1. Water, HPLC grade (*see* **Note 3**).
2. High-purity corn starch as a reference standard.
3. HPLC with refractive index detection (RID): Biorad Aminex HPX-87H column with corresponding guard column, and 0.01 N H_2SO_4 water mobile phase.
4. Starch HPLC standards: water purified using a 0.2 μm filter, high-purity standards D(+)glucose, glycerol, and ethanol. Prepared at concentrations of 0.1 mg/ml, 0.5 mg/ml, 5.0 mg/ml, and 12 mg/ml (*see* **Note 4**).
5. Second set of high-purity standards, as listed above, from a different source (manufacturer or lot), to be used to prepare calibration verification standards (CVS), water purified using a 0.2 μm filter, and prepared at concentration of 3.0 mg/ml (*see* **Note 4**).
6. 15 ml high-clarity polypropylene conical tubes instead of glass test tubes.

2.3 %Solids

"Determination of Total Solids in Biomass and Total Dissolved Solids in Liquid Process Samples."

2.4 Extractives

"Determination of Extractives in Biomass".

Since this LAP was last updated in 2005 we have replaced YSI quantification of nonstructural carbohydrates in the water extractives with HPLC. Additional details are provided below.

1. 7.1.1: Water, HPLC grade (*see* **Note 3**).
2. HPLC with refractive index detection (RID): Shodex sugar SP0810 column with ionic form H^+/CO_3^- deashing guard column, and ultrapure water mobile phase.
3. Nonstructural Carbohydrate HPLC standards: water purified using a 0.2 μm filter, high-purity standards sucrose, D(+)

glucose, and D(+)fructose. Prepared at concentrations of 0.04 mg/ml, 0.01 mg/ml, 1.0 mg/ml, and 4 mg/ml (*see* **Note 4**).

4. Second set of high-purity standards, as listed above, from a different source (manufacturer or lot), to be used to prepare calibration verification standards (CVS), water purified using a 0.2 μm filter, and prepared at concentration of 0.5 mg/ml (*see* **Note 4**).

2.5 Structurals

"Determination of Structural Carbohydrates and Lignin in Biomass"

1. 7.1.1: 72% H_2SO_4 (*see* **Note 5**).
2. 7.2.2: Pressure tubes (*see* **Note 6**).
3. 7.2.4: Filtering crucibles (*see* **Note 7**).

2.6 Ash

"Determination of Ash in Biomass."

2.7 Protein

"Protein Content in Biomass."

3 Methods

Each publicly available LAP has an extensive methods and calculations section which is not reproduced here but can be accessed at: https://www.nrel.gov/bioenergy/biomass-compositional-analysis.html. Please access the below listed documents online under the *Procedures* subheading and refer to the "Procedures" and "Calculations" sections of each technical report. Numbered references from these reports (e.g., 10.3.5.1) are indicated below where *Notes* have been provided or changes to the method have been made.

3.1 Summative Mass Closure for Feedstocks

"Summative Mass Closure," Technical Report, NREL/TP-510-48087, Revised July 2011. This document describes the compilation of all data generated in the separate LAPs into a total mass closure. It also provides guidance for selection of individual LAPs for any relevant feedstock. For Sorghum the mass closure will include ash, starch, protein, extractable materials, glucan, xylan, galactan, arabinan, lignin, acetate. Specific procedures for the determination of these components are detailed below.

3.2 Sample Preparation

1. "Preparation of Samples for Compositional Analysis," *Technical Report* NREL/TP-510-42620, Revised August 2008. This procedure describes drying, size reduction, and representative sampling methods that must be performed prior to analysis for many other constituents (*see* **Note 8**).
2. 10.5: Optional sieving: Omit sieving (*see* **Note 9**).

3.3 Starch Assay

1. Reference AOAC Official Method 996.11, Starch (Total) in Cereal Products. Starch content is determined by measurement of glucose after enzymatic hydrolysis. The AOAC method was adopted from the Megazyme method for determination of total starch content for samples that contain resistant starch (*see* **Note 10**).
2. We incorporate a high-purity corn starch method verification standard (*see* **Note 11**).
3. We incorporate sample blanks to account for free sugars.
4. Samples are capped in between reagent additions (*see* **Note 12**).

3.4 %Solids

1. "Determination of Total Solids in Biomass and Total Dissolved Solids in Liquid Process Samples," Technical Report NRELTP-510-42621, Revised March 2008. This procedure describes methods for the determination of the water content or total solids of a sample. Samples are dried either in a convection oven or by automated moisture analyzer.

3.5 Extractives

1. "Determination of Extractives in Biomass," Technical Report NREL/TP-510-42619, January 2008. This procedure describes aggregate quantification of nonstructural materials in a biomass sample by sequential extraction with water and ethanol. Free-sugars removed during water extraction are quantified by HPLC. Extractives percentages are used to convert components compositions from an extractives-free basis to an as-received basis (*see* **Note 13**).
2. Choose your extraction method: soxhlet or ASE (*see* **Note 14**).
3. *Soxhlet method only*:
   (a) 10.3: Analyze the water extract for sucrose content: required.
   (b) 10.3.5.1: Follow written procedure substituting HPLC analysis for YSI. See Subheading 2 of this chapter for additional details.
4. *Automatic extraction method only:*
   (a) 10.6: Analyze the sample for water extractives (required) and ethanol extractives.
   (b) 10.6.4: Analyze the water extract for sucrose content: required.
   (c) 10.6.4.1: Follow written procedure substituting HPLC analysis for YSI. *See* Subheading 2 of this chapter for additional details.
5. 10.7: Remove solvent from the extractives (*see* **Note 15**).

3.6 Structural

1. "Determination of Structural Carbohydrates and Lignin in Biomass," Revision Date: August 2012 (Version 08-03-2012). This procedure is for measuring structural carbohydrates and lignin content of samples after extraction. This procedure uses a two-step acid hydrolysis to hydrolyze carbohydrates prior to analysis by HPLC. The insoluble portion after hydrolysis is typically identified as lignin.
2. 11. Calculations
   (a) 11.3: Sorghum recommended wavelength for soluble lignin quantification is 320 nm and the recommended absorptivity at this wavelength is 30 l/g cm.
   (b) Accounting for glucose from starch (*see* **Note 16**).

3.7 Ash

1. "Determination of Ash in Biomass," Technical Report NREL/TP-510-42622, January 2008. This procedure describes determination of the residual inorganic content of a sample by oxidation at 550–600 °C.

3.8 Protein

1. "Determination of Protein Content in Biomass," Technical Report NREL/TP-510-42625, Revised May 2008. This document details the calculation of protein factor for converting measured nitrogen content to a protein content by using the ratio of amino acids in the specific sample type (*see* **Note 17**).

4 Notes

1. Sample selection can include grain, sweet, forage, and biomass sorghums. However, these methods have not been optimized for seed head or pure grain. The material to which these methods should be applied is whole plant sorghum with or without the panicle or grain. Anatomical fractions of sorghum can also be analyzed using these methods with the exclusion of pure grain or seed head as previously stated.
2. We do not add sodium azide to any reagents; however, it is present in small quantities in the enzyme preparations. This requires all buffers to be prepared weekly and stored refrigerated.
3. We recommend use of HPLC grade water that has been purified to 18 MΩ-cm for preparation of all reagents.
4. Standards may be purchased at the specified concentrations, which we have found ensures reproducibility between batches and saves significant time for the analysts.
5. Preparation of the 72% sulfuric acid solution from concentrated sulfuric acid is time consuming and highly exothermic. It is

recommended to purchase this reagent at the desired concentration of 72% wt/wt.

6. There are two styles of Teflon that are available for the pressure tubes. We recommend use of the cap with the o-ring seal at the tip of the screw top and not near the base. This prevents the acid and biomass becoming trapped in the threads during analysis. As the pressure tubes undergo repeated heat and cooling cycles the glass can become stressed. We recommend frequent inspection of the tubes for chips, crack, or defects that could lead to failure of pressure vessel. Even with care, pressure tubes have broken during handling. We recommend uses of cut proof gloves while handling the pressure vessel but especially during sealing or opening.
7. We have noticed that over time a buildup of inorganic material begins to clog the porous filtering crucible and affects the filtration of the acid insoluble phase. If this becomes severe enough excessive time under vacuum in the filtration flask may concentrate the acid soluble solution and affect the measured carbohydrates. Crucibles that take longer than typical for filtration should be replaced.
8. Drying must be performed at temperatures not higher than 45 °C to prevent chemically altering the sample. The drying step is critical for the accurate analysis of the samples not only because excess water may affect the acid hydrolysis, but also because a dry sample is more stable to degradation. Samples with significant concentration of free sugars (glucose and sucrose) are very susceptible to microbial growth and loss of measurable free sugars. If the sample begins to show sign of microbial growth, it is likely compromised, and a more aggressive drying technique is recommended. Size reduction to less than 2 mm particle size ensures proper hydrolysis during subsequent procedures.
9. Sieving was introduced to remove very large or very small particles of biomass that would hydrolyze differently than the standard 2 mm particle size. In biomass types for which the procedures were originally developed, the assumption was that all particle sizes were of similar chemical composition. This has not been observed to be true with many herbaceous biomass types and particularly in herbaceous materials which can be high in starch. Because of the nonstructural nature of starch, the starch has been seen to partition differently than the non-starchy material which leads to change in the measured chemical composition due to sieving. For this reason, the sieving step should be omitted for sorghum samples.
10. We have chosen to use the DMSO procedure as recommended by AOAC 996.11 as it has been the most reproducible of the

starch assays in our hands. Accurate determination of starch is crucial to the accurate determination of the cellulose content of the sample. With the standard NREL LAP analysis, the starch is quantified as cellulose during the acid hydrolysis. As starch has a drastically different chemical accessibility when compared to the more recalcitrant cellulose, compositional mis-identification could result in improper treatment during conversion to biofuels. For accurate quantification of starch both a whole sample and an extractives-free sample must be analyzed because the water and ethanol extraction does not effectively remove all of the starch from the sample and the remaining starch would be quantified as cellulose.

11. NREL recommends inclusion of a method verification standard of pure starch to monitor the effectiveness of the enzymatic hydrolysis. Any validation samples with recovery of less than 95% are considered failed and the batch is repeated. The starch standard is prone to thickening during analysis. To prevent thickening the sample should be vortexed vigorously during addition of any reagents.
12. The concentration calculations assume a known volume at the end of the procedure. We have found that tightly capping the vials during analysis prevents evaporation and concentration of measured sugars.
13. Water extraction of the biomass sample is required to remove compounds that may interfere with lignin quantification after acid hydrolysis. Additionally, water extraction serves to remove free-sugars such as sucrose that are present in significant quantities in herbaceous material. If these free sugars remain during the hydrolysis of the cellulose, the free-sugar derived glucose is inseparable from the cellulosic glucose which results in a high bias for cellulose. Additionally, it has been shown that the fructose in the free sucrose is extremely labile to acid hydrolysis and the products are readily measured as pseudo-lignin [11].
14. In our experience samples that are high in starch can clog the ASE extractor. If this is the case reducing the sample size in the extraction cell may help to prevent clogging. Alternately, soxhlet extraction does not seem to have the same clogging issues.
15. Large concentrations of free-sugars and starch can effectively trap water during drying of the water extractives and produce a film that prevents complete drying of the sample. It is important to ensure complete removal of the water prior to extractives weight determination as biases in water extractives values affect many other measurements. Longer drying periods may be used to ensure dryness, but temperatures may not exceed 40 °C ± 2 °C.

16. Starch present in the sample will hydrolyze into glucose and is measured as glucose during HPLC analysis. This glucose from starch and glucose from cellulose are indistinguishable and starch glucose must be quantified separately and mathematically removed from the measured glucan. The starch value used in the correction must be measured on the extractives-free sample as some starch is solubilized during the water extraction.

    Starch correction calculation

    As the glucose measured in the hydrolyzed biomass is derived from both starch and cellulose, the glucan attributable to cellulose must be calculated as follows.

$$\text{Glucan}_{\text{Measured}} = \text{Glucan}_{\text{Cellulose}} + \text{Glucan}_{\text{Starch}}$$

$$\text{Glucan}_{\text{Cellulose}} = \text{Glucan}_{\text{Measured}} - \text{Glucan}_{\text{Starch}}$$

If the starch content has been determined on the extracted material that was used for hydrolysis

$$\text{Starch}_{\text{extracties-free}} = \text{Glucan}_{\text{Starch}}$$

Therefore

$$\text{Glucan}_{\text{Cellulose}} = \text{Glucan}_{\text{Measured}} - \text{Starch}_{\text{Extractives-free}}$$

17. We have not performed this determination for sorghum but instead use the factor of 4.6 which was determined for corn stover.

Acknowledgments

The authors would like to acknowledge Amie Sluiter and David Templeton for their extensive contributions to the development of the original LAPs and Michelle Reed for her contributions to starch method optimization. The authors would also like to thank Jeff Dahlberg for his support of our methods. This work was supported by the U.S. Department of Energy under Contract No. DE-AC36-08GO28308 with the National Renewable Energy Laboratory. Funding provided by USDOE Office of Energy Efficiency and Renewable Energy's BioEnergy Technologies Office.

References

1. Sluiter JB, Ruiz RO, Scarlata CJ, Sluiter AD, Templeton DW (2010) Compositional analysis of lignocellulosic feedstocks. 1. Review and description of methods. J Agric Food Chem 58:9043–9053. https://doi.org/10.1021/jf1008023
2. National Renewable Energy Laboratory (2014) Standard procedures for biomass compositional analysis. http://www.nrel.gov/biomass/analytical_procedures.html

3. Templeton DW, Scarlata CJ, Sluiter JB, Wolfrum EJ (2010) Compositional analysis of lignocellulosic feedstocks. 2. Method uncertainties. J Agric Food Chem 58:9054–9062. https://doi.org/10.1021/jf100807b
4. Rooney WL, Blumenthal J, Bean B, Mullet JE (2007) Designing sorghum as a dedicated bioenergy feedstock. Biofuels Bioprod Biorefin 1:147–157. https://doi.org/10.1002/bbb.15
5. Stefaniak TR, Dahlberg JA, Bean BW, Dighe N, Wolfrum EJ, Rooney WL (2012) Variation in biomass composition components among forage, biomass, sorghum-sudangrass, and sweet sorghum types. Crop Sci 52:1949. https://doi.org/10.2135/cropsci2011.10.0534
6. Dahlberg J, Wolfrum E, Bean B, Rooney WL (2011) Compositional and agronomic evaluation of sorghum biomass as a potential feedstock for renewable fuels. J Biobaased Mater Bioenergy 5:507–513. https://doi.org/10.1166/jbmb.2011.1171
7. Murray SC, Rooney WL, Mitchell SE, Sharma A, Klein PE, Mullet JE et al (2008) Genetic improvement of sorghum as a biofuel feedstock: II. QTL for stem and leaf structural carbohydrates. Crop Sci 48:2180. https://doi.org/10.2135/cropsci2008.01.0068
8. Murray SC, Sharma A, Rooney WL, Klein PE, Mullet JE, Mitchell SE et al (2008) Genetic improvement of sorghum as a biofuel feedstock: I. QTL for stem sugar and grain nonstructural carbohydrates. Crop Sci 48:2165. https://doi.org/10.2135/cropsci2008.01.0016
9. Wolfrum E, Payne C, Stefaniak T, Rooney W, Dighe N, Bean B et al (2013) Multivariate calibration models for sorghum composition using near-infrared spectroscopy. Golden
10. Megazyme International (2016) Megazyme: total starch assay procedure K-TSTA. http://secure.megazyme.com/files/BOOKLET/K-TSTA_1107_DATA.pdf
11. Katahira R, Sluiter JB, Schell DJ, Davis MF (2013) Degradation of carbohydrates during dilute sulfuric acid pretreatment can interfere with lignin measurements in solid residues. J Agric Food Chem 61:3286–3292. https://doi.org/10.1021/jf303727t

Chapter 19

The Role of Sorghum in Renewables and Biofuels

Jeff Dahlberg

Abstract

Sorghum bicolor (L.) Moench is an important annual C_4 cereal crop with unique properties—it can be used in almost all renewable schemes being proposed for renewable fuels and green technologies. In the United States, the grain is currently used as a feedstock in the grain-ethanol process, while in China, the Philippines, and India, sweet sorghums are used in a sugar-to-ethanol process. High-tonnage biomass sorghums are being investigated for their potential use in both cellulosic and lignocellulosic renewables. Other countries have been exploring sorghum's use as a renewable building material and as a potential source of high-value C molecules for the creation of renewable oils and other important industrial chemicals. Sorghum can become a major player in the renewable feedstock industry because of its potential for high-yield production under limited water and inputs, strong research capacities, a well-established seed industry, and a robust history of research on production and cultural practices. The following review highlights various research activities in support of renewables using sorghum as a primary feedstock.

Key words Sorghum, Renewable, Ethanol, Cellulosic, Lignocellulosic, Drought, Low inputs, Adaptability

1 Introduction

Sorghum bicolor (L.) Moench is an important annual C_4 cereal crop with unique properties—it can be used in almost all renewable schemes being proposed for renewable fuels and green technologies. Sorghum is an ancient crop, domesticated in Eastern Africa [1] and selected for various characteristics within a wide range of environments and abiotic stresses. It was the second cereal crop to be sequenced [2] and along with a strong germplasm collection offers a wide range of genetic opportunities to create sorghums tailored to various utilizations and characteristics favorable for renewable uses [3]. In the United States, the grain is currently used as a feedstock in the grain-ethanol process, while in China, the Philippines, and India, sweet sorghums are used in a sugar-to-ethanol process. High-tonnage biomass sorghums are being investigated for their potential use in both cellulosic and lignocellulosic renewables. Other countries have been exploring sorghum's use as

Zuo-Yu Zhao and Jeff Dahlberg (eds.), *Sorghum: Methods and Protocols*, Methods in Molecular Biology, vol. 1931, https://doi.org/10.1007/978-1-4939-9039-9_19,

a renewable building material and as a potential source of high-value C molecules for the creation of renewable oils and other important industrial chemicals. Sorghum can become a major player in the renewable feedstock industry because of its potential for high-yield production under limited water and inputs, strong research capacities, a well-established seed industry, and a robust history of research on production and cultural practices. The following review highlights various research activities in support of renewables using sorghum as a primary feedstock.

2 Renewable Products

Sorghum has been used as a renewable source for a wide range of consumer products ever since its domestication. Photoperiod sensitive sorghums that grow to large heights and produced panicles were utilized as both a grain sorghum and the stems as a building source for fences and other home-building materials in Africa. Broomcorns, a specialty sorghum, were used for the manufacturing of brooms as early as the 1200s in Italy [3], in China around the 1300s [4] and introduced into the United States by Benjamin Franklin in 1757 [5]. Robust broomcorn manufacturing plants (Fig. 1) exist within Europe [3]. Petrini et al. [6] and Rajki-Siklósi [7] reviewed several attempts to use sorghum as a raw material for paper production. Pederson [8] reviewed research that looked at forage sorghum as feedstocks for both paper and cardboard production. Plant stems have been used as a source of fuel for cooking and stems of grassy sorghum have been used to make fishing nets and baskets [9]. Doggett also reported that stems can be crushed, treated with an aqueous extract and leather dyes can be created. The dye was reported to be santalin. Levulinic acid made from grain sorghum have been evaluated as potential deicers and these could be a renewable source of levulinate salts that could replace traditional deicers [10]. Many chemicals that can be produced by oil feedstocks can be manufactured through the use of sugar feedstock. Products such as plastics, solvents, and other industrial solvents can use sugar as the raw feedstocks. As with most every renewable product and manufacturing process, the driving force is prices.

3 Grain Sorghums for Alcohol Production

Sorghum has a long history in its use as a feedstock in alcohol production, primarily as distilled spirits for human consumption. For ethanol production as a renewable fuel, the United States began using sorghum as a cost-effective substitute for corn in regions such as Kansas and Nebraska. Essentially, the grain is ground, processed, and converted using enzymes into fermentable

Fig. 1 Broom manufacturing facility in Serbia, 2010 (picture taken by Jeff Dahlberg)

carbohydrates. In the 1980s, Coble et al. [11] reviewed the utilization of sorghum in ethanol production and concluded that fermentation yields were similar to that of corn. Sweeten et al. [12] evaluated the use of grain sorghum in integrated food and fuel production systems to determine plant operation procedures through use of small scale production facility, the energy requirements and to characterize the stillage from the fermented grain. They also concluded that sorghum was comparable to corn of ethanol production. More recently, several authors have evaluated yield and fermentation efficiency on the performance of feedstocks in ethanol production. They found several key factors in sorghum that impacted both yield and efficiency, such as starch content, starch digestibility, extractable proteins, the number of phenolic compounds, and ratio of amylose to amylopectin [13–15].

Wu et al. [14] reported differences in fermentation efficiency and ethanol yield in 70 sorghum genotypes differing in kernel characteristics. Yan et al. [16] investigated the physicochemical and biochemical properties of field sprouted sorghums on ethanol production and reported that ethanol yields from sprouted grain sorghum were higher than that of non-sprouted sorghums.

Barcelos et al. [17] evaluated starch enzymatic hydrolysis and found good experimental protocols that resulted in conversion efficiencies of 99.3% when using sorghum. They concluded that this had good practical use in Brazil. Ramirez et al. [18] showed that the removal of tannins improved the simultaneous saccharification and ethanol fermentation of high tannin sorghums and could be a method for utilizing these sorghums. There are clear genetic differences in sorghums that could be exploited to enhance these efficiencies and yield; however, the primary trait of importance in the grain-to-ethanol system is total starch production per unit area. As long as sorghum is planted as a rainfed crop or on marginal agricultural lands which limits its yield potential, it will continue to be a small player in the grain to ethanol market.

4 Sweet Sorghums

Sweet sorghums are a specialty sorghum grown specifically for its production of sugary stem juices. In the United States, these have been grown for the commercial production of sorghum syrup or molasses and in more recent years for their interest as a complimentary crop to sugarcane for production of ethanol. The stalks are squeezed to extract the sugars in the juice which are then processed into either a syrup or ethanol. In the 1970, several authors promoted sweet sorghum production as a renewable fuel [19, 20]. Stevens and Holou [21] provide an excellent review of the status of sweet sorghum research as an alternative renewable fuel worldwide. The amount of fermentable sugar is dependent upon the efficiency of extraction equipment, the sugar content of the juice, and the yield. Fermentation is relatively straightforward, the carbohydrates are readily available as sucrose, fructose, and glucose for utilization by the yeast. India (Fig. 2), China, and Brazil have programs working to develop sweet sorghum for ethanol fuel production [22]. The potential for the use of sweet sorghum as a viable biofuel feedstock has been well documented [23–30].

Though sweet sorghums can be grown on most arable lands in the world, limitations to the production of this crop exist. Most of the sweet sorghums utilized within the United States are old sorghum cultivars that were bred and cultivated for syrup production. New hybrids with improved yield production and high juice and sugar concentration are needed. These new hybrids will need to be better managed to optimize both yield and harvest timing to maximize production on a unit scale. As the juice is quickly broken down, it must be either stored or converted to ethanol quickly. Machinery is also a limitation and improved harvesting equipment will be needed to efficiently and quickly harvest mature plants. Stevens and Holou [21] concluded that sugars extracted from juice is still one of the most affordable ways for producing motor fuels from a renewable source.

Fig. 2 Sweet sorghum Ethanol plant outside of Hyderabad, India, 2017 (picture taken by Jeff Dahlberg)

5 Biomass Feedstocks for Renewable Fuels

Much of the research taking place on biomass feedstocks is driven by the one billion ton goal that was established by the U.S. Congress Biomass Research and Development Technical Advisory Committee. Perlack et al. [31] estimated that this could be done through the sustainable use of forestlands and the rest taken from agricultural lands. They projected that approximately 428 million tons could be sustainably harvest through annual crop residues, while another 377 million tons from perennial crops. Each one of these crop systems has its challenges.

Sorghum is a highly productive C_4 annual crop and under ideal conditions sorghum can produce tonnage (Fig. 3) of over 33 T of biomass ac^{-1} at 65% moisture [32]. In this same report, the authors reported that sorghum can produce biomass yields equal to or greater than corn using 33% less water. Work on biomass sorghums

Fig. 3 High-biomass sorghum in demonstration plot at the University of California, Agricultural and Natural Resources, Kearney Agricultural Research and Extension Center in Parlier, CA, 2013 (picture taken by Jeff Dahlberg)

as a potential feedstock is a relatively new area of research. Several authors looked at the potential of lignocellulosic and cellulosic processes, the impact of brown midrib mutations on cellulose, hemicellulose, and lignin content, and structural and nonstructural carbohydrates and their impact on fuel production [24, 33–41].

Several factors impact the composition of plants and this is true of sorghum as well. The type of sorghum, grain, forage, hay, or biomass will influence the types of fuel production strategies that will be used to convert this feedstock into viable fuels. Rooney et al. [37] had reported on yield potential of various sorghum types for renewable fuel production and concluded that sorghum could reliably produce high-tonnage biomass feedstock. Prior to work reported by Dahlberg et al. [42] little work had been done on the compositional characteristics of sorghum feed stock for cellulosic and/or lignocellulosic conversion into fuels. The data reported a tremendous amount of variability in the compositional makeup of sorghum and from this dataset NIR calibration curves were developed and made available for breeding purposes. Genomic technologies have been used since then to predict breeding values that will enhance breeding strategies to improve these biomass feedstocks for various compositional and functional characteristics to improve conversion [43, 44].

6 Conclusions

Sorghum is a uniquely positioned agronomic crop as a feedstock for a wide range of renewable productions. Its genetic diversity, wide range of adaptability, and inherent drought tolerance makes it an ideal annual crop for the production of renewable feedstocks. The grain, hay, stems, and whole plant can be used in a wide variety of renewable products from building material to liquid fuels and chemicals. It is clear that sorghum, through genetic manipulation, can be tailored to most of the renewable strategies being proposed. The availability of robust genomic sequences, NIR technologies to rapidly characterize compositional characteristics, the development of high-throughput phenotyping being supported by various US Department of Energy grants and good commercial breeding companies support the development and rapid deployment of sorghums as a robust feedstock for renewable products. As with all renewable feedstocks, availability, and economics will be the major driving force of a successful renewable strategy in the future.

References

1. Kimber CT (2000) Origins of domesticated sorghum and its early diffusion to India and China. In: Smith CW, Frederiksen RA (eds) Sorghum: origin, history, technology, and production. John Wiley & Sons, New York, NY, pp 3–98
2. Kresovich S, Barbazuk B, Bedell JA, Borrell A, Buell CR, Burke J et al (2005) Towards sequencing the sorghum genome. A U.S. National Science Foundation-sponsored workshop report. Plant Physiol 138:1898–1902
3. Dahlberg J, Berenji J, Sikora V, Latkovic D (2011) Assessing sorghum [*Sorghum bicolor* (L.) Moench] germplasm for new traits: food, fuels & unique uses. Maydica 56(1750):85–92
4. Qingshan L, Dahlberg JA (2001) Chinese sorghum genetic resources. Econ Bot 55:401–425
5. Franklin B (1757) Letter to Mr. Ward from Benjamin Franklin on March 24, 1757. Original letter held in trust with the Nantucket Atheneum, Nantucket, MA
6. Petrini C, Casalboni L, Bazzocchi R, Casamenti R (1993) Il sorgo da fibra. Agricoltura, Luglio-Agosto: 34–36
7. Rajki-Siklósi E (1996) Examination of sorghum fibre and sugar content in Hungary. In: European seminar on sorghum for energy and industry, Toulouse, p 49–52
8. Pedersen JF (1996) Annual forages: new approaches for C-4 forages. In: Janick J (ed) Progress in new crops. American Society of Horticultural Science, Alexandria, VA, pp 246–251
9. Doggett H (1988) Sorghum. John Wiley & Sons, New York, NY

10. Ganjyal G, Fang Q, Hanna M (2007) Freesing points and small-scale deicing tests for salts of levulinic acid made from grain sorghum. Bioresour Technol 98:2814–2818
11. Coble CG, Hiler EA, Sweeten JM, O'Neal HP, Reidenback VC, LePori WH et al (1981) Small scale ethanol production from cereal feedstocks. In: Pomeranz Y, Munck L (eds) Cereals: a renewable resource. American Association of Cereal Chemists, St. Paul, MN, p 611
12. Sweeten JM, Coble CG, Egg RP (1983) The use of grain sorghum in integrated food and fuel production systems. In: Veziroglu TN (ed) Alternative energy sources V. Part D: biomass/hydrocarbons/hydrogen. Elsevier Science Publishers B.V, Amsterdam
13. Wang D, Bean S, McLaren J, Seib P, Madl R, Tuinstra M et al (2008) Grain sorghum is a viable feedstock for ethanol production. J Ind Microbiol Biotechnol 35:313–320
14. Wu X, Zhao R, Bean SR, Seib PA, McLaren JS, Madl RL et al (2007) Factors impacting ethanol production from grain sorghum the dry-grind process. Cereal Chem 84:130–136
15. Zhao R, Bean SR, Ioerger BP, Wang D, Boyle DL (2008) Impact of mashing on sorghum proteins and its relationship to ethanol fermentation. J Agric Food Chem 56:946–953
16. Yan S, Wu X, Dahlberg J, Bean SR, MacRitchie F, Wilson JD et al (2010) Properties of field-sprouted sorghum and its performance in ethanol production. J Cereal Sci 51:374–380
17. Barcelos CA, Maeda RN, Betancur GJV, Pereira N Jr (2011) Ethanol production from sorghum grains [*Sorghum bicolor* (L.) Moench]: evaluation of the enzymatic hydrolysis and the hydrolysate fermentability. Braz J Chem Eng 28:597–604
18. Ramirez MB, Ferrari MD, Lareo C (2016) Fuel ethanol production from commercial grain sorghum cultivars with different tannin content. J Cereal Sci 69:125–131
19. Miller FR, Creelman RA (1980) Sorghum—a new fuel. In: Loden HD, Wilkinson D (eds) Proc. 35th annual corn and sorghum res. conf., Chicago, IL, 9–11 Dec. 1980. American Seed Trade Assoc., Chicago, IL, p 219–232
20. Broadhead D, Freeman K (1980) Stalk and sugar yield of sweet sorghum as affected by spacing. Agron J 72:523–524
21. Stevens G, Holou RAY (2010) Sweet sorghum as a biofuel crop. In: Halford NG, Karp A (eds) Energy crops: energy and environment Series No. 3. Royal Society of Chemistry, London, pp 56–74
22. Reddy B, Ramesh S, Reddy P, Kumar A, Sharma K, Chetty S et al (2006) Sweet sorghum-food, feed, fodder, and fuel crop. International Crops Research Institute for the Semi-Arid Tropics, Patancheru, Andhra Pradesh Bull. 627-2006
23. Alsina E, Valle-Lamboy S, Méndez-Cruz AV (1975) Preliminary evaluation of ten sweet sorghum varieties for sugar production in Puerto Rico. J Agric Univ P R 59:5–14
24. Amaducci S, Monti A, Venturi G (2004) Non-structural carbohydrates and fibre components in sweet and fibre sorghum as affected by low and normal input techniques. J Ind Crop Prod 20:111–118
25. Cundiff JS, Vaughan DH (1987) Sweet sorghum for ethanol industry for the Piedmont. Energy Agric 6:133–140
26. Cundiff JS (1992) Potential reduction of non-structural carbohydrate losses by juice expression prior to ensiling sweet sorghum. Biomass Bioenergy 3:403–410
27. Murray SC, Sharma A, Rooney WL, Klein PE, Mullet JE, Mitchell SE et al (2008) Genetic improvement of sorghum as a biofuel feedstock: I. QTL for stem sugar and grain non-structural carbohydrates. Crop Sci 48:2165–2179
28. Smith GA, Bagby MO, Lewellan RT, Doney DL, Moore PH, Hills FJ et al (1987) Evaluation of sweet sorghum for fermentable sugar production potential. Crop Sci 27:788–793
29. Tew TL, Cobill RM, Richard EP Jr (2008) Evaluation of sweet sorghum and sorghum x sudangrass hybrids as feedstocks for ethanol production. Bioenergy Res 1:147–152
30. Worley JW, Vaughan DH, Cundiff JS (1992) Energy analysis of ethanol production from sweet sorghum. Bioresour Technol:263–271
31. Perlack RD, Wright LL, Turhollow AF, Graham RL, Stokes BJ, Erbach DC (2005) Biomass as feedstocks for a bioenergy and bioproducts industry: the technical feasibility of a billion-ton annual supply. USDA/DOE, DOE/GO-102005-2135. http://www1.eere.energy.gov/biomass/pdfs/final_billionton_vision_report2.pdf (verified 2 Aug 2010)
32. McCollum T, McCuistion K, Bean B (2005) Brown midrib and photoperiod sensitive forage sorghums. In: Plains Nutrition Council Spring Conference, San Antonio, TX, pp 36–46
33. Miller FR, McBee GG, Hons FM, Cothren JT (1989) Biomass production, quantity and quality. In: Hiler E (ed) Advanced concepts in biomass production and biological pretreatment. TAES progress report to Gas Research Institute, College Station, TX

34. Murray S, Rooney WL, Mitchell SE, Sharma CA, Klein PE, Mullet JE et al (2008) Genetic improvement of sorghum as a biofuel feedstock: II. QTL for stem and leaf structural carbohydrates. Crop Sci 48:2180–2193
35. Porter KS, Axtell JD, Lechtenberg VL, Colenbrader VF (1978) Phenotype, Fiber compostion, and in vitro dry matter disapperance of chemically induced brown midrib (bmr) mutants of sorghum. Crop Sci 18:205–208
36. Propheter JL, Staggenborg SA, Wu X, Wang D (2010) Performance of annual and perennial biofuel crops: yield during the first two years. Agron J 102:806–814
37. Rooney WL, Blumenthal J, Bean B, Mullet JE (2007) Society of Chemical Industry and John Wiley & Sons, Ltd., biofuels, bioprod. Bioref 1 (2007):147–157. https://doi.org/10.1002/bbb
38. Saballos A, Vermerris W, Rivera L, Ejeta G (2008) Allelic association, chemical characterization and saccharification properties of brown midrib mutants of sorghum (*Sorghum bicolor* (L.) Moench). Bioenergy Res 1:193–204
39. Sipos B, Réczey J, Somoria Z, Kádár Z, Dienes D, Réczey K (2009) Sweet sorghum as feedstock for ethanol production: enzymatic hydrolysis of steam-pretreated bagasse. Appl Biochem Biotechnol 153:151–162
40. Stanley RL, Dunavin LS (1986) Potential sorghum biomass production in North Florida. In: Smith WH (ed) Biomass energy development. Plenum Press, New York, pp 217–226
41. Vermerris W, Saballos A, Ejeta G, Mosier NS, Ladisch MR, Carpita NC (2007) Molecular breeding to enhance ethanol production from corn and sorghum stover. Crop Sci 47: S143–S153
42. Dahlberg J, Wolfrum E, Bean B, Rooney WL (2011) Compositional and agronomic evaluation of sorghum biomass as a potential feedstock for renewable fuels. J Biobaased Mater Bioenergy 5:507–513
43. Brenton ZW, Cooper EA, Myers MT, Boyles RE, Shakoor N, Zielinski KJ et al (2016) A genomic resource for the development, improvement, and exploitation of sorghum for bioenergy. Genetics 204:21–33. https://doi.org/10.1534/genetics.115.183947
44. De Oliveira AA, Pastina MM, de Souza VF, da Costa Parrella RA, Noda RW, Ferreira Simeone ML et al (2018) Genomic prediction applied to high-biomass sorghum for bioenergy production. Mol Breeding 38:49. https://doi.org/10.1007/s11032-018-0802-5

Index

Zuo-Yu Zhao and Jeff Dahlberg (eds.), *Sorghum: Methods and Protocols*, Methods in Molecular Biology, vol. 1931, https://doi.org/10.1007/978-1-4939-9039-9,

W

X

Y

Z

MIX
Papier aus verantwortungsvollen Quellen
Paper from responsible sources
FSC® C105338

Printed by Books on Demand, Germany